KU-566-980

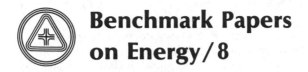

**Benchmark Papers
on Energy/8**

A BENCHMARK® Books Series

ELECTROCHEMICAL, ELECTRICAL, AND MAGNETIC STORAGE OF ENERGY

Edited by

W. V. HASSENZAHL

University of California

Hutchinson Ross Publishing Company

Stroudsburg, Pennsylvania

05436090

Copyright © 1981 by **Hutchinson Ross Publishing Company**
Benchmark Papers on Energy, Volume 8
Library of Congress Catalog Card Number: 81-6476
ISBN: 0-87933-376-6

All rights reserved. No part of this book covered by the copyrights
hereon may be reproduced or transmitted in any form or by any
means—graphic, electronic, or mechanical, including photocopying,
recording, taping, or information storage and retrieval systems—
without written permission of the publisher.

83 82 81 1 2 3 4 5
Manufactured in the United States of America.

LIBRARY OF CONGRESS CATALOGING IN PUBLICATION DATA
Main entry under title:
Electrochemical, electrical, and magnetic storage of energy.
   (Benchmark papers on energy; v. 8)
   Includes bibliographies and indexes.
   1. Energy storage—Addresses, essays, lectures. I. Hassenzahl, W. V.
(William V.), 1940– . II. Series.
TJ165.E43   621.31′26   81–6476
ISBN 0-87933-376-6   AACR2

Distributed world wide by Academic Press,
a subsidiary of Harcourt Brace Jovanovich,
Publishers.

D
621.4
ELE

Benchmark Papers
on Energy

Series Editors: R. Bruce Lindsay, Brown University
Mones E. Hawley, Jack Faucett Associates

Volume

CONTENTS

Contents

PART III: CAPACITORS

SERIES EDITOR'S FOREWORD

The Benchmark Papers on Energy constitute a series of volumes that makes available to the reader in carefully organized form important and seminal articles on the concept of energy, including its historical development, its applications in all fields of science and technology, and its role in civilization in general. This concept is generally admitted to be the most far-reaching idea that the human mind has developed to date, and its fundamental significance for human life and society is everywhere evident.

One group of volumes of the series contains papers bearing primarily on the evolution of the energy concept and its current applications in the various branches of science. Another group of volumes concentrates on the technological and industrial applications of the concept and its socioeconomic implications.

Each volume has been organized and edited by an authority in the area to which it pertains and offers the editor's careful selection of the appropriate seminal papers, that is, those articles that have significantly influenced further development of that phase of the whole subject. In this way every aspect of the concept of energy is placed in proper perspective, and each volume represents an introduction and guide to further work.

Each volume includes an editorial introduction by the volume editor, summarizing the significance of the field being covered. Every article or group of articles is accompanied by editorial commentary, with explanatory notes where necessary. Both an author index and a subject index are provided for ready reference. Articles in languages other than English are either translated or summarized in English. It is the hope of the publisher and editor that these volumes will serve as a working library of the most important scientific, technological, and social literature connected with the idea of energy.

The present volume is one of two edited by Dr. William V. Hassenzahl of the University of California. Their subject is the storage of energy. The purpose of storage is to make energy available at some future time. The concept of serial time, of thinking about time as a series of events as opposed to the unity of "for all time," is an evolutionary step and a basic characteristic in the civilization of mankind. Artifacts for use in the future world are a unique characteristic of human burials. Early examples of cognitive storage of energy include heated rocks, elevated reservoirs, potters' wheels, and animal snares that use bent branches.

The present volumes deal with a wide range of more modern forms of energy storage and are distinguished by the physical sciences with which forms of energy storage are most closely associated—electrochemistry, electricity, magnetism, chemistry, mechanics, and heat. The emphasis of each volume is on current topics of great concern: use of renewable energy sources that are not available continuously, such as solar; conservation of petroleum energy; and reduction of the cost of providing electricity to meet peak power demands. For readers whose basic science needs refreshing, Dr. Hassenzahl has provided an introduction that also facilitates comparisons of the various forms of energy on a common basis. His comments on the reprinted papers and the bibliographies lead the reader to greater depth of specialization. The tables of contents indicate clearly the coverage of each volume.

MONES E. HAWLEY

PREFACE

Effective energy utilization is the basis of today's advanced industrial society. Usually energy is used neither where, nor when, nor in the form in which it is found, thus creating a need for its conversion, transmission, and storage. Perhaps the largest artificial storage systems are the dams and reservoirs that contain enough water to produce electricity sufficient to power the world's largest cities for months. At the other extreme, both in magnitude and kind, almost every living organism stores energy in the form of hydrocarbons to satisfy its own particular variable energy requirements.

Tens of thousands of papers have been written on energy storage. To reduce this topic to a level that is tractable in two volumes, *Electrochemical, Electrical, and Magnetic Storage of Energy* and *Mechanical, Thermal, and Chemical Storage of Energy*, the perspective has been narrowed to the specific and willful storage of energy by use of special devices. This limitation leaves an almost overwhelming subject, which includes batteries, hydroelectric pumped reservoirs, capacitors, magnets, flywheels, hydrogen and certain specially produced chemicals, and thermal energy storage. Fuels such as gasoline are a special topic and not included here.

In selecting articles for inclusion, emphasis was placed on electric utilities' energy storage, which requires far more storage in terms of both capital investment and total stored energy than any other application. A few articles of historical significance set the stage for present developments; some articles outline the basic laws or rules that govern the developments of a given technology, and others describe the present status of the technologies or the materials required for and the material problems associated with these technologies.

The Introduction describes in simple terms the various forms in which energy exists and the mechanisms available for its storage. It is not intended to be a basic science course but should supply a vocabulary for readers who have not had introductory courses in chemistry and physics; should provide a consistent picture of the interrelationships among the various forms of energy; and, in addition, should give readers some feeling for the capabilities, capacities, costs, and sizes of the various energy storage methods.

Bibliographies are provided at the end of each commentary to aid researchers in obtaining more information on each particular energy storage technology.

W. V. HASSENZAHL

CONTENTS BY AUTHOR

ELECTROCHEMICAL, ELECTRICAL, AND MAGNETIC STORAGE OF ENERGY

INTRODUCTION

ENERGY AND ENERGY CONVERSION

The developments and changes of society over the past few millennia, and in particular over the past two hundred years, have been characterized by the substitution of animals and, more recently, mechanical devices for human muscle power. We have witnessed an ever-increasing utilization of energy and diversification of energy sources. These changes have occurred throughout the world but have been most profound in the West.

The developments in agriculture, mining, construction, transportation, and other areas have generated the need for various types of energy that must be deliverable at a high rate or equivalently at a high power level. Frequently the available energy is neither in a convenient form for the planned use nor in a convenient location. For example, petroleum is not naturally found on highways, and chemical energy must be converted into mechanical energy to propel a vehicle. Thus it is often necessary to transform energy from one form to another. Further, because the processes of energy transformation are generally expensive, it has become common to attempt to produce some forms of energy at a constant rate and on a large scale.

Electric utility companies frequently produce excess electrical energy at night and store it in a form that can be converted back into electrical energy for use during the peak load periods, which typically occur during the daylight hours. Of course, larger, more expensive generating plants could be installed and their output cycled, but the use of storage facilities reduces cost and in some cases improves overall system performances. To understand why one type of energy storage is selected over other possiblities, it is necessary to consider the forms in which energy can exist and the relationship among the various forms.

Energy, defined as "the capacity to do work,"* appears in a variety of natural forms, such as the *potential* energy of water in a high lake relative to its final resting place in the ocean, the kinetic energy of the water as it flows from the lake to the sea, the electromagnetic energy in light, and the thermal energy or heat that exists in all matter.

In our society usually we do not use energy directly in these forms; rather it is converted into a more convenient form. For example, the kinetic energy of flowing water can be converted into the rotational kinetic energy of a generator, which in turn is converted into electrical energy. The electrical energy might then be changed back into mechanical energy or used to produce heat, and so forth. People rarely use electrical energy directly, yet nearly 20 percent of all the energy used in the United States is electrical.

Energy is always a positive quantity. One system may have less energy than another, but it can never have zero energy. Thus, it is a change in the energy state or level of one system that may be utilized to effect some change in the energy state or level of another system. Clearly the energy of the driving system must decrease and that of the driven system must increase. Energy is never lost, never destroyed, and never created, yet it can be changed. One of the powerful fundamental laws of physics is the *Law of Energy Conservation*, which states that the total energy of a closed system cannot be changed without some outside influence.

Energy is available for our use in five basic forms:

> Mechanical energy,
> Thermal energy or heat,
> Electrical or electromagnetic energy,
> Chemical energy, and
> Nuclear energy or mass.

Each of these types of energy will be described either in this volume or in the companion volume, *Mechanical, Thermal, and Chemical Storage of Energy*. In almost all cases one form may be transformed into another, although frequently an intermediate step is required.

Although energy may exist in many forms, all may be expressed in the same units. In the new Standard International (SI) set of units the preferred unit of energy is the joule (J). The SI unit

Webster's Third New International Dictionary (Springfield, Mass.: G. C. Merriam Co., 1965).

of distance is the meter (m), 39.37 inches; the unit of time is the second (s); and the unit of mass is the kilogram (kg). The joule is a derived unit given by

$$1 J = 1 \text{ kg m}^2/\text{s}^2. \tag{1}$$

Other units of energy are used for some special applications. The electric utilities use the kilowatt hour (kWh) and the megawatt hour (MWh). These may be expressed in terms of joules:

$$1 \text{ kWh} = 3{,}600{,}000 \text{ J} = 3.6 \times 10^6 \text{ J} \tag{1a}$$
$$1 \text{ MWh} = 3{,}600{,}000{,}000 \text{ J} = 3.6 \times 10^9 \text{ J}.$$

The watt (W), a unit of power, is the rate of using energy:

$$1 \text{ watt} = 1 \text{ joule/second} = 1 \text{ J/s}. \tag{1b}$$

If the power required for a process, or available from a generator, is known, then the total energy required, or available, is equal to the power multiplied by the time of operation. Heat is a form of energy that is usually expressed in calories or kilocalories. One calorie (cal) is defined as the energy required to elevate the temperature of 1g of water 1 Celsius degree and equivalently 1 kilocalorie will raise the temperature of 1 kg of water by 1 Celsius degree. The calorie is now defined in relation to the joule (several types of calories are defined but all are within 1 percent of 4.18 J):

$$1 \text{ cal} = 4.18 \text{ J}. \tag{1c}$$

Another unit of thermal energy is the British thermal unit (Btu), which was originally defined as the heat required to raise the temperature of 1 pound (lb) of water by one Fahrenheit degree.

$$1 \text{ Btu} \approx 252 \text{ cal} \approx 1050 \text{ J}. \tag{1d}$$

Energy can be produced by a force (F) being exerted through a distance (d).

$$E = Fd. \tag{2}$$

In SI units the force is expressed in newtons (N):

$$1 \text{ N} = 1 \text{ J/m} = 1 \text{ kg m/s}^2. \tag{3}$$

3

The unit of voltage in SI is the volt (V) and the unit of current is the ampere (A). Frequently the electric field is used in calculations. The unit is volts/meter (V/m).

Mechanical Energy

Perhaps the energy form that is easiest to comprehend is mechanical energy. Mechanical energy can be the result of a force acting through a distance and may appear as kinetic, potential, strain, or, for a gas, compressional energy.

Kinetic energy is the energy associated with the motion of an object. Linear kinetic energy is given by

$$E = Mv^2/2, \tag{4}$$

where M is the mass (kg) of the object and v is its velocity (m/s). Rotational kinetic energy is given by

$$E = I\omega^2/2, \tag{5}$$

where I is the moment of inertia (kg $\cdot$ m^2) and ω is the angular velocity (s^{-1}).

Potential energy is the energy associated with the position of matter in a gravitational field. (This concept may be extended to other more general fields.) Usually the quantity of interest is the relative potential energy of an object as it moves from one height (h) to another in the gravitational field of the earth:

$$\Delta E = Mg\,\Delta h, \tag{6}$$

where ΔE (J) is the energy available from or required by a change in elevation Δh (m) of the mass M (kg), and g is the gravitational constant at the earth's surface (g = 9.8 m/s^2).

The gravitational constant at the earth's surface is equal to the acceleration of an unsupported object toward the earth. This acceleration will increase the object's velocity by 9.8 m/s (neglecting air resistance) during each second of free fall.

Elastic or strain energy is associated with the stretching or compressing of a solid substance. In the typical one-dimensional case of a spring, the energy is given by

$$E = k\,(\Delta x)^2/2, \tag{7}$$

where k (kg/s²) is the "spring constant" and Δx the change in length of the material. This expression may be generalized to two or three dimensions for solids, and in a modified form it applies to liquids.

Compressional energy of a gas is the energy gained by a gas as it is compressed from one pressure P_i and volume V_i to a higher pressure P_f and a smaller volume V_f. The change in energy depends on the characteristics of the gas being compressed. No simple expression adequately describes this dependence. When most gases are compressed, a significant fraction of the energy goes into heating the gas. If this heat is removed during or after compression, then the relative amount of energy stored in the gas is considerably reduced.

The transformation of one form of mechanical energy into another is generally simple and relatively efficient. For example, potential energy is converted into kinetic energy when water from a reservoir flows downhill.

Thermal Energy

Thermal energy and heat are identical, but they are different from temperature. We are all familiar with temperatures and can distinguish a rather small temperature difference. As the temperature of a substance increases, the energy content also increases. The energy required to heat a volume V of a substance from a temperature T_1 to a temperature T_2 is given by

$$E = C(T_2 - T_1) \cdot V, \qquad (8)$$

where C is the specific heat of a unit volume of that particular substance. A given amount of energy may heat the same weight or volume of other substances to temperatures greater or lower than T_2. The value of C may vary from about 1.0 cal/gm °C (or cal/cm³ °C) for water to 0.0001 cal/gm °C for some materials at very low temperatures.

The energy given off by a material as its temperature is reduced, or taken up by a material as its temperature is increased, or equivalently the energy required to raise the temperature, is called the sensible heat.

Another form of thermal energy, the latent heat, is associated with the changes of state, or phase changes, of a material. For example, energy is required to convert ice into water, to change

5

water into steam, and to melt paraffin wax. The energy required to cause these changes is called the *heat of fusion* at the melting point and the *heat of vaporization* at the boiling point.

Using our example of water, suppose one wishes to evaporate 1 g of ice by converting it to liquid and then heating it until it boils away. First, 80 cal will be required to change ice at 0°C to water at 0°C; then it will take about 100 cal to raise the temperature of the water to 100°C; finally 540 cal will be needed to boil the water, giving a total of 720 cal.

Just as the sensible heat varies from one material to another, so the latent heat also varies a great deal.

It is straightforward to determine the value of the sensible heat for solids and liquids, but the situation is more complicated for gases. If a gas restricted to a certain volume is heated, both the temperature and the pressure will increase. The specific heat observed in this case is called the specific heat at constant volume, C_v. If instead the volume is allowed to vary and the pressure is fixed, the specific heat at constant pressure, C_p, is obtained. The ratio C_p/C_v and the fraction of the heat produced during compression that can be saved significantly affect the storage efficiency.

Electromagnetic Energy

Energy is a characteristic of both electric and magnetic fields and may be transmitted through free space from one place to another through an oscillating combination of the two types of fields. This type of energy transmission is known as radiant energy. In addition to light, it includes radiowaves, microwaves, X rays, and γ radiation.

The storage of electromagnetic energy requires either a capacitor, a magnet, or a microwave cavity. Because it is quite difficult to store large quantities of microwave energy, in which the electric and magnetic fields oscillate, only magnets and capacitors, where the fields are essentially constant, will be considered here.

The energy in a magnetic field is given by

$$E = B^2/2\mu, \tag{9}$$

where B is in tesla (1 tesla = 1 volt s/m^2 = 1 weber/m^2 = 10,000 gauss) and μ is the permeability of the medium. The permeability of a vacuum is $4\pi \times 10^{-7}$ henrys/meter (H/m). The permeabilities of most nonferromagnetic materials (those not attracted to a magnet) are almost identical to the permeability of free space. The energy stored in a magnetic field of 10 tesla (T) is about 40×10^6 J/m³.

The energy of an electric field is given by

$$E = \epsilon \cdot \frac{(\Delta V/\Delta x)^2}{2},$$ (10)

where ϵ is the permittivity of the medium and $\Delta V/\Delta x$ is the field gradient (V/m). In free space, $\epsilon = 8.8 \times 10^{-12}$ farads/meter (F/m). The energy density possible depends on ϵ and the maximum allowable field gradient. Air, which has a permittivity almost equal to that of free space, can withstand a field of about 20×10^6 V/m over short distances. Thus the maximum stored energy in an air capacitor is about 2000 J/m³.

Magnetic fields are produced by the motion of electric charges—for example, the ionized particles in a plasma or the electrons in a current-carrying wire. Electric fields occur when positive and negative charges are separated.

Electromagnetic radiation, including light, X rays, and γ radiation, is produced by the acceleration and deceleration of electric charges, usually electrons.

The most familiar form of man-made electromagnetic energy is that which we receive by power transmission lines from a generator some tens to hundreds of kilometers distant. This form of energy is associated with a varying current in the conductors of the power lines, which is approximately in-phase with the varying voltages between the lines. To transmit the maximum possible power, the current and voltage in an ac power system must increase and decrease together. The energy delivered to a load during a given period (t) is

$$E = V_{av} \cdot I_{av} \cdot t,$$ (11)

where V_{av} and I_{av} are the average voltage and current, respectively. One ampere and 1 volt delivered for one second amount to 1 joule. Electric utility companies try to provide a constant voltage, V_{av}, whereas the current, I_{av}, is controlled by the customers' demand for energy.

Energy is transmitted from the generator to the load at the speed of light, c, 300,000 km/s or 186,000 miles per second. The energy stored in the line at any given time is very small. For a 1,000 MW capacity, 300 km long transmission line, the stored

7

energy (disregarding the inductance and capacitance of the line) is

$$1000 \text{ MW} \cdot \frac{300 \text{ km}}{300,000 \text{ km/s}} = 1 \text{ MJ}. \tag{12}$$

Chemical Energy

Chemical energy is the energy associated with the changing of one or more chemical compounds to form other compounds. The burning or oxidation of fuels is the most familiar use of chemical energy. Fuels, which are composed mainly of hydrogen and carbon, combine with the oxygen in the atmosphere to form various compounds and to give off heat. Perhaps the simplest example of this is the formation of water from hydrogen (H) and oxygen (O). Both hydrogen and oxygen exist as diatomic molecules (two atoms forming a single molecule). This is shown symbolically as H_2 and O_2. Oxygen occurs freely in the atmosphere, and hydrogen can be produced by a variety of chemical processes. At elevated temperatures or in the presence of certain catalysts, these two elements combine to form steam (H_2O) through the process

$$2H_2 + O_2 \rightarrow 2H_2O + \text{Energy}. \tag{13}$$

The energy released for each 18 g of water produced is 57,700 cal; equivalently, 1.21×10^8 J is released for each kg of hydrogen consumed. At 25°C the total energy released in this process is 1.34×10^7 J per kg of water produced. This energy is called the heat of formation of water. A similar energetic relationship exists for every other chemical reaction. In some reactions, a single compound decomposes and energy is released. For example, nitroglycerin [really trinitroglycerol $C_3H_5(NO_3)_3$] releases 6.8×10^6 J/kg when it decomposes. The energy available in the formation of 1 kg of water is greater than the energy released when 1 kg of nitroglycerin decomposes, but the nitroglycerin decomposition is much faster. Very rapid decomposition is the characteristic that defines explosives.

The energy associated with chemical reactions is another manifestation of electromagnetic energy. Chemical bonds are the result of the electric fields that exist between negatively charged electrons and the positive nuclei of atoms. It is the changing of these fields during chemical reactions that leads to the release of energy or requires external energy input.

We have seen that chemical energy may be available in the

8

form of fuels such as hydrogen, alcohol, gasoline, or fuel oil or in compounds such as nitroglycerin. It may also be in the form of two or more compounds stored in a container, such as a battery. In a lead acid storage battery, the charging reaction is given by

$$2PbSO_4 + Energy + 2H_2O \rightarrow PbO_2 + Pb + 2H_2SO_4. \qquad (14)$$

During charging, lead, lead oxide, and sulfuric acid, which have higher heats of formation, are produced from lead sulfate and water. The energy stored in a battery is typically referred to as electrochemical energy because in a battery cell, the electrical energy induces chemical reactions at the plates and is thereby converted back to electrical energy. Similarly, during discharge, the chemical energy is converted back to electrical energy.

Nuclear Energy

The energy associated with nuclear reactions is perhaps the most fundamental form of energy. It is in fact the simplest in concept, but it is not well understood because it is so remote from our day-to-day experiences. This form of energy is associated with the different mass of the original and final nucleii involved in a nuclear reaction. Thus, when a uranium atom breaks up in the process called fission, the neutrons and atoms that are produced have a total mass that is less than the original mass of the uranium atom. The excess mass has gone into the energy of motion of the resulting neutrons and atoms and possibly γ rays and neutrinos, particles produced during a certain type of nuclear reaction that are difficult to detect because they interact very weakly with other forms of matter and energy. Thus:

$U^{235} \rightarrow$ fission products + neutrons + γ rays + neutrinos + energy.

The energy available in any reaction can be determined by weighing the original and final particles and then applying Einstein's mass-energy relation to the mass difference:

$$E = \Delta M c^2. \qquad (15)$$

The fission of 1 kg of uranium will yield 8×10^{13} J, which is equal to the heat from burning 2 to 3×10^6 kg of coal. Although uranium may be a very valuable material as a fuel, it is not suitable for energy storage because it cannot be created from other nucleii except

9

at the temperatures and pressures typical of the centers of some of the more massive stars.

Nuclear energy is also available from fusion, the opposite of fission. In fusion two nucleii combine to form one or more different nucleii, of which at least one is larger than the larger of the original nucleii. Thus, deuterium (D) plus tritium (T) combine to form helium (He) and a neutron (n) that subsequently decays into a hydrogen atom:

$$D^2 + T^3 \rightarrow He^4 + n \quad + \text{Energy.} \tag{16}$$
$$\llcorner\, H^1$$

This process is a source of energy in some stars. It is also being considered as a source of energy for electrical power reactors. One kg of a mixture of 2/5 deuterium and 3/5 tritium will yield 3.38×10^{14} J, which is four times as much as the energy available through the fission of 1 kg of U^{235} and is equal to the energy available from burning about 10^7 kg of coal.

Unfortunately, for the fusion reaction to occur, the initial elements, deuterium and tritium, must be raised to very high temperatures so that thermal motion will overcome the electrostatic repulsion between the two nucleii and allow them to meet and fuse. There are great hopes for fusion as a power source of the future and research is being carried out in this area, but the technical difficulties are great.

Although this form of energy is available, it is almost impossible to reverse the reaction and produce tritium and deuterium from He^4 and a neutron, or two deuterium atoms from He^4. Thus, because the conversion to other forms is difficult, nuclear energy does not appear to be suitable for energy storage.

It is useful to look for a thread that brings all of these energy forms together. It is in this perspective that the elegance of Einstein's mass-energy relationship may be understood. Through very careful weighing of the constituents of any reaction, we would find that the mass of the products is less than that of the original components when energy is produced in a reaction and that the final mass is greater than the initial mass when energy is required to make a reaction proceed. Thus, equation 15 applies not only to nuclear reactions but also to mechanical, electrical, and chemical processes. Although it would be impossible to measure with conventional techniques, the mass of a storage battery is greater in the charged state than in the discharged state. The difference for a typical automotive ignition battery would be about 10^{-9} kg out of a total mass of about 20 kg.

ENERGY STORAGE

Energy storage devices can be classified and categorized in several different ways. Each of the parts in this book and in the companion book, *Mechanical, Thermal, and Chemical Storage of Energy*, describes the storage of one particular kind of energy. Although articles are included on small-scale applications of energy ergy storage in the parts on specific technologies, one major, large-scale application for energy storage—meeting the diurnal load variations of electric power systems—is the primary application addressed in both volumes.

To understand the magnitude of this storage requirement, we can compare the stored energy in all the automotive starting, lighting, and ignition (SLI) batteries in the United States to the energy stored in one large pumped hydroelectric plant. (A pumped-hydro plant is a hydroelectric power plant that can be filled by operating electrically driven pumps. The reservoir may or may not have a natural water source in addition to the pumps.) A standard automotive SLI battery stores 1 to 2 MJ when new and typically discharges 0.1 to 10 percent during a starting operation. In round numbers, there are about 10^8 of these batteries in the United States today. If the average storage capacity is 1 MJ, then the total capacity is about 10^{14} J. In comparison, the Ludington, Michigan, pumped-hydro facility has a capacity of about 4×10^{13} J and can be discharged in six to eight hours. That one large pumped-hydro storage plant has 40 percent of the storage capacity of all the automotive SLI batteries in service gives an idea of the magnitude of the need for energy storage by the electric utility companies.

This storage requirement comes about because, on a yearly basis, the U.S. electric utility industry transmits power to its customers at a rate equivalent to only about 60 percent of its generating capacity. Aside from some downtime necessary for plant maintenance and standby equipment for emergencies, the unused 40 percent of capacity represents essentially wasted capital investment. This situation arises because the demand for power varies periodically on a daily, weekly, and seasonal basis and varies randomly during time periods of seconds to tens of minutes. Load variations are such that on a yearly basis, the minimum and peak power requirements for most electric utilities differ by a factor of two or more, and even on a daily basis the minimum load averages only 60 to 70 percent of the peak load.

The electric power industry must be prepared to meet the peak power demands, and generally this is done through a combination of three (or more) power sources: the base-load generation (typically about 45 percent of the peak power) is furnished by relatively efficient, large fossil-fueled or nuclear generators that operate around the clock at or near full throttle; the intermediate load (typically about 40 percent of peak power) is met by older, less efficient plants that are cycled in and out of service as the power demand ranges from 45 to 85 percent of the peak demands; and finally, the peak 15 percent of the power is usually delivered by gas turbines. The base-load units supply about 70 percent of the total system energy at the lowest delivered cost, while the intermediate cycling plants furnish about 25 percent of the energy at significantly higher cost. The remaining 5 percent is derived from the peaking units, which, although relatively inexpensive in terms of capital investment, require costly and scarce special fuels (such as JP-4 or number 2 fuel oil), operate at low thermal efficiencies, and necessitate high expenditures for maintenance.

Delays in deliveries of base-load plants have forced utility companies to substitute gas turbine peaking units for the generators on order. These turbines may deliver up to 15 percent of the total energy of some systems. To add to the power industry's problems, the partially successful drive to reduce overall power consumption between 1974 and 1979 often has not simultaneously reduced peak demand. The net result is that an even larger percentage of the total system power must be derived from the more expensive sources.

If we assume that it is not possible to control the variable load, then an ideal solution to this problem would be to have base-load plants provide all of the system's energy requirements by operating at full capacity (corresponding to the annual average load) and to have energy-storage systems that could absorb excess energy generated during periods of less-than-average demand and deliver it back to the system during periods of greater-than-average demand. These energy-storage systems should be highly efficient, nonpolluting, easily sited, reliable, long-lived, and easily and inexpensively maintained, and they should have a capital cost ($/kW) lower than the peak- or intermediate-load generating plants they replace. To date only pumped-hydro installations such as the one at Ludington have been used for this application.

One area where energy storage will aid the utilities is in capital investment. Partly because of the load variations, the electric utilities are the most capital intensive of all major industries. This would not be a serious problem if the total, yearly demand for electric power were not increasing, because the capital investment would then be relatively constant, with a few old or worn-out generation facilities being replaced each year. However, electric energy usage doubles every decade or so. Thus, the electric utility industry is expected to install about 400,000 MW of generating capacity at a total cost of $120 billion between 1977 and 1987. If inexpensive energy storage were available today, however, part of this investment in generating capacity could be postponed.

The potential savings associated with the utilization of energy storage systems on electric power systems are in fact much greater than just those in the area of capital investments. Let us assume that 10 percent of all generating capacity is in the form of gas turbines and that 5 percent of the total electrical energy used in the United States is generated by these units. Then about 8×10^8 barrels of number 2 fuel oil are consumed each year by these devices on utility systems. Since it should be possible for energy-storage systems to replace a large fraction of the gas turbine capacity (50 percent substitution should be possible) the potential saving is 4×10^8 barrels of oil per year. Of course some other fuel such as coal or uranium would have to be consumed to generate the electrical energy that is to be stored. However, coal and nuclear fuels are considerably more abundant than oil and natural gas.

Another potentially large-scale application of energy storage is for vehicular propulsion. It has been estimated that hydrogen, ethanol, batteries, and flywheels could replace 70 to 80 percent of the petroleum, including oil derivatives, now used for transportation. Three major applications of these storage technologies in the transportation sector are: battery-, hydrogen-, and ethanol-powered automobiles, buses, and trucks; flywheels for power augmentation and regenerative braking; and liquid hydrogen-powered aircraft. The rapid integration of storage technologies into the electric utility and transportation sectors could extend for many years the availability to humanity of the earth's very limited petroleum supplies.

The types of energy storage available are determined by the forms that energy can assume. The value of a storage technology for a particular application depends on the ease with which the stored energy may be used or converted into another useful form.

The energy in a small dry-cell flashlight battery is converted to heat and then to light quite effectively, and, though low efficiency and high cost remove these cells from consideration for electric power system applications, they have been of great value for a variety of uses. Thus the type of storage must fit the end use of the energy.

Continuing with this example of a dry-cell, primary battery, it must have a long shelf life; that is, it must retain a charge for an extended period, it must be relatively inexpensive, and it must deliver energy for a period of hours once discharge has started. Though rechargeable batteries are now widely available, this capability does not appear to be a primary consideration for most consumers because several hundred million dry-cell flashlight batteries are sold in the United States each year. For some applications, efficiency (energy withdrawn divided by energy deposited) is a primary criterion. In particular, storage units for electric utility systems must have an efficiency of about 60 percent or greater.

There are several characteristics of an energy-storage device that make it effective for a particular application. Cost ($/J) is perhaps the most important factor. However, for some applications, such as satellites and space probes, reliability and lifetime are the primary considerations, and cost is of little concern because the cost of the system is certain to be only a small fraction of the total mission cost.

Energy storage density (J/m³) and specific energy storage (J/kg) may be important considerations for other applications. For example, most car owners require a vehicle that will travel 300 km or more without refueling. A quantity of gasoline, hydrogen, or alcohol sufficient to supply the energy required to move a 2,000-kg vehicle this distance occupies a modest volume, (less than 0.1 m³) and has a modest weight (less than 100 kg). But lead acid batteries or steel flywheels for the same purpose would be considerably larger and more massive.

Another characteristic of the storage device is the area or, for large installations, land requirements. This parameter is not of great concern for a vehicle or a flashlight, but it may become extremely important if the energy-storage system is to be part of an electric utility system and is to be installed in or near a metropolitan area where real estate costs are a serious consideration.

Thus several characteristics affect the usefulness and effectiveness of a storage system. The characteristics mentioned above and a few others, with their appropriate units indicated, are:

unit cost ($/J),
efficiency (%),

energy density (J/m³),
power density (W/m³),
specific energy (J/kg),
specific power (W/kg),
charged shelf life (years or seconds depending on application),
discharge time(s),
land requirements (J/m²),
useful life (years or cycles),
reliability (mean time between failures, mean time to repair),
and
safety (probability of an injury per year).

Each of the various types of energy may be stored once a suitable storage medium has been found.

Mechanical Energy Storage

Mechanical energy may be stored as the kinetic energy of linear or rotational motion, as the potential energy in an elevated object, as the compressional or strain energy of an elastic material, or as the compressional energy in a gas. It would be very difficult to store large quantities of energy in linear motions because one would have to chase after the storage medium continually. However, it is rather simple to store rotational kinetic energy. In fact, the potter's wheel, perhaps the first form of energy storage used by man, was developed several thousand years ago and is still being used.

The stored energy in a flywheel is given by

$$E = \frac{1}{2}I\omega^2,$$
(17)

where I is the moment of inertia of the rotating mass and ω is the angular velocity. A tensile stress or load is created in the structural material of the flywheel as it spins. This stress, which increases as the angular velocity increases, limits the amount of energy that can be stored in a given mass of material. Materials with very high tensile strengths can store more energy per unit volume than can materials with low tensile strengths. It can be shown that there is a relationship between the mass M, the stored energy E, the strength σ, and the density ρ of a material under load.

$$M = \kappa \cdot \frac{E\rho}{\sigma},$$
(18)

15

where κ is a constant that is greater than or equal to 1 and is determined solely by the geometry of the system. This relationship is very fundamental and applies to all forms of energy storage in which the energy is contained by a structural element that is put in tension as the energy is stored.

For flywheels that are not tapered near the rim, κ is greater than two ($\kappa \geq 2$) because the loads are almost always in the tangential direction. This type of stress is called a hoop stress. From equation 18, it is straightforward to get a rough estimate of the cost of a unit from the cost factors ($/kg) used by industries that fabricate various devices. For example, a fabricated cost of $3/kg may be used for steel with the relatively high tensile strength of 30,000 pounds per square inch (psi). In SI units tensile strengths and pressures are given in pascals (Pa). One psi is equivalent to 6,900 pascals; thus 30,000 psi is 207,000,000 Pa, or about 2×10^8 Pa. If the flywheel container and the motor drive and other parts double the cost, then a rough estimate of the storage cost is 2.2×10^3 J/$ or 4.6×10^{-4} $/J, which is independent of the storage capacity of the device.

An estimate of the energy stored per unit volume, E/V (J/m^3) can be determined by substituting the value of E/M from equation 18:

$$\frac{E}{V} = \frac{E\rho}{M} = \frac{\sigma}{\kappa}. \tag{18a}$$

The energy-storage density in a steel flywheel with $\kappa = 2$ and a 2×10^8 Pa (30,000 psi) working strength is 10^8 J/m^3, and about 5×10^7 J/m^3 for a complete system.

Potential energy storage is most often utilized in the form of pumped, hydroelectric storage. Energy generated in an electric power grid during periods of low energy demand is converted into potential energy by pumping water from a lower reservoir into a higher one. The storage density depends on the elevation difference of the two reservoirs and is about 10^4 J/m^3 for each meter of head. For a 100 m height difference for the two reservoirs, the energy density is 10^6 J/m^3 and the specific energy is 10^3 J/kg.

Although the energy density and specific energy of pumped hydro is relatively small, the cost is low, which makes it attractive to electric utilities. For example, the Ludington, Michigan, pumped-hydro facility, which was constructed between 1968 and 1974, cost $350 million for about 14,000 MWh of storage and 2,000 MW of power capacity. This facility occupies about 2.5 square miles (640 hectares) of land. The storage cost was 1.4×10^5 J/$ or $6.9 \times$

10^{-6} \$/J, about one-seventieth the cost of the flywheels described above.

The storage of energy directly as strain in a material such as a spring has been used for many years in specific applications such as watches and clocks. Because of the small size of watch springs, expensive materials with working stresses of 300,000 psi (2×10^9 Pa) are used. The energy density and specific energy are quite small, however. For example, at the same stress level, the energy density in a spring is smaller than the energy density in a flywheel by a factor equal to the strain , ϵ, which is usually 0.01 or less.

An advantage of storing "strain energy" is that it has a long shelf life. Springs have been used for years in mine shafts on protective devices. When a cable fails the springs release and snap a stopping mechanism into a position to limit the descent of the cable car.

Energy storage in the form of compressed gas has been used since the late 1800s for special applications such as mining. Because massive containment vessels are required for compressed gases, it is not practical to store large quantities of energy in this form unless the vessel is very cheap, such as an underground cavern. The structural material is then the rock surrounding the cavern. This type of energy storage appears to be practical for the electric utility diurnal storage requirement in some locations.

Thermal Energy Storage

Thermal energy may be stored by elevating the temperature of a substance (increasing its sensible heat), by changing the phase of a substance (increasing its latent heat), or by a combination of the two. Both forms will see extended applications as new energy technologies are developed. Energy stored in sensible heat appears very promising for high-temperature storage of large quantities of energy at fossil-fired power plants. Oil will likely be used as the medium for this type of storage.

Many new homes are being equipped with solar heating and cooling systems that require diurnal energy storage. Materials being considered for this application include water, two- to four-inch diameter rocks, and phase-change materials such as Glauber's salt ($Na_2SO_4 \cdot 10H_2O$). These materials are all readily available, and water and rocks are obviously quite cheap, though the containment system may not be.

For a typical house in the high mountains and plateaus of the southwestern United States, where there is adequate sunshine in winter but the nighttime temperatures are quite low, the energy storage requirements are about 15 Btu/°F/ft$_h$² (in SI units 2.9 × 10⁵J/°C/m$_h$²) where ft$_h$² (m$_h$²) is the heated floor space and the °F (°C) is the average temperature difference inside versus outside. For a house with an area of 2000 ft$_h$ (~200 m$_h$²) the volume of water or rocks required to provide most of the heating for one day, with the outside temperature at 0°F will be 13 m³ and 42 m³, respectively. The cost of the water would be trivial, and the required 70 tons of rocks would cost only $700 to $3,000 depending on availability, giving an energy storage cost for the rocks alone of about 1.2 × 10⁶ J/$ or 8.0 × 10⁻⁷ $/J. Unfortunately the cost of storage tanks, heat exchangers, and the space inside the building is considerable. These must be weighed against the ever-increasing cost of other heating methods such as natural gas, fuel oil, and electricity.

The cost of the oil for thermal storage at central power plants is based on the present cost of fuel storage vessels and heat exchangers. The storage density, which depends on the temperature swing, nominally between 530°F and 250°F (230°C and 121°C), is 4 × 10⁸ J/m³. At present fuel oil costs, this amounts to 2 × 10⁻⁶ $/J. Other costs, however, raise this estimate considerably. It has been calculated that additional equipment and oil treatment will bring this cost up to 1.2 × 10⁵ J/$ or 8.4 × 10⁻⁶ $/J for a 2.6-GWh thermal storage plant on a nuclear boiling water reactor.

Thermal storage has also been considered for vehicle propulsion, but the cost and weight of the storage medium required for travel over distances greater than about 100 km make it an unlikely candidate compared to batteries and hydrogen.

Electrical or Electromagnetic Energy Storage

Energy may be stored in the magnetic field of a coil or in the electric field of a capacitor. The energy storage density in a capacitor is given by

$$E/V = \frac{\epsilon(\Delta V/\Delta x)^2}{2}$$

and is about 60 J/m³ for air. Many materials have higher dielectric constants and can also withstand higher electric fields than air. Some of these, such as mica, glass, and oil-impregnated paper, have been used for capacitors for fifty to seventy years; others, such as Mylar, Teflon, titanium dioxide, and various titanates have been used only recently. The energy storage densities currently possible in small capacitors using barium titanate are about 5×10^5 J/m³, whereas the theoretical limit is 10^7 J/m³.

The most recent studies of large capacitive energy-storage systems have shown the oil-impregnated paper is the cheapest, most reliable dielectric. Costs of 10^{-2} $/J have been achieved on large, 10 to 20 kV, 0 to 10 MJ systems. The single advantage of capacitive storage has been its rapid discharge capacity or equivalently high-power density. Most capacitors can be almost completely discharged in 1 to 10 microseconds. The energy density of capacitors is too low and the cost too high for most energy-storage applications.

Energy storage in magnetic fields was not economical until the recent, practical demonstrations of superconductivity. The development of materials having no measurable resistance to the flow of electric currents has increased the charged shelf life of practical magnets from a few seconds, which was the maximum possible ten to fifteen years ago, to days or weeks. Some small superconducting coils have been charged and have remained in the charged state for years without a detectable change in current or field.

Superconducting magnets capable of storing up to 9×10^8 J have been constructed for special nuclear physics applications. Some small magnets have been used for pulsed energy storage, but to date no large magnets have been constructed specifically to store energy for long periods.

The energy-storage density in a magnetic field of five tesla, which is easily achieved with many commercially available superconductors, is 10^7 J/m³. In a very large magnet, such as those proposed for the diurnal energy-storage application, the superconducting coil occupies only a few percent of the field volume. As a result, the energy density relative to the superconductor volume may be as high as 10^9 J/m³.

The cost of large superconducting magnetic energy-storage (SMES) units proposed for diurnal load leveling will not be proportional to size. The unit cost ($/J) of larger coils will be less than the unit cost of smaller coils. It appears that with present technology SMES units storing more than 10^3 MWh will be competitive with other storage devices on electric utility networks. A SMES unit having the capacity of the Ludington pumped-hydro plant would require less than one square mile of land, one-third as much as Ludington; would have about the same initial cost; would have an efficiency of 95 percent rather than 70 percent and could be more easily sited. Smaller storage units, for example, for automotive propulsion, will not be economical with present technology. The advantages of this type of storage are efficiency, cost, fast response time, and, depending on environmental restrictions, decreased land requirements.

Most magnets, large inductors, and transformers use iron. Iron, however, is a very poor storage medium for magnetic energy. Its advantage is that a moderately high (up to about 2 T) magnetic field can be obtained with a relatively small driving force, usually the induction produced by a small current. Unfortunately this effect, which may provide a field multiplication of 1000 or more, saturates as the internal field approaches 2T. This enhancement then decreases as the external field increases, and, at 2T, the external and internal fields are approximately equal. At a field of 5 T iron would increase the stored energy in a given volume by 10 percent at most, or about 1 MJ/m^3. Storage of energy in the magnetic field of iron costing \$3/kg would be about 2×10^{-2} \$/J, 10^4 times more expensive than pumped hydro and other economical forms of energy storage.

The third form of electromagnetic energy, radiant energy, which includes light, radiowaves, and X rays is very difficult to store. It is possible, however, to store microwave energy for a short time in a cavity with superconducting walls. At present the energy density is a fraction of a percent of that possible with other techniques, the storage efficiency is only a few percent, and the equipment required to provide the appropriate environment is costly and complicated.

Chemical Energy Storage

Energy may be stored in systems composed of one or more chemical compounds that release or absorb energy when they react to form other compounds. The most familiar chemical energy-storage device is the battery. Energy stored in batteries is frequently referred to as electrochemical energy because chemical reactions in the battery are caused by electrical energy and subsequently produce electrical energy.

It is useful to speak of storage efficiency only for secondary or rechargeable batteries. Primary batteries, such as the dry-cell flashlight battery, are fabricated in a charged state and are discharged only once. The energy-storage efficiency of batteries depends on the battery design and type. The batteries for vehicle propulsion and utility energy storage are expected to have an efficiency of 70 or 80 percent for deep discharge.

Another important form of chemical energy storage is possible by using hydrogen. Hydrogen does not exist as a free element on the earth. It is, however, extremely abundant as a constituent of water, which may be decomposed by electrical or thermal processes to yield hydrogen and oxygen. In the past electrolysis has been used, with a relatively low efficiency, to produce hydrogen for special applications. If hydrogen is to be used for energy storage, then the total efficiency for the combined processes of hydrogen production, storage, and use must be increased to 60 or 70 percent. Assuming this is possible, there are several methods of storing hydrogen: as a liquid at a temperature of 20 K, as a compressed gas in cylinders or underground, or as a compound that is easily formed and decomposed, such as a metal hydride. Each of these has advantages and disadvantages. If hydrogen is to be used as a fuel for vehicles, then the specific energy should be as great as possible, and liquid hydrogen in a dewar is probably the best solution. A dewar is a device that uses a vacuum to insulate thermally cryogenic systems at temperatures below 77 K. The same principle is used in the vacuum or thermos bottle. The energy density is about 8.6×10^9 J/m^3 as compared to 3.6×10^{10} J/m^3 for gasoline. But the specific energy is 1.21×10^8 J/kg, which, even when the weight of a dewar is included, is greater than that of gasoline, 4.8×10^7 J/kg. If maximum efficiency is desired and weight is not a consideration, then a metal hydride storage system may be used. This type of storage has been proposed for electric utility diurnal storage. The energy-storage density of hydrogen in the hydride is approximately equal to the storage density of the hydrogen as a liquid. The storage of hydrogen as a compressed gas is not promising because the affinity of hydrogen for metals, which makes metallic hydride storage so effective, leads to hydrogen embrittlement in steel tanks containing high-pressure hydrogen.

An advantage of hydrogen over some other energy-storage methods is that it is easily transportable and can be used as a fuel some distance away from its production site. For example, part of the hydrogen produced by the excess capacity available at night could be stored in metal hydrides and used in fuel cells for power generation during the peak-load periods, part could be liquefied and used for automotive or airplane propulsion, and part could be used to augment natural gas for residential or commercial use.

Another form of chemical energy storage is ethanol, which

Table 1 A Comparison of Energy-Storage Technologies

Energy Storage Technology	Cost[a] ($/J)	Efficiency (%)	Energy Storage Density (J/m³)	Max Power[b] Density (W/m³)	Specific Energy (J/kg)	Max Specific Power[b] (W/kg)	Charged Shelf Life	Typical[c] Discharge Time (s)	Useful[d] Life (cycles or years)	Applications[e]
Mechanical energy										
Flywheels	$4 \cdot 10^{-4}$	70–90	$5 \cdot 10^7$	$5 \cdot 10^6$	$6 \cdot 10^3$	$6 \cdot 10^2$	min–days	$1 - 10^3$	20 yrs	RB/SA
Strain energy	$4 \cdot 10^{-2}$	50–90	10^6	10^5	$2 \cdot 10^2$	$2 \cdot 10^1$	years	$10 - 10^6$	20 yrs	SA
Compressed gas[f]	10^{-5}	60–70	10^7	10^3	—	—	weeks	$10^4 - 10^5$	20 yrs	EU
Pumped hydro storage[g]	$7 \cdot 10^{-6}$	70	10^6	10^2	10^3	10^{-1}	weeks	$10^4 - 10^5$	20 yrs	EU
Thermal energy										
Sensible heat										
High temp oil	10^{-5}	70–80	10^8	10^4	10^5	10	days	$10^4 - 10^5$	20 yrs	EU
Low temp-rocks	$8 \cdot 10^{-6}$	90	$3 \cdot 10^7$	$3 \cdot 10^3$	10^4	1	days	$10^4 - 10^5$	20 yrs	RH
Latent heat										
Various salts	10^{-5}	80	$3 \cdot 10^8$	$3 \cdot 10^4$	10^5	10	weeks	$10^4 - 10^5$	$10^2 - 10^3$ cycles	RH
Electromagnetic Energy										
Capacitors[h]	10^{-2}	90	10^5	10^{11}	$3 \cdot 10^2$	$3 \cdot 10^8$	days	$10^{-6} - 10^{-3}$	10^4 cycles	SA
Magnets/coils[i]	10^{-5}	90	10^8	10^4	10^6	10^2	days	$10^{-3} - 10^5$	20 yrs	EU/SA
Chemical Energy										
Batteries[j]										
Pb acid	$2 \cdot 10^{-5}$	60–80	$5 \cdot 10^7$	$5 \cdot 10^3$	10^5	10	weeks	$10 - 10^4$	10^3 cycles	ES/SLI/AP
Other[k]	10^{-5}	65–80	10^4	10^4	$5 \cdot 10^5$	—	weeks	$10 - 10^5$	$10^3 - 10^5$ cycles	EU/SA/AP
Hydrogen[l]	$2 \cdot 10^{-5}$	30–70	$9 \cdot 10^9$	—	10^8	—	days/mos	$10^4 - 10^5$	20 yrs	EU/SA/AP
Ethanol		60–80	$2 \cdot 10^{10}$	—	$3 \cdot 10^7$	—	days/mos	$10^4 - 10^5$	20 yrs	EU/SA/AP

Table notes:

[a] The cost of energy storage is given for the entire system, including equipment required to convert the energy from and/or into a useable form. Actual costs may vary by a factor of two or three from the value given, which is a mean. The cost in $/kWh is obtained by multiplying by 3600.

[b] Specific power and power density are maxima based on minimum possible discharge times, except for those technologies where the device is for the electric utility diurnal storage application.

[c] Minimum discharge times are usually for partial discharges. For specific technologies, smaller units can be discharged faster than larger units.

[d] For economic considerations, a lifetime of twenty years is equivalent to an infinite life. There is no indication that the various devices given a twenty year life expectancy in the table will wear out in twenty years.

[e] Abbreviations are given in the table for various applications:
EU, electric utility diurnal storage;
ES, emergency service—for electric power system failure, and so on;
SLI, automotive–starting, lighting, and ignition;
RB, regenerative braking;
RH, residential heating;
SA, special applications; and
AP, automotive propulsion.

[f] Data for compressed gas are based on the plant installed at Huntorf, West Germany.

[g] Data for pumped-hydro storage are based on the facility at Ludington, Michigan.

[h] Data are for oil-impregnated-paper capacitive energy-storage systems.

[i] Most data are for a 10^4 MWh superconducting coil constructed underground and used for diurnal storage on an electric power system. Energy density is based on excavated volume.

[j] Most data are based on diurnal energy-storage application. Automotive propulsion batteries will have higher power rating.

[k] Several advanced battery systems have been proposed for automotive propulsion and electric utility applications.

[l] Storage efficiency depends on type of hydrogen storage, and storage cost depends on the medium used. Liquid hydrogen storage is less efficient but cheaper than metal hydride storage. Power and storage capacities are almost unrelated for hydrogen storage.

may be produced through biological processes or from coal or other hydrocarbons. The energy density and specific energy for ethanol are 2.2×10^{10} J/m^3 and 2.7×10^7 J/kg. Ethanol and a related alcohol, methanol, may be substituted for gasoline in internal combustion engines. Only partial substitution may be possible for existing engines, and complete substitution may require engine modifications or new engines. Methane can be produced by similar chemical processes. It can be transported with ease, can be added to natural gas supplies, and can be used in a chemical heat pipe, which makes use of the reversible reaction, $CH_4 + H_2O \leftrightarrow CO + 3H_2$. This reaction provides a convenient method of transporting low-grade heat but is not an effective means of storing energy.

Any reversible chemical reaction in which all of the reactants are solids or liquids may be considered for storing energy. In general the driving force for the reaction will be heat or electrical energy and, when the reaction is reversed, the output will again be heat or electrical energy.

Nuclear or Mass Energy Storage

Because it is difficult to produce specific nuclear species or isotopes in a controlled way and then to decompose them and produce other forms of energy, it is not practical to use nuclear energy as a form of energy storage.

A Comparison of Energy-Storage Technologies

The various types of energy storage are compared in table 1 on the basis of the characteristics that affect the usefulness and effectiveness of a storage system. An attempt has been made to make this table as complete as possible at this time. Two characteristics, reliability and safety, have been left out because they are not true technical characteristics of the storage technologies but instead depend on manufacturing technique and quality control. All of the data in the table are taken from the articles herein and in *Mechanical, Thermal, and Chemical Storage of Energy* or from the articles listed in the bibliographies at the end of each of the Editor's Comments.

BIBLIOGRAPHY

Brown, H. L., Director. 1972. *Effective Energy Utilization Symposium.* Drexel University, Philadelphia.

Chalmers, B. 1963. *Energy.* Academic Press, New York.

Ford, K. W., G. I. Rocklin, R. H. Socolow, D. L. Hartley, D. H. Hardesty, M. Lapp, J. Dooker, F. Bryer, S. M. Berman, and S. D. Silverstein, eds. 1975. *Efficient Use of Energy.* American Institute of Physics, New York.

Garvey, G. 1972. *Energy, Ecology, Economy.* W. W. Norton and Co., New York.

Guyol, N. B. 1969. *The World Electric Power Industry.* University of California Press, Berkeley, Calif.

Hutchinson, F. W. 1957. *Thermodynamics of Heat-Power Systems.* Addison-Wesley Publ. Co., Reading, Mass.

Krenz, J. H. 1976. *Energy Conservation and Utilization.* Allyn and Bacon, Boston.

McMullan, J. T., R. Morgan, and R. B. Murray. 1976. *Energy Resources and Supply.* Wiley, London.

U.S. Dept. Defense and Calif. Inst. Tech. 1958. *Seminar on Advanced Energy Sources and Conversion Techniques.* Electro-Optical Systems, Pasadena, Calif.

Walsh, E. M. 1967. *Energy Conservation: Electrochemical, Direct, Nuclear.* The Ronald Press. New York.

Part I

ELECTROCHEMICAL ENERGY STORAGE

Editor's Comments
on Papers 1 Through 17

Batteries are the most familiar and most common of all energy-storage devices. They are available in a variety of sizes and capacities and are used in almost every household in the industrialized countries of the world.

The earliest description of a battery is in a report to the Royal Society of London in 1800 by Volta. He described the electricity produced by a stack of silver and zinc plates separated by pasteboard soaked in water and other fluids.

In 1859 the lead-acid battery was first developed by Planté (1860, 1883) in France. His cells consisted of two sheets of lead separated by strips of rubber, rolled into a spiral and immersed in a dilute sulfuric acid solution. At that time the only source of electric currents was other batteries so the utility of this type of battery, even though it had a high current capacity, was not apparent.

Good descriptions of the history of battery development and battery technology circa 1950 are given in the books *Storage Batteries* and *Primary Batteries* by G. W. Vinal (1951, 1955). For many years batteries were mere laboratory curiosities and then, for a while, they were used only as scientific tools for research in other areas. It was with the advent of widespread, large-scale applications of electricity, beginning with the telegraph and then the telephone, electric lights, and electric-powered vehicles, that batteries finally found practical application.

A battery consists of one or more cells that store energy in two or more chemicals. On demand, a battery can be discharged to produce electricity. There are two types of batteries.

A **primary battery** is fabricated by bringing together separately formed chemical components and then assembling the battery in a charged state. During discharge, these components are irreversibly changed as chemical energy is converted to electrical energy. In general, primary cells cannot be recharged. Typical examples of a primary battery are the dry cells and alkaline cells used in flashlights, transistor radios, and similar devices.

A **storage** or **secondary battery** consists of several chemical and elemental materials that are reversibly changed during charge and discharge. A storage battery may be assembled in either the charged or discharged state. The most frequently found battery of this type is the lead-(sulfuric) acid storage battery.

The U.S. retail market for batteries is about $1 billion per year. Of this market about half is for automotive applications and nearly 80 percent is storage batteries.

The most familiar battery application is starting, lighting, and ignition for automobiles and the small batteries used to energize flashlights and small household appliances. Another major use of lead-acid storage batteries is for traction in lift trucks, small carts, and other industrial applications. Any list of battery applications would be very long. It would include hearing aids; electric toys; watches; clocks; power for missiles and satellites; propulsion for submarines and electric vehicles; emergency lighting; and cardiac pacemakers. Some of these applications have very stringent requirements in terms of specific energy, energy density, or reliability, whereas for other uses, cost may be the major factor. As with almost every other technology there is a trade-off between cost on the one hand and various special requirements such as reliability or energy density on the other hand.

In some applications batteries must be cycled many times during their life, and in others they need be used only once but must be available with extremely high reliability. The depth of discharge of the battery during each cycle may be 1 or 99 percent of the total stored energy, and it may take one second or thousands of hours. Because of these differences in performance requirements and in the types of materials available for batteries, the variations in the different types of batteries alone are almost as extensive as the variations in all the other forms of energy storage together.

The number of chemical reactions possible for batteries is nearly infinite. It is only when the requirements of inexpensive and safe materials, reasonable temperatures, ease of fabrication, and reliability, among others, are imposed that many reactions are eliminated. Even then there are several hundred possible combinations of materials that have been used in batteries or are being considered for future battery systems. Let us examine a few of these possible reactants and the approach that has been used to select the most effective materials.

A cell generally consists of two conducting electrodes, one being positive relative to the other, and an electrolyte. One electrode material must be an electron donor and the other an electron receiver. The choice of materials for the electrodes and the electrolyte is based on the potential differences associated with the reactions that occur at the two electrodes. The procedure in battery development is to explore the various types of electron donors and receivers, to select components that will meet, at least theoretically, the technical requirements, and to develop economical batteries from them.

The most effective electron donors are the alkali metals, such as lithium and sodium, and the next most effective are the alkaline-earth metals, including beryllium and magnesium. The most effective electron acceptors are fluorine, chlorine, oxygen, and sulfur.

Table 1 Chemical Reactions, Potentials, and Specific Energies of Several Batteries

Battery	Chemical Reaction	Nominal Cell Voltage (V)	Commercial Energy Density (Wh/lb)
Lead acid	$Pb + PbO_2 + 2H_2SO_4$ $\rightleftarrows 2PbSO_4 + 2H_2O$	2.2	7–15
Leclanché (dry cell)	$2MnO_2 + 2NH_4Cl + Zn$ $\rightarrow ZnCl_2 \cdot 2NH_3 + H_2O + Mn_2O_3$	1.58[a]	5–40
Nickel-cadmium	$Cd + 2NiOOH + 2H_2O$ $\rightleftarrows Cd(OH)_2 + 2Ni(OH)_2$	1.25	10–20
Edison (nickel-iron)	$Fe + 2NiOOH + 2H_2O$ $\rightleftarrows Fe(OH)_2 + 2Ni(OH)_2$	1.2	8–14
Lithium-iron sulfide[b]	$4Li + 3FeS_2$[c] $\rightleftarrows Li_4Fe_2S_5 + FeS$	2.1	50–60
Sodium-sulfur[b]	$2Na + 3S \longrightarrow Na_2S_3$[d]	1.8	50–60

[a] Cell voltage depends on pH of electrolyte.
[b] These two developmental batteries will operate at 400° to 500°C.
[c] One of several possible reactions.
[d] Several intermediate reactions occur.

Several materials and reactions used in batteries are given in table 1. Both primary and storage batteries are included. This table is not complete but is representative of a few of the batteries in use today or being considered for future applications.

Because chemical reactions involve only the outermost electron shell, on the basis of weight, two obvious choices for battery constituents would be lithium and fluorine. Unfortunately these materials are extremely reactive, and fluorine is not at all suitable for battery use. Lithium does appear promising for several future applications. On the other hand, because of their high density, common battery constituents such as lead, nickel, and mercury are not the most logical choices for battery components. The main reasons cells made of these materials are widely used are that many battery applications are not extremely demanding in terms of energy density or specific energy, and the lead-acid battery and the carbon-zinc dry cell are simple and easily fabricated.

This is not to say that the lead-acid battery is thoroughly understood. In fact, although the net reactants and the relative potentials are known (see table 1), in private the experts still debate the sequence of the detailed chemical reactions at the electrodes. Because of the incomplete understanding of battery chemistry, the development of batteries has been less a result of research than of engineering application.

The characteristics of one type of heavy-duty, lead-acid battery available today are given in table 2 and the construction is shown in figure 1 (pages **34** and **35**).

Table 2 Sizing To 1.55V Final Voltage, kW per Cell for Indicated Time

Telcell[a] Number		5 Minutes	15 Minutes	30 Minutes
UPX 11	7½"	2.638	1.782	1.251
15	7½"	3.368	2.417	1.752
19	7½"	3.911	3.033	2.252
UPX 23	10½"	4.908	3.756	2.625
27	10½"	5.448	4.187	3.026
31	10½"	5.864	4.540	3.405

[a] Registered trademark of Globe Union Inc.

The battery most used today is the lead-acid battery. Paper 1 gives a brief description of lead-acid battery development at the beginning of the twentieth century and describes some of the early applications of "accumulators," as they were then called.

Around the turn of the century (1895–1910), the great interest in electric-powered vehicles stimulated a rapid development of lead-acid and a host of other storage batteries. The status of lead-acid batteries for electric automobiles is described in Papers 2A and B. That the development of batteries was very rapid during this period may be seen in Paper 3. This paper was presented a scant six years after Thomas Edison began experimenting with nickel/iron batteries, and yet it describes the construction of several hundred cells that were in operation in electric vehicles at the Edison factory. Paper 4 focuses on the present status of the continuing development of the lead-acid battery.

Primary batteries for applications such as flashlights were also being developed during this period. However, it is interesting to look at some later articles that examine the improvements in technology by the mid-1930s when batteries were used extensively in radios and when there were scientific investigations of the detailed reactions within various cells. Paper 5 presents data showing a threefold improvement in the capacity of Leclanché or dry cells from 1910 to 1934. Paper 6, which was presented at the same meeting as Paper 5, looks at some of the controversy at that time regarding the nature of the reactions that occur within the dry cell and the effects of the different sources of materials on the voltages produced.

Paper 7 reports on the early developments of the cadmium nickel battery, mainly in Europe, and discusses potential applications for automotive starting, lighting and ignition. Because of its light weight, this type of battery is extensively used today for starting aircraft engines, but it has not penetrated the automotive market because of its high cost. The materials, processes, designs, performance characteristics, and applications of a variety of state-of-the-art alkaline storage batteries are covered briefly in Paper 8. Paper 9 describes some simple procedures for improving the performance of cells that have gone through many cycles of partial discharge. Paper 10 is a collection of engineering data from a battery manufacturer on a variety of primary battery types now available. The data are included because generally they are not published in technical papers today but are essential to understanding the present battery technology.

One of the major applications of batteries in the future will be for electric vehicles. Some of the batteries being evaluated and

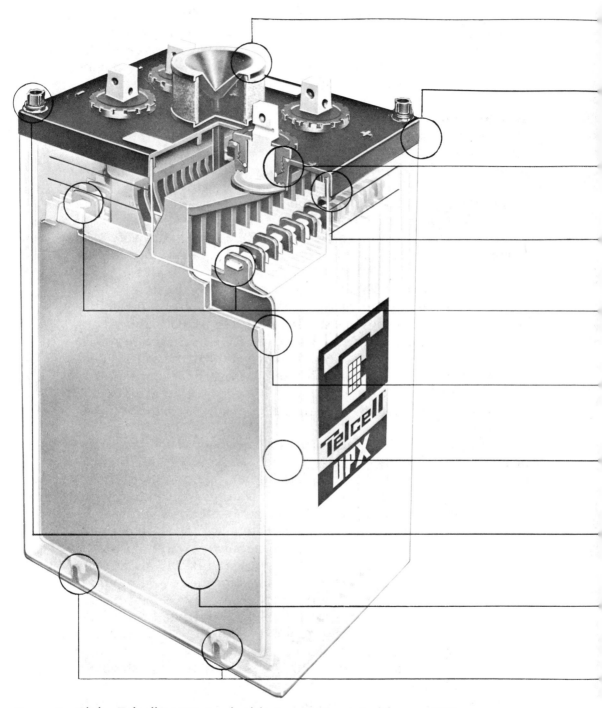

Figure 1 Globe Telcell® UPX standard features. (*Reprinted from a 1975 engineering brochure; copyright © 1975 Globe Battery Division of Globe Union, Inc.*)

1 Spark Arresting Ceramic Vent & Filling Funnel permitting routine maintenance while in operation.

2 Case & Cover of high impact Thermoplastic 100% annealed to relieve subsequent stress & strain of normal service . . . Polycarbonate available.

3 Dynamic Post/Cover seal is a triple seal providing for typical post movement while maintaining highest degree of seal integrity.

4 Tongue & Groove case-to-cover seal provides a triple sealing surface.

5 Positive Plates suspended with two top supports providing for unrestrained, normal growth patterns in both vertical & horizontal planes while evenly distributing the load on the negative plates.

6 Integrated glass fiber reinforces and strengthens the positive active material for increased durability in cycling and higher performance throughout life.

7 PVC separator for reduced resistance and increased plate protection.

8 Electrolyte withdrawal/maintenance tubes for ease of accessibility.

9 Computer Assisted Grid Design with central termination for reduced resistance, maximum efficiency and strength.

10 Negative plates with positive locating feet provide maximum element support & alignment integrity.

developed for this purpose are described in Paper 11. It is interesting to compare the battery characteristics in table II of this paper to the batteries available around 1900 (see Papers 2A and 2B) and to note how little improvement has been made in three-quarters of a century. Details of a new type of electric-vehicle battery are given in Paper 12. The advantage of this type of battery is the high specific energy and power that result from using the oxygen in air much as conventional internal combustion engines. A comparison of the many existing and proposed batteries is given in figure 3 of this paper.

One of the promising areas of battery research involves the use of high temperatures and molten salt electrolytes. The present status of high-temperature batteries is given in Paper 13. This type of battery is not a new idea. Investigation into the effectiveness of molten electrolytes was started late in the nineteenth century. The results of one early study are discussed in Paper 14. Many of the problems with materials in caustic electrolytes at high temperatures and the difficulties of separating the chemical potentials from thermoelectric effects are brought out in this paper.

There have been several biological applications of batteries. Perhaps the most demanding is the use of batteries in implanted cardiac pacemakers. A battery for this purpose is the subject of Paper 15.

As the exploration and utilization of the oceans continue, there will be many uses for batteries under water. Several attempts have been made to use seawater as an electrolyte in a battery. Paper 16 addresses the development and characteristics of a low-power, long-life, undersea battery.

Paper 17 reviews a very special class of battery that has been used extensively for artillery, ordnance, and nuclear weapons. This type of battery has a charged shelf life of several years but can deliver energy only for a few minutes once it has been activated.

REFERENCES

Planté, G., 1860, Nouvelle pile secondaire d'une grand puissance, *C. R.* **50**:640.

Planté, G., 1883, *Recherches sur l'electricite*, Gauthier-Villars, Paris, p. 29.

Vinal, G. W., 1951. *Primary Batteries*, Wiley, New York.

Vinal, G. W., 1955, *Storage Batteries*, 4th ed, Wiley, New York.

Volta, A., 1800, On the Electricity Excited by the Mere Contact of Conducting Substances of Different Kinds (in French), *Roy. Soc. (London) Philos. Trans.* **90**:403.

BIBLIOGRAPHY

Beccu, K. D. 1974. *Storage System for Secondary Energy.* Sesssion of the Association of German Engineers Held at the Technical University of Stuttgart.

Brown, J. T., and J. H. Cronin. 1974. Battery Systems for Peaking Power Generation. *Intersoc. Energy Convers. Eng. Conf., 9th, Proc.,* pp. 903–911.

Cairns, E. J., and R. K. Steunenberg, 1973. High-Temperature Batteries. In *Progress in High Temperature Physics and Chemistry,* vol. 5, ed. C. A. Rouse. Pergamon, New York, pp. 63–125.

Chilenskas, A. A., J. E. Battles, P. A. Nelson, and R. O. Irvins. 1974. Lithium/Sulfur Batteries for Marine Application. *Intersoc. Energy Convers. Eng. Conf., 9th, Proc.,* pp. 654–664.

Cohn, E. M. 1967. *Outline for Electrochemical Space Power Sources.* An AGARD/NATO lecture in the series XXVII Energy Sources for Space Power. NTIS No. N68-14818.

El-Badry, Y. Z., and J. Zemkoski. 1974. The Potential for Rechargeable Storage Batteries in Electric Power Systems. *Intersoc. Energy Convers. Eng. Conf., 9th, Proc.,* pp. 896–904.

Falk, S. U., and A. J. Salkind. 1969. *Alkaline Storage Batteries.* Wiley, New York.

Garrett, A. B. 1957. *Batteries of Today.* Research Press, Dayton, Ohio.

Hanneman, R. E., H. Vakil, and R. H. Wentdorf, Jr. 1974. Closed Loop Chemical Systems for Energy Transmission, Conversion, and Storage. *Intersoc. Energy Convers. Eng. Conf., 9th, Proc.,* pp. 435–441.

Jasinski, R. 1967. *High Energy Batteries.* Plenum Press, New York.

Lyndon, L. 1911. *Storage Battery Engineering.* McGraw-Hill, New York.

Mantel, C. L. 1971. *Battery and Energy Systems.* McGraw-Hill, New York.

Nelson, P. A. 1974. The Lithium/Sulfur Battery. Paper presented at the Great Lakes Power Club Winter Meeting.

Park, B. 1893. *The Voltaic Cell.* Wiley, New York.

Proceedings of the Symposium on Batteries for Traction and Propulsion, Columbus, Ohio. 1972.

Proceedings of the Symposium and Workshop on Advanced Battery Research and Design, Argonne National Laboratory. 1976.

Proceedings of the Tenth Annual Battery Research and Development Conference, Power Sources Division, Signal Corps Engineering Laboratory, Fort Monmouth, N. J. 1956.

Specifications for Dry Cells and Batteries. American National Standard, New York, 1972.

Thaller, L. H. 1974 Electrically Rechargeable Redox Flow Cell. *Intersoc. Energy Convers. Eng. Conf., 9th, Proc.,* pp. 924–931.

Treadwell, A. 1906. *The Storage Battery.* Macmillan, New York.

[*Editor's Note:* The following journals and proceedings have detailed technical information on current battery development: *Journal of the Electrochemical Society, Journal of Chemical Physics,* and *Proceedings of the Intersociety Energy Conversion Engineering Conference* (IECEC), held annually, published by the American Society of Mechanical Engineers.]

1

Reprinted from *Am. Electrochem. Soc. Trans.* 3:159–168 (1903)

AN HISTORICAL REVIEW OF THE STORAGE BATTERY.

BY H. B. COHO.

The vast strides that have been made in the use of the storage battery and the many advantages claimed by the various patentees as to the durability and efficiency of their cells naturally leads to the question as to what improvement, if any, is to be found in the storage cells now employed over those of former days.

There has been a doubt in the minds of many engineers as to the desirability of a secondary receiver, but up to within a very few years it was indeed a very courageous engineer who dared to recommend to his clients the use of an accumulator.

The field of research along the lines of storage battery development is of particular interest to the electrochemist, as both the chemical and electrical sciences enter largely into his calculations, and there is probably no field wherein the reward would be greater could certain elements or combinations of elements be found that would be capable of doubling the output of the present lead cell without increasing the weight or bulk of the battery as a whole.

The consulting engineer is called upon more frequently perhaps to pass upon new batteries than upon almost any other subject, and as it is important that any meritorious discovery be carefully investigated, great care has to be given to this subject.

As a matter of fact, one of the easiest propositions in the storage battery field is to build a cell of light weight and to get a great output for a short time, but when that same battery is called upon to give its output under the varying conditions met in an automobile, for instance, the result is generally disastrous to the light lead cell.

The amount of money which has been lost in the endeavor to force upon the market secondary batteries having little or no intrinsic value, to say nothing of the heartaches to honest inventors from unfulfilled dreams, makes the history of the storage battery's development almost tragic, and now that the Brush patent has expired there will probably be a new growth to be cut down by the ruthless test of durability, and this, therefore, seems to be a fitting time to review briefly what has been done and to give short descriptions of some of the more important experiments covering the discovery of secondary currents, as well as to describe a few of the more important types of cell.

The first attempt in the storage of electricity was that of Musschenbroek in 1745. He only succeeded in storing it in a bottle for ten, twenty, thirty or more seconds, and his discovery was of no practical value. It, however, led to his discovery of the principle upon which his invention of the Leyden jar is based.

In 1783 Volta invented his condenser, based upon the same principle, while in 1800 he invented the "Crown of Cups" voltaic pile, a battery consisting of two different metals and a liquid.

Humboldt in 1796 observed the phenomenon of what later became known as "polarization." He noticed the action of a plate of silver brought into contact with a moistened plate of zinc, and the action between other metals when in the presence of water. Humboldt directed attention to the decomposition of water into its constituent elements, and he created the phrase "decomposition of water by galvanic action."

The year 1801 saw the inception of what might be termed the scientific era in the history of the storage battery, which terminated with the advent of Plante's first practical cell. It was in this year that Gautherot, a French scientist, discovered that platinum or silver electrodes which had been used in the decomposition of a saline solution by the current of a voltaic cell, after being disconnected from the battery itself, were capable of giving an electric current lasting but a few moments. Carlisle and Nicholson experimenting with the electrolysis of water are said to have "unwittingly constructed an apparatus vaguely resembling an accumulator."

Erman in 1801 found that the electrode connected with the positive pole of a voltaic battery, after being disconnected from the cell, will return a positive current, and he is said to have discovered the principle of polarization.

In 1801 Ritter observed the same facts. He succeeded in decomposing water by means of the polarized electrodes and demonstrated the inversion of the currents. In 1802 he made the first secondary battery. This battery consisted of a pile composed of disks made of copper plates, separated by a liquid. The liquid became charged when the current of a voltaic battery was passed through it, and produced all the characteristics of a primary battery.

Gautherot's discovery was confirmed by Ritter in 1803, except that Ritter used gold wire and made a secondary battery by superposing a series of pieces of gold between circular pieces of cloth which had been moistened with a saline solution. Ritter is said to have always used a saline solution instead of acidulated water.

In his experiments he found that copper was not the best kind of metal for secondary batteries, and he tried bismuth, brass, gold, iron, lead, platinum, silver, tin and zinc. He had the most success with carburet of iron and peroxide of manganese, and he obtained no results with lead, tin or zinc. It is interesting to note here his reason for not employing lead in his secondary batteries. In the saline solution which he always used, lead chloride formed on the negative electrode, and this being but very slightly soluble, as well as a very poor conductor, the primary current itself was not strong as compared to that produced in acidulated water.

It was thought at the time that the secondary currents were due to the recombination of the acids and bases yielded during the decomposition of the salts by the current of a voltaic battery. In 1826 De la Rive noticed similar secondary currents produced by a voltameter composed of platinum plates immersed in acidulated water, or even in distilled water. The absence of salts to be decomposed and the fact that the metal was not oxidized led to the belief that the secondary currents were produced by a physical rather than a chemical action, and the electrodes were said to be "polarized."

Becquerel, Marianini and Volta have shown that Ritter's secondary currents had a chemical origin. Becquerel proved that the current produced by a secondary battery was due to the reciprocal action of an acid and its base. He immersed one of the platinum plates in an acid and the other in a basic solution, and these were united by a moist conducting material. The plate immersed in the acid solution became the positive pole, while the plate in the basic solution became the negative pole. This led to the detection of the direction of the current of the secondary battery, and this direction was such that the plate in connection with the positive pole of the primary cell was actually the positive pole of the secondary battery.

Nobili's experiment is next recorded in 1826. He is said to have deposited peroxide of lead on lead plates by means of a primary current.

Schoenbein in 1837 showed that lead plates prepared with peroxide of lead were capable of giving a current similar to that of a primary battery.

Grove in 1842 made his famous gas battery with platinum strips, which, having previously been used for the decomposition of acidulated water by the current, would, when the battery was disconnected, cause the recombination of the hydrogen and oxygen, thus producing an inverse current.

Faraday observed in the electrolysis of a solution of lead acetate that peroxide of lead was formed at the positive electrode, while metallic lead was produced at the negative pole. He found that the peroxide was a very good conductor and observed its ready ability to yield its oxygen.

Grove discovered in 1843 that plates covered with the oxides of metals gave better results than metal plates. Siemens used carbon plates in the same year, and Wheatstone made a secondary battery of platinum plates covered with the peroxide of lead and potassium amalgam, but they both found that the peroxide of lead appeared to give the most satisfactory results for these purposes.

In 1853 Siemens used carbon plates covered with lead oxide, which was converted into peroxide by the current, but like its predecessors it belonged to the scientific era.

Plante in 1856 began his investigations with secondary batteries and in 1859 he discovered that lead was the best metal for them. In March, 1860, he constructed the first practical accumulator, which was based upon the property of lead as a powerful depolarizer, a fact that Faraday, Wheatstone, De la Rue, Niaudet, Sinsteden, and others observed before him, yet it remained for Plante to perceive the possibility of giving the secondary battery the power, capacity, permanence, and efficiency necessary to a practical accumulator, and his discovery brought with it the dawn of the practical era in the history of the storage battery.

Plante's cell consisted of two coiled sheets of lead separated from each other by heavy cloth. These plates were immersed in dilute sulphuric acid. Upon sending a current through the liquid from one plate to the other some twenty or thirty times and frequently reversing the same, the positive plate became coated with lead peroxide, while the negative plate became "spongy," a process of "forming" which rendered the plates active.

In 1861 Kirchoff made batteries consisting of alternate plates of spongy lead and peroxide of lead.

Percival in 1866 experimented with metal plates covered with crushed coke and powdered lead in a solution of dilute sulphuric acid. It was for this battery that the first United States patent was granted for a storage battery. In 1869 he used a positive plate of amalgamated zinc and a negative electrode of lead.

It is recorded that Faure in 1872 attempted to obtain a secondary battery, and he is said to have improved his battery by perforating the plates and alloying the lead of the supports with antimony.

D'Arsonval in 1879 tried carbon plates covered with an oxide of lead and manganese, and with plates of zinc in a solution of zinc sulphate.

Plante's cell was practical in that it was powerful and utilizable, but the time required to form the plates was a factor that undoubtedly retarded its commercial adoption and one that received the attention of many eminent scientific men. Plante himself continued his researches until 1879. In this same year Metzger devised a means of applying the active material by a mechanical process which dispensed with Plante's slow-forming process.

In 1878 Faure conceived the idea of his secondary battery and discovered spongy electrolytic lead obtained from insoluble oxides, its value as a positive material and its practical manufacture. He constructed a battery with its elements made entirely of lead sulphate.

Reynier in 1880 invented his regenerating alkaline battery, but soon abandoned it and took up the Faure accumulator, with the inventor of which Reynier worked for many years.

In 1881 Faure took out a patent covering a method of getting the reducible plate by means of preparing the active material from oxides or salts mechanically applied to the surface of the plates, thus doing away with the great number of reversals required in the Plante cell. This saved power and reduced the time element, of which so much of either was required to form the plates of the Plante cell. A difficulty experienced in making the Faure or "pasted" cells was to keep the coatings on the plates, and heavy cloth was employed to facilitate this.

Swan and Sellon made the next improvement in the Faure cell. They introduced the substance which formed the coatings into the interstices in the lead.

After Faure obtained his patent in 1881, inventors in the United States, England, Germany, France and Italy and in other countries patented so-called improvements on the Plante and Faure types of cell. Confining research for the data for this paper entirely to the litigation in this country, owing to limited time and space, the difficulty in deciding the matter of the patent rights of the numerous claimants can be readily conceived when it is stated that in the patent office one decision alone is said to have been reversed twelve times before the case was finally decided.

In 1882 Gladstone and Tribe were the first to declare that sulphuric acid had a direct action upon the electrodes, and they also made the statement that "the principal, if not the only, function of the hydrogen of the water is that of reducing the lead compounds."

In 1885 a patent was issued to Brush by the United States Courts, after long litigation, for the mechanical application of oxides of lead to a supporting grid of conducting material, as it was decided that Brush's invention antedated Faure's application for patent. Brush received other decisions in his favor in this country, while a compromise was effected before decision was rendered in the Brush-Julien litigation.

Since Plante's time Reynier's lead-zinc cell, Sutton's copper-lead cell and the Thomson-Houston copper-zinc cell have been devised, while another type of cell that appeared after Reynier's cell was the alkaline-zincate secondary battery, which had one zinc electrode and an alkaline electrolyte. The Desmazure, Edison-Lalande, Lalande-Chaperon and the Waddell-Entz storage batteries are of this latter class.

Nickel was used in accumulators by Dun in 1885, and porous

nickel electrodes were used by Desmazure in 1887.   Michalowski in 1889 discovered a new method of making nickel electrodes.

Pollak in 1896 experimented with iron compounds, while Krieger invented a cell using nickel.

In 1898 Jungner made cells, using copper and cadmium-silver electrodes.

The chloride battery consisted of plates cast of a fused mixture of chloride of lead which had a rigid metallic rim cast about them. When immersed in a solution of chloride of zinc or a similar substance in connection with a piece of metallic zinc, the chlorine in the zinc chloride and in the lead chloride combines with the zinc plate and forms zinc chloride.   Thus prepared the plates are ready for use in the same way that Plante plates are when "formed."

In his first cells Edison used cadmium and copper in an alkaline electrolyte.

A patent was issued to Edison on February 4, 1902, for a storage battery consisting of sheet iron or steel grids, nickel-plated, to prevent the electrolyte of potassium hydrate from acting on the same.   The grids have a series of openings into each of which fits a pocket made of thin crucible-steel with perforations having internal burrs.   These pockets contain briquettes of finely divided cadmium, which is used as the oxidizable element and an oxide of nickel or cobalt which is used as the depolarizer.   The plates are placed alternately, and between them are placed rubber partitions.

In the foregoing historical sketch it will be seen that the main points intended to be brought out are the various metals, compounds and electrolytes that were used in experiments by the leading physicists in the past, and it is hoped that this will prove of service to prospective inventors.

The history of storage battery installations has been covered by a number of writers, and the following are only intended to give a few of the early uses to which the storage battery was put in so far as it was possible to get definite information on the subject at this time.

In 1882 one of the first storage battery installations was made on a car belonging to the Pennsylvania Railroad Company for the purpose of train-lighting, but this experiment did not result satisfactorily.   The cells for this purpose were imported from France.

44

One of the first isolated plant installations was made with electric power storage cells imported from England, in Baltimore in 1883. The battery was used for incandescent lighting in several stores in the vicinity where it was installed.

The first storage battery installation in a central station is supposed to have been in Phillipsburg, Pa., in 1885.

In 1886 an electric boat "Volta" made a trip in the English Channel, and the trials of the submarine boat "Nautilus" took place, both boats being equipped with Elwell-Parker cells.

Applications of storage batteries became quite general about this time, and while a good many of the earlier installations were failures, yet the demand caused the manufacturers to improve their cells to meet conditions.

In his specification for an invention, Faure, in 1887, claimed an improvement "in the application of secondary batteries or accumulators preferably carried by a vehicle or upon one or a train of vehicles to supply electrical energy to one or more electro-magnetic motors, also carried by or upon said vehicle, or one or more of a train of vehicles, combined with means for varying the amount of energy supplied by the battery." The Commissioner of Patents decided that the combination was not new and therefore not patentable, and attributed the commercial success of the alleged invention to the Faure battery alone, and not to the combination. The Supreme Court of the District of Columbia affirmed this decision, the case having been appealed.

The first important installation of modern storage batteries of German manufacture was made by the Edison Electric Illuminating Company, of Boston, in 1894, and in 1896 they installed their first American cell.

As is well known, while comparatively little improvement has been made in the manufacture of the secondary battery, yet the necessity for its use from engineering and commercial standpoints is plainly evinced by the many purposes for which it has proved itself practical and for which it is being constantly demanded.

As indicative of the amount of business done in storage batteries, the Electric Storage Battery Company, of Philadelphia, has installed up to date about 85 central station plants having a capacity of 120,000-kilowatt-hours; 225 railway batteries having a capacity of 160,000-kilowatt-hours; and 500 isolated plant batteries having a capacity of 30,000-kilowatt-hours.

It might be well to mention here that theoretically there is no difference between a secondary battery and a primary battery, as in both the chemical energy becomes transformed into electrical energy. If peroxide of lead prepared by electrolysis is placed around a cylinder of lead and put into acidulated water with another cylinder which has spongy lead similarly prepared placed about it, the cell would ordinarily be called a primary cell, while if a reversed current is passed through a Daniell cell, the cell would become a storage battery. Other well-known primary cells, if treated in the proper way, could be made secondary batteries. It is entirely a matter of a reverse current, and it is in this that the lead peroxide battery has up to the present time demonstrated its practicability over other cells.

One of the most recent inventions is in a patent granted to Edison on March 3, 1903, covering a modification in his alkaline storage battery. The electrode is composed of oxide of cobalt mixed with metallic mercury or with metallic mercury and copper or silver, by which a great proportion of the cobalt oxide is kept in electrical contact with the electrode. Mercury, copper or silver are added by preference to the oxide of cobalt for the purpose of maintaining contact between the active materials, but any insoluble conducting material, preferably in flake form, such as flake graphite, can be used for this purpose.

DISCUSSION.

DR. H. S. CARHART: Mr. Chairman, I should like to add a small item to the historical part of this subject. I happened to be present in 1881 at the Paris Exposition of Electricity as one of the International Jurors for the United States, and during that time I became acquainted with M. Planté and was invited to go to his laboratory to see his storage cells. I went in company, I remember, with Prof. Barker. During that visit, I remember with very great distinctness, Planté told us that he had some time previously tried the application of active material to his plates and found that he could not get as good results as he got with the formed plates. He did not publish this; he did not apply for a patent; I think he never obtained any patent; and in the public mind has not had credit for the application of active material to the plates. That is the point that I wish to add to this paper.

DR. N. S. KEITH: I desire to add a little to the historical interest in this matter by stating that I was one of the early litigants with Brush before the Patent Office. I received a patent, after extensive efforts in that line, for a storage battery made by depositing peroxide of lead upon the anode in a now well-known way, and using those plates as the positive and negative by the reduction of the lead upon the one plate to metallic lead—the oxide to metallic lead upon the one plate—and retaining the peroxide deposited by electrolysis upon the other. It was shown by the testimony that I made my storage battery in that way in the year 1877 or 1878 here in New York City.

Reprinted from *Electrochem. Ind.* **1**:148–151 (1902)

AMERICAN AUTOMOBILE BATTERIES, I

THE SPERRY BATTERY.

In the Sperry battery, which is manufactured by the National Battery Co., Buffalo, N. Y., the grid, as shown by Fig. 1, is made of thin, pure sheet lead, which is corrugated horizontally. In the bottom of the hollows, formed by the corrugations, numbers of small trapezoidal holes are punched. The punch also cuts diagonally across the trapezoid, and, instead of making a clean hole, fins are left which project on either side of the plate. The whole plate, when punched, resembles a corrugated grater, except that the fins are longer and triangular in shape. These projections are then spread wider apart, so that a pressure on the plate will cause them to bend down in a direction away from the holes from which the fins project.

The material which is to become active is spread in the form of a powder on both sides of the grid, filling up the corrugations and being in sufficient quantity to make a flat plate of usual thickness after press-welding. The whole is then subjected to pressure—about 1,000 pounds per square inch of surface—which forms it into a solid mass. The material is bound to the grid by being pressed together into a continuous mass, which is on both sides of the grid and is welded through the multitude of small holes, thereby riveting itself in the grid. Additional hold is furnished by the fins, which bend down and are clinched over the active material, and thus retain the material at the surface.

The retention by the fins and riveting under pressure of the powder into a solid mass cannot be successfully accomplished with every form of material to become active, but the material used in the Sperry battery welds up strongly in the press, and, after the chemical formation, becomes hard like soapstone. This active material consists of 80 to 85 per cent. of finely-divided pure lead, obtained by dissolving a precipitation, to which is added 15 to 20 per cent. of lead oxide. These are thoroughly mixed, and to the mass thus formed is added about one-twenty-fifth of a compound of alkaline salts and other ingredients.

The alkaline salts and other ingredients are themselves inert, and their function is twofold:

First. They decompose or dissolve when the plates are subjected to the acid bath, just before being subjected to the electrolytic forming process, leaving numerous minute cells throughout the mass of material upon the plates, which cells permit the electrolyte to thoroughly penetrate the mass of material upon the plates during the time they are subjected to the forming process, and also permit the expansion of this

however, that trouble from disintegration and dropping away of active material is obviated. There are a number of these batteries that have been in use for over two years, and the positive plates are hard, solid and durable.

The corrugation of the grid horizontally allows proper expansion, and relieves the plate of any tendency to buckle. As an additional precaution, however, the Sperry pyroxylin envelope is used, which is applied to the outside of the plate and covers it, passing down one side, under the bottom of the plate, and up the other side. It is evident that if any active material should be loosened by any extraordinary circumstance, it would not be able to drop off the plate, being held in position by the outside coating. If a particle of material should, by any chance, drop off, it is retained in the pyroxylin envelope and will not be able to short-circuit the battery.

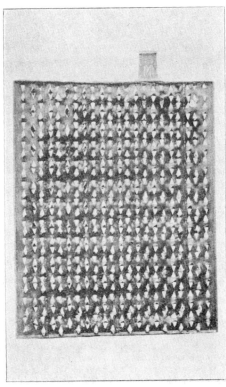

FIG. I.—GRID, SPERRY BATTERY.

mass of material during the expansion which takes place in the material while the plates are subjected to the forming process, and thereby prevents the active material from crowding off or disintegrating from the grid or support and seriously injuring the plates.

Second. They have the peculiar property of causing the mass of material to harden, not only during the time that the plates are subjected to the forming process, but also during the use of the plates in a storage battery. In other words, this mixture accelerates the forming of the plates by increasing the penetration of the electrolyte, and also prevents serious injury to the plates that otherwise would be caused by the expansion and contraction of the mass of material; and, as above stated, also increases the hardening condition of the mass of material during the use of the plates.

The composition of the active material used, its method of application and the formation of the Sperry battery are such,

FIG. 2.—PYROXYLIN ENVELOPE.

This covering, or pyroxylin envelope (as shown by Fig. 2), is made of open-mesh cotton cloth, such as "cheese cloth," which is chemically treated, forming a cellulose nitrate, which is termed "pyroxylin," and which is of the same general character as gun cotton. The addition of a small quantity of nitrobenzol renders it inert.

The jars in which the elements are placed are of the hard-rubber variety, but different from the usual hard-rubber jars in that they have a series of ribs, capped with soft rubber, extending across the bottom of the cells. On these soft, resilient ribs rest the plates, which are in this way relieved from excessive shock or jar. The separators are corrugated, perforated, ribbed, hard rubber.

Fig. 3 shows a curve of discharge of the Sperry battery taken from 44 cells in series. These cells weigh 1,005 pounds

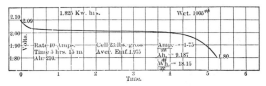

FIG. 3.—SPERRY BATTERY DISCHARGE.

gross, or 23 pounds each. The discharge rate was nearly 40 ampères, and the average voltage 1.975. The curve shows the unusual results of 1.75 ampères, 9,187 ampère hours and 18.15 watt hours per pound gross weight at 5¼-hour discharge, giving, at this rate of discharge, 40 pounds per horse-power hour, corresponding to an energy sufficient to raise its own weight against gravity at sea level about 9 miles.

This is equivalent to 2.65 ampères, 7.95 ampère hours, 5.03 watts and 15.09 watt hours per pound at the three-hour rate. There are several sizes of these cells manufactured, the principal ones being 2 9-10 x 5¾ x 10 inches, and weighing 15 pounds, gross; 2 x 7¼ x 10½ inches, and weighing 18 pounds, gross, and 3 x 7¼ x 10½ inches, weighing 23 pounds.

A great number of life tests have been made on this battery, both in laboratories and on vehicles and in practical work.

One vehicle made 7,770 miles with the batteries at a loss of 28 per cent. of the original capacity when the test was discontinued. Several of the Sperry batteries have been in operation actively for over two years, and showed splendid results. The negative plates were apparently as good as when the battery was made, and a test conducted under the general supervision of Prof. John W. Langley, of the Case School of Applied Science, shows that the battery carried through a uniform rate of discharge of 1 ampère per pound until the carriage had run the equivalent of 3,060 miles. This test shows the capacity of the battery for the last ten discharges to be 10 per cent. in excess of the first ten discharges. The battery had not revealed any indication of falling off in capacity, and the forty-fourth discharge of this battery, taking it at 1 ampère per pound of complete working cell, is a mean potential difference throughout the curve of over 2 volts, namely, 2.014, as shown by the following fifteen-minute readings of potential difference and positive and negative cadmiums:

Time.	Voltage between Plates.	Voltage between Positive Plate and Cadmium.	Voltage between Negative Plate and Cadmium.
12.	2.27	2.37	.13
.15	2.12	2.22	.13
.30	2.12	2.22	.13
.45	2.10	2.20	.13
1.	2.10	2.20	.13
.15	?	2.20	.13
.30	2.0	2.19	.13
.45	2.07	2.18	.13
2.	2.05	2.16	.13
.15	2.03	2.15	.13
.30	2.03	2.15	.13
.45	2.03	2.15	.13
3.	2.03	2.15	.13
.15	2.00	2.13	.13
.30	2.00	2.12	.13
.45	1.98	2.10	.13
4.	1.98	2.08	.13
.15	1.97	2.07	.15
.30	1.93	2.03	.15
.45	1.93	2.03	.15
5.	1.90	2.03	.15
.15	1.87	2.02	.16
.30	1.83	2.00	.17
.45	1.80	1.97	.17

The company also manufacture a battery for automobile purposes, known as the "National." This battery is a combination of Sperry negative and Planté positive plates. The negative plates are 3-16 inch in thickness, and the positive plates ¼ inch in thickness. The former are made as explained in the description of the Sperry battery. The latter are made from chemically pure sheet lead, rolled to the desired thickness, cut into strips or ribbons and passed through machines of special design, which knurl and weave a straight and corrugated strip alternately. The ribbons are then placed in a form and subjected to hydraulic pressure, fitting them to the frame or grid, which is cast from an alloy of antimony and lead in

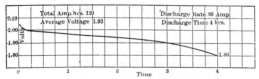

FIG. 4.—SPERRY BATTERY DISCHARGE CURVE.

two pieces. They are next placed in the grid, the center supports and ends of which are burned together, making the plate one piece.

The plates are electrochemically formed by the usual Planté method. Fig. 4 shows the discharge of 24 cells of this battery

after 80 discharges, and the battery shows no sign of deterioration. Fig. 5 is the "National" positive plate.

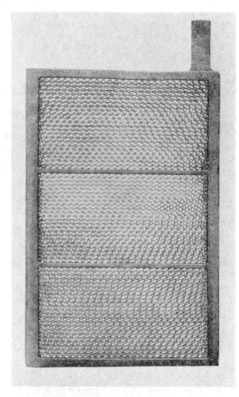

FIG. 5.—NATIONAL POSITIVE PLATE.

The extremely large surface secured by this method of construction makes it possible to obtain the capacity desired with a thin, hard, coherent layer of active material, and permits a free flow of the electrolyte through the plate, allowing it to withstand rapid charging and discharging without serious injury.

WESTERN STORAGE BATTERY.

The more completely the active material covers the conducting plate or grid, the more enduring is the plate, more especially the positive electrode. For this reason the grid used in the Western storage battery, made by the Western Storage Battery Co. for the National Vehicle Co., Indianapolis, Ind., has a conducting diaphragm of thin, perforated lead running through the center of the plate, just midway between the surface of the outer frame of the plate; then, horizontally across this diaphragm are cast thin lines of lead, the outer edges of which are turned down, so that a slight shoulder is made, under which the oxide paste, to become active, is forced by hand pressure, thus confining the paste against outward pressure, which takes place in the process of forming the positive plates. When the plate is thoroughly pasted, small points of lead only show on the surface. Thus the forming process, in which the electrical current tends to follow the more prominent lead as its easiest line of conduction, is forced to seek the main conductor in the center of the plate, passing through the oxide paste to reach it, and producing the necessary chemical changes on its way. This tends to form the electrodes more evenly. The plates made in this way are light, but stiff, and, being cast of an alloy of lead and antimony, are less rapidly decomposed by the constant action of electrolysis in charging and discharging than soft lead would be. A plate, for example, of a surface area ot 4⅞ x 7¾ weighs 11 ounces, and will hold, when properly pasted, about the same weight of lead oxide paste when dry. Seven of these plates, when formed and charged, have a capacity of 68 to 70 ampère hours when discharged at the high rate of 20 ampères per hour.

In order to give more endurance than is usual in a light battery, the positive electrodes are protected from washing, which the motion of the vehicle gives to the battery electrolyte, by a thin pad of spun glass, made for this purpose. It is held against the two surfaces of the positive plate by perforated hard rubber, on one side of which are several perpendicular ribs, about 1-16 inch thick and ⅛ inch high. This gives free circulation of the liquid and gas along the negative plate, and also holds the very resilient glass pad firmly against the positive plate. The combination of spun glass and hard rubber for separators and insulators has proved an improvement, since there is no chemical action in this material, as there would be in any form of wood or cellulose, which chemical action, set up by electrolysis, is always detrimental to the durability of the lead grid. The following data are the results of a test of a 40-cell battery, each cell weighing 14½ pounds, complete. This battery was made up for the National Vehicle Co., for their road wagon, designated in the list as R 2 battery, and consists of seven plates, three positives and four negatives, the plates having each a surface area of 4⅞ x 7¾ inches. The specific gravity of the electrolyte was 1.250 fully charged, and 1.117 at the end of the discharge:

Time.	Ampères.	Volts.
12.	20	82
1.	20	81
1.30	20	80
2.	20	79.5
2.30	20	78.5
3.	20	77
3.30	20	75
3.45	20	68

This gives, for a 3¾-hour discharge, 75 ampère hours, with an average power of, approximately, 1,600 watts. At 68 volts the discharge is cut off and recharging at once commenced, since the drop to 68 volts for the series of 40 cells shows 1.6 volts to each cell, which is as low as any battery could be discharged.

The next size of battery, made for stanhope vehicles and park traps, contains the same number of plates, the surface area of which is 5½ x 8¼ inches, and the discharge rate is 25 ampères, instead of 20, for 3¾ hours, which gives 94 ampère hours. Forty-four of these cells, after making a trip over country roads for three days, of about 65 miles each day, charging in this time three times, made a run of 118 miles the day after charging on the asphalt streets of Indianapolis, taking about eleven hours for the test. In other respects, this battery does not differ outwardly from other batteries in general use.

SMITH STORAGE BATTERY.

The tray storage battery, invented by Malcom O. Smith, and made by the Smith Storage Battery Co., Binghamton, N. Y., differs in its form widely from the batteries made by other manufacturers, although it involves no new principle. The main new point is that the jars are dispensed with, and the lead plates themselves are used for holding the electrolyte. The gridded plates are formed into oblong trays, with pointed bottoms, the ends also receding from the top to the bottom, so as to permit of nesting.

Any desired number of trays can be nested on top of each other, perforated separators of sheets of hard rubber or other suitable material keeping the trays from contact with each other. When nested, there is sufficient space between the trays to hold the electrolyte, which is as much in contact with the bottom of the tray above it as with the under one, which holds it. This permits of an equal chemical action upon the upper surface of one tray and the under side of the one above it.

In preparing the plates, the under side of each tray is chemically treated to become oxide, and the upper side peroxide of lead. This makes the upper side of each tray the positive side, and the under one the negative side, of what in other batteries would be a single cell.

The addition of another tray to the nest is simply the same as adding another cell to an ordinary battery. Of course, the bottom of the lower tray and the inside of the upper one are plain, and are not in contact with the electrolyte. A battery of 11 trays is practically the same as an ordinary one with 10 cells. The voltage of the battery will be twice the number of trays minus one.

One of the battery terminals, the positive, is attached to the bottom side of the bottom tray, and the other, the negative, to the top of the upper one. For the use of the entire battery, there is no connection between the trays except the electrolyte, the use of "jumpers" between the plates, as in ordinary batteries, being unnecessary.

Any combination of series-multiples or multiple series connections can be obtained that are possible with any battery. One terminal can be applied to the top of any tray of the nest, the other to the bottom of any other one. The voltage will always be twice as much as the difference between the number of trays. Between the top of the second tray and the bottom of the fourth one, the e. m. f. is 4 volts; the top of the third and the bottom of the ninth, 12 volts, etc.

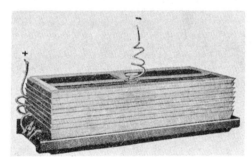

SMITH STORAGE BATTERY.

One of the advantages claimed for this type of battery is its extremely low internal resistance. There are no lugs, soldered joints or loose connections, with connecting wires or strips from plate to plate, to offer resistance. At the latter part of a discharge, the current does not have to come from the bottom of the plate through the already discharged upper end. The discharge is practically continuous from the entire surface of every plate.

This results in another advantage, which is the rapid or uneven manner in which it can be discharged without fear of buckling. Inasmuch as the discharge is from the entire surface of all of the plates, it is claimed to make little difference whether it is accomplished in two or three hours or two or three days. Response is made to sudden changes of load with slight variation in voltage. This feature of even response to variable loads is of particular advantage for automobiles, where the demand varies so suddenly.

The form of the battery is claimed to preclude possibility of the deposit of active material to form a short circuit between the plates, as frequently happens with jar batteries; such deposits come from the positive plates. In these batteries the positive face of the tray is up, so that there is no tendency for loosening of material; and, should that happen, it simply rests in the bottom of the tray, where it can cause no trouble.

Dispensing with all jars, either of glass or hard rubber, decreases the weight of the battery considerably, and reduces the cost. It is possible to nest an entire battery of any number of cells up to 20 or more.

[To be continued.]

Reprinted from *Electrochem. Ind.* **1**:293–294 (1902)

AMERICAN AUTOMOBILE BATTERIES, II

THE MADIGIN BATTERY.

This battery is only manufactured at present in Canada, although large quantities are being exported to the United States and foreign countries notwithstanding the heavy tariff duties on this class of goods. It is manufactured there under Madigin patents by the Croftan Storage Battery Co., Toronto, Canada, which are the only firm manufacturing storage batteries in the Dominion of Canada.

The construction of the plate in this battery is shown by figures No. 1 and No. 2. Figure No. 1 shows the grid, which, as will be noticed, is made of one thin piece of sheet lead $\frac{1}{32}$ inch thick. This sheet is highly perforated with over 11,000 holes to the square foot, and is horizontally corrugated so as to present parallel ridges about $\frac{1}{8}$ inch apart. The positive plates are corrugated to such a depth as to render the plate $\frac{1}{8}$ inch thick when finished, and the negative plate $\frac{3}{32}$ inch thick. In the process of manufacture a rim and lug are burned on at the operation so as to form a border of $\frac{3}{16}$ inch all around the plate. Fig. 2 shows a positive plate fully pasted and formed, and, as will be noticed, the grid is almost entirely covered by the active material, with the exception of the small protuber-ances of the corrugations which present themselves on the outside surface of the plate. In this manner the grid in itself is almost entirely protected from dangerous electrolytic action, which always causes rapid depreciation and disintegration.

The positive grid of their standard automobile type weighs but 10 oz., and when pasted and fully formed will give a discharge of 24 ampère hours at a five-hour rate. The positive plates before pasting are coated with a thin coherent layer of lead peroxide by electrolysis, and are then removed from the solution and pasted with usual oxides, with the addition of a chemical (the nature of which they withhold), which renders the active material extremely hard and yet highly porous. After pasting the plates are subjected to the regular forming process.

In assembling the elements the positive plates are enveloped with a layer of pure fibrous cellulose, which is chemically treated so as to render it neutral and at the same time to guard against disintegration. The plates are then packed closely together, with the cellulose envelope separating each, and are then connected or burned together in the usual manner.

It will thus be seen that this method of insulation gives a complete perfectly-porous insulating wall between the plates, and if the cell is only half full of electrolyte (which often happens when batteries do not receive proper attention) the entire surface of the plates will remain in action, owing to the juxtaposition of the insulating material and the capillary attraction resulting in the use of this method of insulation. It also prevents any active material from possibly becoming dis-

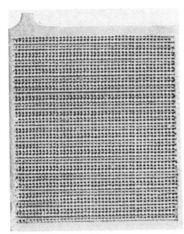

FIG. 1.—GRID UNFORMED.

engaged from the grid, and, furthermore, eliminates all washing action of the electrolyte on the plates, which has heretofore been one of the greatest difficulties to be contended with in automobile practice, owing to the usual excessive vibration while in operation.

From the construction employed it must be noticed that notwithstanding the extreme lightness, as well as the extraordinary elasticity which has been attained, durability has not been sacrificed, but on the contrary is always in evidence. The battery requires but a small portion of the space usually demanded by storage batteries, owing to the compact nature

FIG. 2.—GRID FORMED.

of the elements and the small but sufficient quantity of the electrolyte employed. As an illustration, their 13-plate cell occupies but a space of 2⅝ inches x 6¼ inches x 9 inches high and weighs but 18 pounds. This cell has a capacity of 132 ampère hours when discharged at the unusually high rate of 35 ampères. This feature is due very much to the perfect contact of the active material with the grid, as well as the

extremely low resistance of the grid itself, which admits of discharges much higher than this amount without depreciation. The practice, however, which they employ is to furnish batteries of such capacity as to give from six to eight hours' operation at normal load. In this manner a mileage of 80 miles per charge is easily secured by ordinary vehicles with a battery weighing less than has usually been employed.

This firm also manufactures a patent removable and non-

FIG. 3.—FLEXIBLE REMOVABLE CONNECTION.

corrodible connection for automobile batteries, which is flexible and has a large number of advantages over the usual method. By its adoption any cell or set of cells may be removed or replaced instantly without the necessity of sending the entire set of batteries to the factory to be welded together. This is a valuable feature, when batteries are required to be washed at regular intervals. Fig. 3 shows a cut of this connection as applied to two cells, and also shows its construction. It is manufactured in sizes adaptable for all existing American batteries.

CLARE STORAGE BATTERY.

The special feature of the Clare battery, built by the Consolidated Storage Batteries Co., of New Jersey, is the use of earthenware receptacles for holding the active material, constructed as follows: Plates are made of earthenware, of sufficient porosity to allow the electrolyte to pass freely through them, and at the same time free from impurities, which would make the substance a conductor.

Each plate is divided on one side into many small, square windows by narrow ribs, the active material being pressed into these windows; the other side of each plate has parallel ribs in one direction only. Each plate also has a heavy rib on one or two of its edges, extending slightly further from the plate than the smaller ribs which form the windows containing the active material.

When two such plates are put together, so that the windows face each other, there is thus a small space left between them opposite windows, and into this space a thin sheet of pure rolled lead, about 1-32 inch in thickness, is forced, so that it comes into intimate contact with the active material in each window. This sheet of lead serves only for collecting the current from the active material, and not as support.

The heavy ribs at the edges of the plates are cemented together by an acid-proof cement. A pair of such earthenware plates, joined together in this way, forms a receptacle of rigid structure, which represents a single positive or negative element of a battery

The principal advantages claimed for this construction are lightness and durability of the plates, with sufficient porosity to allow the electrolyte to pass freely through the sides of the receptacles; the internal resistance is very low, and the active material is so firmly held in the receptacles that it cannot fall out. The receptacles are also sufficiently elastic to allow for all expansion of the active material.

Concerning weight and capacity, the manufacturers state that their batteries made for automobile work give 5 ampère hours per pound of cell complete.

3

Reprinted from *Am. Electrochem. Soc. Trans.* **6**:135–151 (1904)

THE PRESENT STATUS OF THE EDISON STORAGE BATTERY.

BY DR. A. E. KENNELLY AND MR. S. E. WHITING, *Harvard University.*

The following paper is presented on behalf of Mr. Thos. A. Edison, as a brief synopsis of the results obtained with his storage battery up to the present date.

HISTORY.

About 1898, Mr. Edison began seriously to investigate alkaline storage cells. After trying a very great number of combinations, the nickel-iron couple became fairly well developed in 1900.

The first type of cell put into service in vehicles was named Type C. This was an 18-plate cell (9 of iron and 9 of nickel), weighing complete 17 1/2 lbs. (7.95 kilos). At 20 amperes discharge (9.5 hours) to 0.75 volt the average potential difference was 1.23 volts and the quantity 191 amp-hours representing 234.9 watt-hours or 13.4 watt-hours per lb. (29.6 watt-hours per kilo). At 120 amperes discharge (1.4 hours) to 0.75 volt, the average potential difference was 1.03 volts and the quantity 171 amp-hours, representing 176 watt-hours or 10.06 watt-hours per lb. Several hundred of these cells were made up and are still in use in vehicles at the factory.

This Type C cell was not regarded as entirely satisfactory because it did not promptly recover its full normal voltage after heavy rates of temporary discharge. This seemed to indicate an inadequate circulation of electrolyte in the cell and pointed to a reduced current density; i. e., to a larger plate-surface for the same mass and volume of active material. The improvement required new tools and months of delay, but was nevertheless undertaken, and the new cell developed was named Type D.

The type of D cell was first completed in May, 1903. It was a 28-plate cell (14+ and 14—) weighing 17.8 lbs. (8.05 kilos) and occupying the same space as the 18 plates of Type C, thus providing an increase of about 50 per cent in active surface with a corresponding reduction in briquette thickness, to the improvement of circulation.

Tests of the Type D cell were made not only at the Edison laboratory but also by several independent experimenters both in this country and in Europe. These tests will be referred to later. The factory tests of Type D showed at 20 amperes discharge (8.45 hours) to 0.75 volt, the average potential difference of 1.28 volts and the quantity 169 amp-hours, representing 216.5 watt-hours or 12.18 watt-hours per lb. (26.9 watt-hours per kilo). At 120 amperes (1.38 hours), to 0.75 volt, the average potential difference was 1.17 volts and the quantity 165 amp-hours, representing 193 watt-hours or 10.85 watt-hours per lb. (24 watt-hours per kilo). The corresponding internal resistance is 0.0013 ohm.

The change from Type C to Type D had succeeded in enabling the cell promptly to return to its normal e.m.f. after temporary heavy discharges, and improved the output under heavy discharge from 10.06 to 10.85 watt-hours per lb. (22.2 to 24 watt-hours per kilo). On the other hand, the output at low rates of discharge had somewhat suffered by the change, and the new cell was also found to be somewhat unbalanced in the individual capacities of the + and — plates, there being an excessive capacity in the iron plates.

A further modification was, therefore, made in the cell, resulting in the production of Type E, the latest type.

Types C, D and E have the same height, length and general external appearance and differ externally only in their breadth. Type E is the cell now manufactured and placed on the market in three sizes, which differ only in the number of assembled plates. The cell of Type E tested by the author's and here considered is E-18, having 12 nickel plates and 6 iron plates. The weight of the E-18 cell complete is 12.66 lbs. (5.75 kilos). The normal rated delivery of these cells is 114 amp-hours, which is practically the same at either the 30-ampere or 120-ampere discharge rate. At the 30-ampere rate (3.8 house), the mean potential difference to 0.75 volt is 1.234 volts, representing an output of 141 watt-hours or 11.1 watt-hours per lb. (24.6 watt-hours per kilo). At the 120-ampere rate (0.95 hour), the mean potential difference to 0.75 volt is 1.04 volts, representing an output of 118.5 watt-hours or 9.38 watt-hours per lb. (20.7 watt-hours per

FIG. 1.— PARTS OF EDISON STORAGE BATTERY.

FIG. 2.— EDISON STORAGE BATTERY.

kilo). Although the above are the normal rated capacities, the full capacity of the cell is found to be about 140 amp-hours, which makes the total output at 30 amperes, 173 watt-hours and at 120 amperes 145 watt-hours; or the specific output 13.7 watt-hours per lb. (30.2 watt-hours per kilo) and 11.5 watt-hours per lb. (25.4 watt-hours per kilo) at the two rates respectively.

CONSTRUCTION OF TYPE E.

The construction of the cell is shown in Fig. 1 which exhibits the various parts both separate and assembled. The containing cell, as also the grids and pockets, are made of thin sheet-steel plated by a special process. The jar completely encloses the element and the lid is soldered permanently in place. There are four openings in the steel top. Two of these are insulated by rubber bushings through which the terminal posts project. The third opening is a filler-hole with gas-tight hinge-stopper. The fourth opening is a gas-vent provided with a valve and gauze screen to prevent escape of entrained liquid.

The insulation between the walls of the can and the enclosed element is all of hard rubber. On the bottom is the four-barred grating, seen on the right-hand side of the figure, and 0.4 in. (10.2 mm) deep. On the ends are ladder-like frames giving about 0.11 in. (2.8 mm) clearance and grooved to hold the edges of the plates. On the sides are solid sheets 0.014 in. (0.56 mm) thick. It has been found necessary to subject all the rubber insulation to a special chemical treatment to prevent a foaming action on the alkaline electrolyte.

Between the plates are threaded four-cornered rods of rubber about 0.1 in. (2.5 mm) thick and spaced 0.57 in. (14.5 mm) apart. The distance between opposed plate surfaces is about 0.04 in. (1 mm).

Plates of like polarity are bolted to a horizontal-bar at the top provided with spacing washers and joined to the vertical terminal post.

For use in vehicles the cells are grouped in wooden trays (Fig. 2) containing 3, 4 or 6 cells each. Connections between cells are ingeniously made by taper lugs and each jar is insulated from its neighbors and from the base by spacing blocks and a rubber pad, respectively.

The grid itself, Fig. 3, and the pockets containing the active materials are identical for both positive and negative electrodes. The grid is 9.25 in. high (235 mm)— excluding the lug — 4.75 in. (121 mm) wide and 0.015 in. (0.38 mm) thick. It contains 24 rectangular holes, each 2.95 in. (.75 mm) high and 0.5 in. (12.7 mm) wide. Into these holes are fitted the pockets illustrated in Fig. 4. These are made up of strips of 0.003 in. (0.076 mm) steel, having flanged edges that telescope together to form the pocket or container of the active material. These pockets have each about 5000 perforations.

The active material of the positive plate is an oxide of nickel in finely-divided form commingled with a conducting substance such as flake graphite, in order to improve the conductivity of the mass. The active material of the negative plate is finely divided iron similarly commingled with a conducting substance. The electrolyte is a 20 per cent aqueous solution of potassium hydrate.

In assembling, the pockets are filled with active material, closed and inserted in the holes of the grid. The plate is then subjected to a pressure of 150 tons for the purpose,— first, of flanging the pockets over the holes in the grid, in order to lock them firmly in position; and second, of corrugating the surface of the pocket, in such a manner as to provide adequate elastic movement of the envelope in view of the contraction and expansion of the contents.

The weight of the active material in its initial condition and including the conducting material is about 3.2 grammes per pocket for the nickel and 4.6 grammes for the iron plate. This represents a total quantity of 922 grammes positive active material, and 662 grammes of negative active material in an E-18 cell. The weight of electrolyte per cell is 3.1 lbs. (1.40 kilos) at a normal density of 1.190, which represents about 25 per cent of the total weight of the cell.

The cell is at present manufactured in three sizes stated in the following Table:

FIGS. 3 AND 4.— DETAILS OF EDISON STORAGE BATTERY.

Type.	Positive plates.	Negative plates.	Capacity in ampere-hours.		Normal charging current.	Normal charging time.	Weight complete.		Dimensions over all (excluding trays).							
			Normal.	Maximum.	Amp.	Hours.	Lbs.	Kilos.	Length.		Breadth.		Height.		Volume.	
									Ins	Cms.	Ins.	Cms.	Ins.	Cms.	Cu. ins.	Cu. cm.
E-18....	12	6	110	140	40	3.75+	12.5	5.66	5.	13.0	2.6	6.6	13.2	33.5	175	2,874
E-27....	18	9	165	210	60	3.75+	17.5	7.95	5.1	13.0	......	......	13.2	33.5	......	......
E-45....	30	15	275	350	100	3.75+	30.0	13.6	5.1	13.0	......	......	13.2	33.5	......	......

60

CHEMICAL THEORY.

The Edison cell is of the oxygen-lift type. That is to say, the process of charging consists in driving oxygen electrolytically from the negative to the positive plate. During discharge the oxygen leaves the positive plate and enters the negative plate. The chemical actions in the cell have not, as yet, been completely investigated.[*] The following conditions may, however, be accepted provisionally as forming a working theory:

Condition.	Positive plate.	Electrolyte.	Negative plate.
Charged.......	NiO_2 NiO_2	KOH H_2O KOH	Fe
Discharging ..	NiO_2 $\overset{+}{K}$ NiO_2 $\overset{+}{K}$ H_2O		$\overset{-}{HO}$ $\overset{-}{HO}$ Fe
" ..	$+Ni_2O_3$ KOH KOH	H_2O	FeO—
Discharged...	Ni_2O_3	KOH H_2O KOH	FeO

The cycle represented in the above table shows that during discharge the electrolyte divides into potassium cations and hydroxyl anions, the former being directed toward the positive plate and the latter toward the negative plate. On arriving at these plates the ions give up their respective charges. At the positive plate, the potassium robs the nickel oxide of a portion of its oxygen and, in combination with the water present, forms new molecules of potassium hydrate, the original electrolyte. At the negative plate the hydroxyl ions deliver oxygen to the ion and form water. Thus the electrolyte tends to become concentrated in the pores of the positive plate, and attenuated in the pores of the negative plate. Diffusion ultimately destroys this difference of concentration and leaves the electrolyte in its original condition, since at any instant the total quantity of water and of potassium hydroxide (including the pores of both plates), remains the same.

[*] E. F. Roeber. *Transactions* A. E. S. Vol. 1, 1902. M. V. Schoop, *Electrochemical Industry,* July and August, 1904.

It would appear from the form of e.m.f. curve during complete discharge, as shown in Fig. 5, that after the cell has become almost entirely discharged other stages of oxidization develop, and further investigation may show that the outline of the chemical cycle above presented is very incomplete. Whatever the complete cycle may be, it is, in all probability, however, of the type indicated.

It would seem that neither the plates of nickeled steel nor the active materials within them are subject to local or chemical corrosion, or solution, in the electrolyte of potassium hydroxide. For this reason the cell is chemically stable either in the charged, discharged or any intermediate condition.

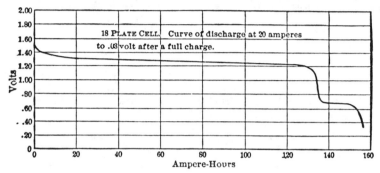

Fig. 5.— P. D. CURVE FOR E. 18 CELL.

ELECTRICAL THEORY.

Electromotive Force.

Various observations both of the authors and other observers show that the discharge curve of e.m.f. of the Type E cell possesses the following peculiarities:

(1) An initial period of rapid descent occupying about 10 per cent of the delivery period.

(2) A nearly steady gradient of gradual descent continuing until within about 10 per cent of the end of the whole delivery period.

(3) A final period of rapid descent occupying the last 10 per cent of the delivery period (assumed as stopping at a potential difference of 0.75 volt).

A fair sample of this curve is represented by the heavy line *a b c d* in Fig. 6, where *ab* is the initial descent, *bc* the nearly uniform descent, and *cd* the final rapid descent. The initial e.m.f. at *a* is influenced by the previous history of the cell, more particularly by the time which has elapsed since the close of the preceding charge. The range of initial e.m.f. is between 1.35 and 1.65 volts. It would seem that this e.m.f. depends upon the amount of gases remaining after the charge, occluded in the pores of the cell.

The e.m.f. at and near the end of the discharge is also somewhat variable, depending in some measure upon the rate and nature of the discharge. It clearly accompanies the exhaustion of the active materials.

Between the points *b* and *c,* the gradient of descent is substantially uniform and approximately independent of the discharge rate within the usual working limits. This diminution of e.m.f. may possibly be due to the increasing thickness of the layer of effete material formed by the progressive exhaustion of the active material.

If the curve *a b c d* be integrated, the mean ordinate is found closely to approach the actual ordinate at *C* or the point of half delivery. This means that a straight line *A C D* may be chosen nearly coinciding with the steady gradient of the actual curve *bc.* This is shown in a dotted line *A C D* in the figure. The area above this line *A a b* at the beginning is approximately equal to the area below the line *c D d* at the end. These areas represent, therefore, sensibly equal and opposite quantities of electrical energy, with respect to the curve *a b c d.* Consequently, we may for practical purposes replace the actual curve *a b c d* by the equivalent straight line *A C D,* which is represented by the formula

$$e = 1.37 - \frac{x \times 0.14}{100} \text{ volts}$$

where *x* is the percentage already yielded, of the full delivery. The mean discharge e.m.f. is, therefore, 1.30 volts. This approximation greatly simplifies practical computation with the Edison battery.

Internal Resistance.

It has been observed both by the authors and by other experimenters that the internal resistance of an Edison Type E cell is substantially constant during the main working portion of the discharge corresponding to *bc* in Fig. 6. Thus in the E-18 cell, the internal resistance at ordinary temperatures is about 0.0022 ohm. This would correspond to 0.0278 ohm in a Type E similar cell weighing 1 lb.; or to 0.0126 ohm in a cell weighing 1 kilo. The internal resistance is slightly less at the beginning of the discharge, but becomes considerably greater near the end of the discharge. Since, however, this rise in resistance occurs only during a small portion near the end of the delivery, its effect in the total discharge may be neglected for most practical purposes.

The internal resistance does not vary greatly with the discharge rate within the usual working limits of 0 — 150 amperes. In other words, the drop of pressure in the cell is approximately proportional to the discharging current strength.

Power.

The maximum amount of power obtainable in the external circuit of any cell is $\frac{e^2}{4\,r}$ watts; where e is the actual e.m.f. and r the actual internal resistance of the cell. Under these conditions the cell will work at 50 per cent efficiency, with equal internal and external resistance. In other words, it is impossible to obtain from any cell a greater amount of external power than corresponds to the above expression, which represents the maximum power of the cell. Taking the mean e.m.f. of the Edison Type E-18 cell as 1.3 volts and the internal resistance as 0.0022 ohm, the maximum power is 192 watts or about 1/4 horse-power or 15.15 watts per lb. (33.4 watts per kilo) at a current strength of 296 amperes, corresponding to a discharge time of less than one-half hour. We have taken a current averaging 250 amperes from such a cell on approximate short-circuit (through heavy leads and an ammeter) for a period of 30 minutes.

The practical rate of power delivery at 150 amperes is $\{\,1.3 - (150 \times 0.0022)\,\}\,150 = 145.5$ watts, or 11.5 watts per lb. (25.4 watts per kilo). This is on a par with the full-load output of good dynamo machines.

Energy.

The energy liberated in an Edison E-18 cell is 1.3 volts $\times$ 141 amp-hours $= 183.3$ watt-hours $= 14.48$ watt-hours per lb. (31.9 watt-hours per kilo) and this liberation of energy is constant to a first approximation for all rates of discharge within the working limits. The amount of this liberated energy which is delivered in the external circuit depends only on the electrical efficiency of the circuit, or on the drop of pressure in the cell. Thus, at 90 per cent electrical efficiency, or 10 per cent internal drop (corresponding to about 60 amperes in the E-18 cell), the externally delivered energy would be $183.3 \times 0.9 = 165$ watt-hours.

It can be shown that if any Edison Type E battery, in any grouping of cells, be discharged at the rate of p watts per unit mass, reckoned at the mean voltage of discharge, the electrical efficiency,

η, of the battery circuit will be $\eta = \dfrac{1 + \sqrt{1-n}}{2}$, where $n = \dfrac{p}{P}$ and P is

the maximum power per unit mass or $\dfrac{e^2}{4mr}$, where m is the mass of a cell.

Thus if a battery of 68 Edison Type E-18 cells be discharged at 120 amperes or 124.3 average external watts, the total external power will be $68 \times 124.3 = 8454$ watts, and the total weight being 861 lbs., the power rate per lb. taken from the battery is 9.83 watts. The maximum power rate is, however, 15.15 watts per lb. as above; so that the ratio $n = \dfrac{9.85}{15.15} = 0.65$. The electrical efficiency of the battery circuit is, therefore,

$$\eta = \frac{1 + \sqrt{1 - 0.65}}{2} = 0.796$$

or 79.6 per cent. The delivery per lb. of battery will, therefore, be $14.5 \times 0.796 = 11.5$ watt-hours per lb., and the entire battery will deliver 9900 watt-hours to the external circuit, when discharged to 0.75 volt P. D. per cell. The same formula applies to *all* batteries capable of maintaining definite values of e and r under load.

CHARGING CONDITIONS.

Electromotive Force.

The charging e.m.f. of an Edison cell is approximately 1.6 volts (somewhat greater at the outset) until about 60 per cent of the charge has been stored, when it rises to about 1.75 volts, simultaneously with increased evolution of gases. This rise is, therefore, apparently connected with gaseous polarization. To the e.m.f. of the cell must be added the drop in internal resistance, in order to find the potential difference at charging terminals. The mean e.m.f. of the cell during charge may, therefore, be taken as roughly 1 2/3 volts. The resistance of the cell during charge is approximately the same as during discharge, being greater at the outset of the charge, corresponding to the condition of resistance at the end of the discharge.

Gassing.

During the charge of Edison cells, bubbles of gas are liberated at both plates, oxygen at the anode or positive plate, and hydrogen at the cathode or negative plate. The collected gases form an explosive mixture. In the ordinary charging process there is a distinct evolution of gas at the outset which reaches its maximum in about a quarter of an hour. The gassing then decreases to a lower steady value which continues until about 60 per cent of the charge, after which it rapidly rises to a steady much higher value than at first. The electrical energy expended in the liberation of these gases is practically all lost.

To replace the water thus lost by decomposition during charging, a small quantity of distilled water has occasionally to be added to the cells through a filler conveniently adapted to this purpose.

Efficiency.

Since the mean e.m.f. of discharge is approximately 1.3 volts, and the mean e.m.f. of charge approximately 1.67 volts, it follows that the superior limit of watt-hour efficiency in an Edison cell is $\frac{1.3}{1.67}$ or 78 per cent. In practice it is always less than this, due to internal drop.

The efficiency may best be examined first in relation to the voltage of charge and discharge, and second, in relation to the electric quantity charged and discharged.

1. If there were no drop of pressure due to internal resistance, either in charging or in discharging, and all the electricity charged in the cell (as expressed in coulombs or amp-hours) were discharged, the amp-hour efficiency of the cell being thus 100 per cent, the watt-hour efficiency would be 78 per cent or thereabouts. This figure can, in fact, be approached by employing very low rates of charging and discharging; i. e., by taking many hours to each process. On the other hand, the more rapid the charge and discharge, the greater the IR drop in the cell, and the less the efficiency, even on the assumption of 100 per cent amp-hour efficiency. Thus at 60-amperes charging and discharging rates. the potential difference in charge would average about 1.8 volts in the E-18 cell and the potential difference in discharge would average about 1.17

volts to the 0.75 volt limit, representing a watt-hour efficiency of 65 per cent, with 100 per cent amp-hour efficiency.

2. The Edison cell would manifestly possess an amp-hour efficiency of 100 per cent if no gases or irreversible substances were generated in its cycle of chemical action. Thus if a certain number of grammes of ion were reduced and a certain number of grammes of nickel were oxydized by one ampere-hour of charging current, and no other action took place, then on discharge the reconversion of these masses of active material would develop the complete ampere-hour of electricity. On the other hand, every gramme of hydrogen (or the equivalent mass of oxygen) liberated and discharged from the negative plate during charge absorbs 26.8 amp-hours of electricity, which is not returned to the circuit during discharge. The amp-hour efficiency of the Edison cell is, therefore, determined by the amount of gas escaping during the charge. This in turn depends upon the rate of charge, or charging-current strength, since the greater this strength the greater the drop in internal resistance, and the sooner the plates are brought to that difference of potential at which the water is rapidly decomposed.

At high rates of charge, then, the watt-hour efficiency of the cell must fall off, not only because of the voltage drop, but also on account of the reduction of amp-hour efficiency in gasification.

Nevertheless, in many cases, and particularly in automobile work, this sacrifice of energy is amply warranted by the convenience and saving of time effected by rapid charge. The normal charging time of Edison E cells is about four hours. In cyclical charges and discharges of four hours each, the amp-hour efficiency is about 75 per cent and the voltage ratio 70 per cent, so that the watt-hour efficiency may be taken in round numbers as 50 per cent ($0.75 \times 0.7 = 0.525$). It is possible, at a still greater sacrifice of efficiency, to charge the cell at fully double the above normal rate.

TESTS.

Laboratory Tests.

Tests of the various types of Edison cell have been made in great numbers and in great variety at the Edison laboratories. Tests have also been made in Europe on Type D cells and published near the end of the year 1903 by Finzi and Soldati at Milan (Associazione Elettrotecnica Italiana, 12th October, 1903), by M. E. Hospitalier at the Central Laboratory of Electricity in Paris (*L'Industrie Elec-*

trique, pp. 493–497, November, 1903), and by Mr. W. Hibbert and Dr. J. A. Fleming, in London (*Journal* of the Institution of Electrical Engineers, Vol. 33, No. 165, April, 1904, pp. 203–238). The authors also have conducted at Harvard University a number of tests on cells of Types D. and E. All of these tests are fairly concordant and in conformity with the data given in this paper.

The essential results of these various tests are represented in the accompanying chart, Fig. 7. This chart gives the output of the Edison cell (per unit mass) at varying power rates of discharge. Curves I, II and III represent the laboratory tests obtained for Type D by the Laboratoire Central d'Électricité, Mr. Hibbert and M. Hospitalier, respectively. Curve V represents the corresponding values for the E-18 cell on normal-rated capacity and delivery. Curve IV represents, on the other hand, the corresponding values

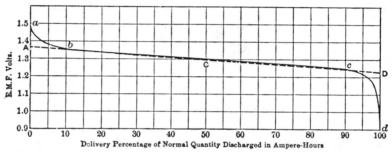

FIG. 6.—ACTUAL AND EQUIVALENT E.M.F. CURVES DURING DELIVERY.

for the E-18 cell, when operated under laboratory conditions of maximum capacity and delivery. In practice, the output may be expected to fall between these two curves, IV and V.

Although the authors' tests of the Type E cell have been confined to the E-18 12 1/2 lb. cell, factory tests of the larger sizes of this latest show about 12 per cent greater maximum output per unit mass than corresponds to Curve V. This may be accounted for by the lesser proportion of dead weight (solution, jar and connections) than is presented in the E-18 cell.

Automobile Tests.

Much experience has been attained in actual service from the Edison battery on automobiles. This experience has led to the elimination of some of the minor practical difficulties that any new battery always encounters. The successive improvements in the battery have, in fact, been due to the perception of defects discovered in actual vehicle practice.

Mr. H. M. Wilson of the Boston Edison Illuminating Company reported some observations on an Edison automobile battery of 68 D-28 cells regularly employed in a single-seat service wagon of the Boston company carrying two persons. The weight of the battery was 1260 lbs. (570 kgs). In four rows of 17 cells each, it occupied a space (including trays and lugs) of about 6 ft. x 2 ft., and was about 1 1/3 ft. high, thus having an over-all volume of 16 cu. ft. (0.45 cu. meter). The weight of the vehicle complete was about 3150 lbs. (1425 kgs). During 22 working days of regular urban service in October and November, 1903, the battery received 1380 amp-hours at 110 volts, or 151.4 kw-hours; and delivered at terminals 68.85 kw-hours at an average potential difference of 85 volts. The amp-hour efficiency was 58 per cent, the volt efficiency, 77 per cent, and the watt-hour efficiency, 45 per cent. The distance run was 324 miles, or 14.7 miles daily, and the energy consumed per mile was 0.47 kw-hour. These results are not only much surpassed by similar batteries of type E, but they are also stated to be exceeded by later batteries of the type D here referred to. Consequently, these results may be regarded as conservative.

DURABILITY AND DEPRECIATION.

It would seem that the Edison cell is so durable that no electrical depreciation is discernible in the cells during the three years' total experience of the practical construction of the battery. No mechanical corrosion of the plates or pockets has been discernible during that time and no depreciation seems to have yet occurred in the active material, judging from the capacity tests of cells which are stated to be as great at the present time as they were when the cells were first constructed. The authors have recently confirmed this observation in the case of Type D cells that have run nearly 3000 miles in an automobile of the Edison Company in Boston. The capacity of these cells was found to be equal to that of new cells of the same type. Two of these cells were opened for examination and the sediment in them collected and dried. The dry sediment weighed 3.9 and 7.1 grammes respectively, probably less than one-third of 1 per cent of the active material in the grids. In fact, a new cell freshly set up will show about this quantity of material washed by the solution from the external surface of the plates. No signs of depreciation or corrosion appeared on any of the plates or connections.

Effect of Rest.

According to observations at the factory, Edison cells lose 15 per cent of their amp-hour charge in eight weeks of idleness. Another test showed 11 per cent of loss by standing one week. Hibbert's tests on Type D showed 9 per cent loss in 48 hours and 27 per cent in 26 days. Hospitalier's tests on Type D showed less than 10 per cent loss after 24 days. A test of the authors on four Type E-18 cells gave, after 26 days' idleness, 100 amp-hours per cell. This represents a loss of 9 per cent of the normal-rated capacity of the cells or 28.6 per cent of the maximum-rated capacity. At a corresponding charge the delivery within 24 hours would have been about 140 amp-hours.

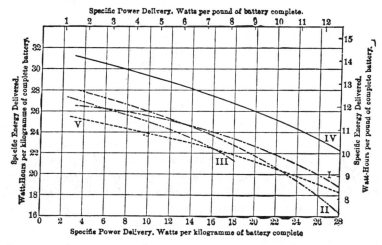

Fig. 7.—Curves showing useful energy delivered in discharging to a P.D. of 0.75 volt at varying power-rates of delivery.

All of the tests show that the cells are remarkably immune from deleterious effects due to careless treatment. Cells have been allowed to dry out, have been permanently short-circuited or even charged in the wrong direction. These cells have shown a full restoration of their capacity after a prolonged restoring normal charge.

CONCLUSIONS.

The Edison nickel-steel alkaline storage cell, in its large output at heavy discharge rates, its low depreciation in capacity and its durability under severe and adverse treatment, approaches the capabilities of a piece of mechanical apparatus more nearly than is ordinarily credited to electrolytic structures. For this reason it is specially adapted to automobile service, where the treatment is abnormally arduous and severe.

The authors desire to express their indebtedness to Mr. Edison for the cells which he placed at their disposal for test as well as facilities for becoming acquainted with the facts concerning their manufacture. Also to his assistant, Mr. R. A. Fliess, of the Edison Storage Battery Company, who very kindly placed an invaluable collection of experimental data at our disposition. We are also indebted to Mr. C. L. Edgar, president of the Boston Edison Illuminating Company, for the use of his battery in tests and information concerning the same from his assistants.

DISCUSSION.

Prof. C. F. BURGESS: I should like to refer to some of the tests which Messrs. Almond and Davidson, in the department of applied electrochemistry of the University of Wisconsin, have carried out on the Edison cell during the past year. Their results corroborate, to a considerable extent, the figures given in the paper just presented. The tests were made on two type " E-27 " cells taken from an automobile in service in Madison, and submitted to 20 charges and discharges. The charging rate was 6 hours, and the rate of discharge varied in different runs from 1 to 6 hours. In no case was a watt-hour efficiency obtained which materially exceeded 50 per cent, even on the lower rates of charge and discharge. It seems possible, however, to obtain a slightly higher efficiency, by decreasing the amount of charge given to the battery.

Tests were made on the loss of charge while standing on open circuit, and it was found that standing fully charged during 30 days and then discharged at a five-hour rate, the loss of ampere-hours was 25, or 19.5 per cent. In regard to gas evolution, some interesting observations were made. The total volume of gas on the charge and discharge was measured on several runs. It was found that with a 22-ampere charging rate there was .59 of a cubic foot of gas liberated during the charge, as measured by an ordinary gas meter. At a 35-ampere charging rate the amount of gas liberated was .63 of a cubic foot, while at the rate of 175 armatures—a very high rate—the total amount of gas liberated was 1.59 cubic feet, thus showing where a good deal of the energy goes. It was found that during discharge at a low rate there was about .01 of a foot of gas liberated, and at the highest rates of discharge gas was not liberated in sufficient quantities to be measured. I would like to ask the authors of the paper if they made any observations upon the amount of gas liberated during the charge or discharge of the cell as confirming the above mentioned measurements?

Dr. A. E. KENNELLY: We made no measurements of the amount of gas liberated in the cells.

The gist of our paper is that the Edison cell appears to possess a definite energy-storage capacity per unit weight (in the measurements detailed in our paper about 32 watt-hours per kilo), of which any desired portion may be delivered to the external circuit by suitably proportioning the external resistance, or, in other words, by assigning the electrical efficiency of discharge. Expressing the same result in another way, the energy delivered to the entire discharging circuit (both internal and external) seems to be practically the same within all working ranges, and the externally delivered energy only varies as the external resistance forms a larger or smaller portion of the total resistance in the circuit. The externally delivered energy will be 60, 75, or 90 per cent. of the total charge according as the external resistance is 60, 75, or 90 per cent of the total resistance in the discharging circuit, at least to a first approximation sufficient for most practical purposes.

This nearly constant release of energy into the discharging circuit and the hardihood of the cell under severe usage are, perhaps, the most salient points in the cell's behavior.

Mr. S. S. SADTLER: I would like to ask Messrs. Kennelly and Whiting whether they noticed any trouble due to the alkaline solution taking up the carbonic acid from the air. I should think there would be a loss of efficiency in that way.

Mr. S. E. WHITING: Owing to the construction of the cell, which is virtually sealed hermetically, the carbonic acid of the air has very little opportunity to enter the cell.

The Chairman then invited Dr. J. W. Richards to read his paper.

4

Reprinted from pages A-34–A-48 of *Symp. Workshop Advanced Battery Research and Design, 1976, Proc.*, Chicago Section, The Electrochemical Society and Argonne National Laboratory, 1976, 422pp.

THE LEAD-ACID BATTERY

A. C. Simon
Naval Research Laboratory, Washington, D.C.

S. M. Caulder
International Lead Zinc Research Organization
Research Associate at NRL

ABSTRACT

The lead-acid battery has shown a slow but steady development over the years of its history. Most of its improvements have been obtained through engineering development and there has been relatively little basic research. At a period when there is the possibility of two major markets for batteries, those of energy storage and electric vehicles, one should not too soon dismiss this battery from consideration. Although at present marginal for both of these applications, there is the distinct possibility that with the proper research the life and energy density of this battery can be increased to the point where it may look a lot more attractive than at present. Some of the problem areas and possible solutions are discussed.

INTRODUCTION

At the present time the lead-acid battery appears to be in a very auspicious position. Despite temporary set-backs caused by the present state of the economy, the starting, lighting and ignition battery market has shown a steady increase that promises to continue with population growth. There is also a growing market for specialty batteries for self-powered tools, recreational vehicles, toys and various other self-powered equipment. With serious consideration now being given to batteries as a means of energy storage for the electric power generating industry and also as the main power source for electric service vehicles, we are faced with the realization that, at least for the near future, the lead-acid battery is the only feasible contender for these services.

On the other hand there are a number of factors that could seriously alter the present favorable position of this battery. There are alternate battery systems now being developed that promise to have much higher energy densities than does the lead-acid battery. There are constantly changing conditions in the materials supply situation that may drastically change the availability or cost of various battery components. Environmental protective regulations are causing increasingly difficult problems in product waste disposal. Energy shortages not only could greatly curtail automobile usage, with attendant disaster to the SLI market, but could also increase the cost of production of such items as containers, separators and other battery components. Even now, changes in grid alloy composition, required to meet the demand for maintenance-free batteries, are making it increasingly difficult to use recycled lead in production.

73

In order to meet these problems a great deal more reliance must be placed upon research efforts than has been the case in the past. Some of this research and development effort is currently underway, some is in the planning stage, and there are some areas where research will be needed that as yet have not even been explored.

CURRENT STATE OF THE ART

A.   Underline Underline

The most significant recent developments in lead-acid battery engineering have been the introduction of light weight, high temperature and impact resistant plastic containers; the manufacturing of sealed and maintenance-free batteries and the development of lighter weight electrodes and connectors.

Attempts at discrediting the lead-acid battery usually concentrate on the low energy density of this system. Until recently the container was a large factor in the overall weight of the battery. The development of plastic containers has made possible a package that not only looks better but is also smaller and of less weight than its predecessors.

The use of thermo-plastics such as polyethylene, polypropylene, polyvinylchloride and polycarbonate is rapidly becoming standard in automotive and stationary batteries. These plastics may be transparent or translucent (for easy monitoring of electrolyte level) or highly colored to increase their sales appeal. The decrease in weight not only permits a greater energy density but, since a decreased wall thickness is required, a greater internal volume, thus allowing for a greater number of plates and giving an increase in capacity.

In addition, the self extinguishing flammable plastics such as polyvinyl chloride reduce the fire hazard and the semi-flexible nature of some of the plastics greatly reduces the possibility of breakage. Along with these advantages have come better methods of sealing the tops and terminals so that leakage problems have been greatly reduced, if not eliminated.

A relatively new and fast growing field is that of specialty batteries such as those used to provide power for emergency lighting, burglar and fire alarms, portable medical instruments, toys, handyman tools, etc. The first maintenance-free, non-spill batteries were developed for such applications. These batteries either contain a gel electrolyte or an electrolyte-containing separator, and the grids are usually of a lead calcium alloy or pure lead to reduce gassing and its accompanying problems.

In all such batteries care must be taken to limit the end of charge voltage. They are not truly sealed since they provide a Bunsen type valve for venting in an emergency, such as overcharging. Most such batteries are designed for low to medium discharge rates. They have a linear charge loss of about 3 to 4% a month, so that their shelf life is long as compared with other lead-acid batteries.

A different approach to obtaining maintenance-free batteries is to use recombination catalysts such as platinum or palladium alloys. These proprietory alloys are incorporated into the electrodes or placed in specially designed caps so that the hydrogen and oxygen that are formed during electrolyte decomposition are recombined and no loss of water occurs.

Recently there has been a demand for maintenance-free SLI batteries. The manufacturers of SLI batteries in this country have on the market or under development long life maintenance-free SLI batteries, the success of which is dependent principally upon the use of lead-calcium alloys in the grid, high purity active material and a somewhat larger than usual electrolyte reservoir. By the use of lead-calcium, rather than lead-antimony, grid alloys the deposition of antimony on the negative is avoided, so that very little water undergoes decomposition. Most modern car batteries, during car use, tend to be maintained at almost constant potential by the alternator and voltage regulation systems. This makes possible the use of lead-calcium grids, which normally do not stand up well to charge-discharge cycling. Thus maintenance-free service can be obtained with these grids without loss of battery life. Although maintenance-free batteries for car use appear to be sealed, investigation shows that venting is provided.

Maintenance-free batteries are also being tested in Europe. The approach used there, however, has been to develop low antimony-lead alloys (less than 3% Sb) which are apparently satisfactory. The much lower rate of antimony transfer to the negative, using these low antimony grids, reduces self-discharge to acceptable levels.

A great deal of unusable battery weight resides in connectors and grids, and the industry in the past few years have been highly successful in reducing this toward a minimum. By the use of cell connectors that pass through the cell wall rather than out of the top of the cell a considerable weight reduction has been achieved as well as lower electrical resistance. Studies of grid design have recently shown relationships between grid resistance and grid design that have made possible the design of grids that achieve the maximum support with the least electrical resistance and minimum weight consistent with the type of service for which they are intended.

Three considerations limit the possible weight reduction of the grid. One is the need for sufficient grid conductors to be present and properly distributed to supply the necessary electrical conductivity at specified rates of charge and discharge. A second consideration is the necessity for sufficient grid material to support the active material throughout the steps of manufacture and in subsequent use. Finally, it is necessary that the grid be of sufficient thickness to provide for the two above considerations throughout the required service life, since during this period there is a continuous reduction of grid thickness by a corrosion process.

No corrosion-free grid, that is economically feasible, has as yet been produced and grids continue to be made of lead or lead alloys. At present lead-antimony and lead-calcium appear to be the most feasible alloys although research is continuing in this area to develop other alloys with equal or greater corrosion resistance.

75

There has been considerable interest in the use of lead plated aluminum as a grid material since it would improve conductivity and provide a lighter grid. Most attempts to use this or other plated metals have been unsuccessful although Sweden's Axel Johnson Institute for Industrial Research claims to have developed lead coated aluminum grids that have been successful in tests and they are now offering this process to battery manufacturers. In this country, claims have been made for an undercoat process that makes the lead coated aluminum impervious to attack even if the lead plating is not perfect. This latter process is now being tested by the Electric Power Research Institute. One hopes that these claims will be proved valid, since such a lead plated aluminum grid would not only increase the energy density and reduce internal resistance but could possibly also reduce costs of manufacture.

A battery development claimed to give extremely long life is the computer designed battery of Bell Telephone Laboratories. This battery uses pure lead grids of a conical construction. This is supposed to be the best compromise between grid growth and active material expansion, and thus best suited to retain the active material pellets throughout life. The active material paste is made from chemically prepared tetrabasic lead sulfate, and does not require the curing steps of conventional paste formulations. A very long life is claimed for this battery under the emergency stand-by, float potential operation required by the telephone industry.

B. Environmental Pollution Considerations

As the populace gradually becomes aware that the environment is becoming poisoned by society's own waste products, there is a growing hue and cry to control all manufacturing processes to the point where there are no harmful environmental pollutants. Since, in many cases, there are no reliable figures as to what constitutes harmful pollution for a given product there has no doubt been a tendency to impose unrealistic limits.

In this respect lead is no exception. The battery manufacturers are finding it increasingly difficult and expensive to comply with restrictions on pollution being enforced by Federal and local agencies. This applies equally to the lead content in the waste water from battery plants, to the lead content in the air surrounding such plants and to the lead content in the workers within the plant. Similar and perhaps worse problems of lead control will be faced by the primary and secondary lead smelters. It is also probable that arsenic and antimony will face similar regulation as their properties become better known to the public.

Manufacturing dry charged batteries contributes greatly to lead bearing waste water. It has been necessary in the past to wash the plates after formation to remove excess acid. Considerable effort is now being directed toward eliminating this washing. Gould and Globe have produced, by slightly different procedures, batteries that eliminate the washing procedure. These batteries can be shipped and stored "dry" and activated when desired by the addition of electrolyte, although the plates in such batteries are not really dry but contain 20 to 40% of the acid that would normally be present in a wet battery. The elimination of the washing step saves up to 40 gallons of wash water per battery and gas savings also occur from not having to dry the battery plates. The cost of such processes are substantially lower than for the drying ovens, washing equipment, water clarification, etc. required for dry charging.

Although the public has now become accustomed to dry charged batteries an alternate route is to return entirely to the shipment of wet charged batteries. Some manufacturers are considering this option and are experimenting with methods to reduce loss of charge and to minimize corrosion during storage.

A final point is the increasing hazards faced by the manufacturer in the marketing of his product. The formation of the Consumers Products Safety Commission and applicable regulations such as the Hazardous Substances Act are examples of the trend in government today to protect the rights and safety of consumers while increasing the responsibilities of the manufacturers.

According to the Kierney Management Consultants the product liability claims in 1972 totaled 12.5 billion dollars, which represented a 100-fold increase from 1965. It is foreseen that, at this rate, the 1975 claims may reach 50 billion dollars. Because of possible mishaps with the acid electrolyte as well as the possibility of explosion, electrical burns, etc., the lead-acid battery can well be considered hazardous. This applies particularly when the many new uses to which this battery is being applied are considered. Therefore research and development into providing a hazardproof battery must also be considered by manufacturers. It has truly become an age in which environmental impacts and environmental protection costs must be considered in every step of manufacture and distribution.

We mention these burdens on the manufacturer, not as an indictment of the lead-acid battery, but only to show the increasing headaches that all manufacturers must face. It is also well to consider what such legislation and restrictions will mean to manufacturers of sodium-sulfur, lithium-chlorine, lithium sulfide, or similar high energy density batteries that contain highly reactive, extremely dangerous materials. When one is tempted to paint too optimistic a future for such batteries these points should be kept in mind.

C.   Separators

Separators are an essential part of battery construction, and properly built separators contribute greatly to both battery life and power. Needless to say, the separators serve the multiple purposes of preventing electrical contact of plates of opposite polarity, restraining the movement and bridging of sediment particles, providing needed electrolyte space at the surface of the plate and assisting in the retention of the active material. There is also evidence that they reduce self-discharge by restraining the migration of antimony to the negative plate. At the same time they have the detrimental effect of increasing the internal resistance of the battery by interfering with current flow through the solution.

There are several essential requirements that must be met by successful separators. The first is mechanical strength sufficient both for the handling during assembly and in subsequent life service. Next is an ability to withstand both the corrosive effects of the acid electrolyte and the redox conditions generated during battery operation. Third, the separator is required to have low electrical resistance and uniform and controlled pore size. Fourth, the separator must not contain soluble impurities, such as chloride or metals, that would interfere with cell performance or produce side reactions. Finally, a demonstrated ability to inhibit the transfer of antimony is desirable when lead-antimony grid alloys are used.

At the present time many different separator materials are available, of different porosity and thickness, and separators can be tailored to specific uses of the battery. The original wood separators were first replaced by microporous rubber separators which are produced by compounding sheet rubber formulations with a silica hydrogel and subsequently driving off the water to develop the necessary porosity. This separator material has pore dimensions in the order of one micron and is relatively pure. Unfortunately it is fairly brittle so that it normally is not produced with a rib thickness of less than .020". It has now been largely replaced by other types, but is used in some applications.

Cellulose separators have been in use for a long time and are currently still in use. Cellulose mats of suitable pore size and corrosion resistance are made up and impregnated with phenolic resin and cured. These phenolic impregnated cellulose separators can be produced in corrugated form or with extruded ribs on a flat sheet. While the porosity of these separators may be as high as 60%, they are brittle and the mechanical strength and corrosion resistance are only fair. Pore size is not particularly uniform and averages about 15-25 microns.

A major improvement in separators is provided by the microporous polyethylene separator, which is compounded of a mixture of silica, ultra high molecular weight polyethylene and a plasticizing oil. These components are extruded as a continuous sheet which is then treated to extract most of the oil, thus developing the required porosity. The material produced in this way has extreme toughness combined with flexibility and has excellent corrosion resistance and mechanical strength. The pores are much finer than in other separator material, averaging about .05 micron. This material can also be made into very thin separators because of its excellent mechanical strength and flexibility.

Polyethylene separators are also made from 1-5 micron fibers produced by blowing from the melt. The filaments are thinned and aligned by a high velocity hot air blast. The resultant mat is then sintered by pressure and heat into a continuous structure. Although this material has the same mechanical strength as the microporous polyethylene described above it has a larger pore size, averaging about 15 microns. Since it is made without the use of fillers or binders its purity is exceptional.

Sintered polyvinyl chloride was a pioneer polymeric material for separator use. Its manufacture involves forming the required shapes from finely divided PVC powder and then sintering the particles. Over the years this product has been improved somewhat, principally by the development of finer grades of PVC powder and by reducing the separator thickness. This separator also appears to be undergoing replacement by materials of greater flexibility, mechanical strength and porosity, although recent developments have brought the pore size down to about .05 microns and considerably increased the flexibility.

Fiberglass separators are also made. These are made from glass fibers that are subsequently bonded by heat and pressure. They may contain an intermediate layer of silica powder and latex coating to reduce pore size and inhibit antimony transfer. These separators are tough and flexible enough for most applications. Fiberglass is usually found as an active material retaining layer in partnership with one of the above mentioned separators.

For high power applications low electrical resistance can be obtained by reducing rib thickness and using material of high porosity. Such advantages are usually accomplished at the expense of corrosion resistance and mechanical strength. A primary consideration, in addition to those mentioned above, is cost. This may in many cases determine which of the above types of separator is chosen, irrespective of other advantages. In this area, also, energy shortages have had an effect on cost, and safety and health have become a source of concern.

As to the future, consideration is being given to thinner separators, with lower electrical resistance. Sealed separators or wrapped separators would prevent edge shorting and probably greatly extend life, as well as retaining sediment. For this purpose flexibility and sealability would be paramount among desirable properties. The increasing use of maintenance-free batteries will require separators of exceptional purity. When lead-calcium grids are used their tendency to cause active material softening will best be controlled by enveloping the plates with separator material of great flexibility and very small pore size.

D.   Batteries for Vehicle Propulsion

There is no battery now available or in prospect that can drive a vehicle with the performance and range that can be achieved with an internal combustion engine. There is no compact, packaged, electrochemical energy system that can compete in weight, portability or economy with gasoline.

However, the time may be fast approaching when the fossil fuels are either not available or become so much more expensive that the public will gladly accept vehicles of the limited range and speed offered by battery power. Increasing atmospheric pollution from internal combustion engines is also a factor that will act to promote the use of all-electric vehicles or hybrids.

Electric vehicles using lead-acid batteries, such as fork lifts, trucks and similar commercial applications have been able to successfully compete with the internal combustion engine for a number of years because of economical service and the non-polluting characteristics that electric vehicles offer. This type of battery use will undoubtedly become more popular in the future. In fact, a number of recent studies here and abroad have shown that, for various types of stop and go vehicles, such as milk vans, mail trucks and buses, the battery propelled vehicles show much lower operating and maintenance costs over their service life than do their internal combustion engine counterparts.

Thus, for industrial vehicle applications, there is no doubt that the lead-acid battery has proven practical. Recent improvements in charger design, which permit fast charging yet prevent overcharging and allow a minimum of supervision have done much not only to extend the life but improve the popularity of such vehicles. At the same time there have been continuing improvements in cases, grids and separators that have increased the dependability and life of the batteries used in such applications. In this type of application, performance and range are secondary to economic or special environmental consideration and the lead-acid battery has proven itself acceptable.

The case is quite different with automobiles used in normal family pursuits. Here economic considerations have been consistently sacrificed to the goal of ever increasing performance and it will take extensive consumer education to gain acceptance for electric automobiles, unless energy shortages make their use mandatory.

Even in this case, however, the low energy density lead-acid battery is the only feasible electric storage system currently available and one might say that, for automobile operation, this battery is not inspiring. The electric car designer is faced with a vicious cycle. Increasing the battery size to obtain increased performance adds sufficient additional weight to largely defeat the effort, not to mention the mounting cost. According to Dr. Paul D. Agarwal of the G. M. Research Laboratory, to equal the heat energy stored in a 20-gallon tank of gasoline would require 15,000 lbs of batteries. In comparison, a sub compact internal engine car with a full 20-gallon tank need not weigh more than 2200 pounds.

However, the Electric Vehicle Council has compiled figures that indicate that most existing automobiles usually run short trips and stay under 35 miles per hour. Considering our large urban population this is probably correct. It would seem especially true in the case of second cars in a family and there are about 25 million such multicar households in the U.S.

While it would appear that an electric car must be designed with all elements keyed to the propulsion system there have been numerous conversions of existing cars to battery propulsion. While these demonstrate the feasibility they must be considered as short term intermediates to the true electric automobile.

The Fiat X1/23, for example, has a curb weight of about 1800 pounds, uses nine 12-volt batteries and has a 65 mile range and a 40 mph top speed. The CGE-Gregoire is powered by eight 12-volt batteries, has a curb weight of about 2000 pounds and claims a 60 mile range with a 55 mph top speed.

An example of a car specifically designed for electric operation is the British Enfield 8000. This two passenger vehicle has a maximum speed of 40 mph, with a 62 mile range under peak traffic conditions through the heart of London. The vehicle uses four 12-volt 110 Ah batteries, providing power through a series/parallel control system. The curb weight is 1800 lbs.

Although not generally realized, there has been a steady gain in energy
density by the lead-acid battery over the past years. According to figures
presented to the 5th International Lead Conference in Paris, November, 1974
by the Chloride Group, Inc., England, the typical energy density at the
5 hr rate has increased from 20.9 Wh/kg (9.5 Wh/lb) in 1920 to 29.2 Wh/kg
(13.2 Wh/lb) in 1970. Current U. S. traction batteries have energy
densities of 31.5 Wh/kg (14.2 Wh/lb) while batteries in certain Japanese
vehicles have energy densities of 36 Wh/kg (16.32 Wh/lb). The Japanese
have set a goal of 60 Wh/kg (27.2 Wh/lb) but so far have achieved only
48 Wh/kg (21.7 Wh/lb) with a life of only 150 charge-discharge cycles.
While waiting for high energy density batteries it is well to ponder the
following.

The higher energy density batteries that are promised for the future
can no doubt greatly improve these performance figures. But in order for
electric vehicles to be successful the vehicle itself must be specifically
designed for batteries and there must be improvement in motor design, control
mechanisms, power transmission and vehicle design. These concepts can be
tested with the lead-acid battery.

Serious development on electric vehicles should therefore start now,
using the lead-acid battery as a basis. There is a waiting second car
market for electrics that will increase as gasoline becomes more scarce
and expensive. As we approach the convenience and economy of mass production of electric vehicles, with attractive financing programs for the consumer and nationwide sales and service this market should continue to grow,
even without super batteries.

The sodium-sulfur, the lithium chlorine, the lithium-sulfide systems
all have a very attractive theoretical energy density. But practical
compromises to obtain a working cell have in each case greatly reduced the
actual energy density below the theoretical value, just as is the case in
the lead-acid battery. In each case the high temperatures required for
operation would make intermittent operation difficult. Various operational
problems have also made it difficult in each case to produce batteries of
extended service life, even under laboratory conditions. In the case of
the sodium-sulfur, a limited power density would almost certainly require
operation in parallel with another power source (hybrid application).

The zinc-air battery gives performance superior to the lead-acid but
rechargeability remains one of the major problems. Other systems that
might be considered require circulating pumps, heat exchangers and other
supplemental equipment that reduce their actual energy density and add
considerably to the complexity of operation.

The high temperature systems make claims of being very inexpensive,
because of the abundance and low cost of their components. This may be
more than offset in actual production by the high costs associated with
the rigid quality control that such batteries will require throughout
all steps in fabrication and assembly.

On the basis of these considerations we believe that if electric automobiles are produced in quantity, a large number of them will be lead-acid
battery powered for a long time.

E.   Batteries for Energy Storage and Peak Shaving

The steadily increasing demands for electrical power and the fact that there are peak periods of electrical demand much larger than the average load are a matter of concern to the suppliers of electrical power. These peak demands can be met by a large capital outlay for new equipment but this is uneconomical because of the long periods during which such equipment would remain idle. Electric utilities therefore use their most efficient and economically obtained power for base loading, while using less efficiently generated power for intermediate and peak energy demand periods. For example, plants using nuclear fuel or water power might meet peak demand with less efficient and more costly fossil fuel sources of power.

It is obvious that methods of storing excess power generated during off-peak periods to be used at periods of peak demand offers an economical means of fully utilizing generating equipment and avoiding crisis in electrical supply. The practicality of this has been proven in the past by systems using pumped hydroelectric storage. However, there are relatively few areas where conditions are suitable for this type of storage. One practical alternative system is that using secondary batteries to store energy. This idea is attractive because batteries would allow a very efficient energy transfer and the modular construction, non-polluting character of the installation would have very little, if any, effect on local environments. Consequently, such load leveling systems could be distributed throughout the utility network at sites near load centers or wherever their use would result in savings in transmission line costs. Since the batteries are modular units they could be efficiently scaled to whatever size local conditions required, and easily modified for changing requirements. In addition, as has been well demonstrated, batteries have additional advantages such as surge suppression, voltage regulation, filtering, etc., that would not be obtainable from other methods of storage.

As a point of fact, such energy storage plans were originally based on the use of high energy density batteries that are not yet available. However, as plans progressed it became evident that lead-acid batteries are capable of meeting this need now and possibly even in the more distant future when such high energy batteries do appear.

At the present time there are two design studies that illustrate this point. The first is that sponsored by the Energy Research and Development Administration and the Electric Power Research Institute that proposes to build a test facility (BEST) for testing various batteries for applicability to energy storage. As presently conceived this test facility will have three bays, each capable of testing a 1MW, 10MWh battery, with a fourth bay containing the control equipment, computer facilities and test instruments. At the present time the lead-acid battery is the only system available for test and plans are being formulated to obtain suitable lead-acid batteries for this facility.

The second design study involves the construction of a 20 MW, 200 MWh lead-acid battery energy storage demonstration plant, sponsored by the Energy Research and Development Administration.

The feasibility studies have brought out the fact that the principal
problem at present is that of producing these batteries at a cost that will
be attractive to the electric utilities. However, with careful planning,
it appears that this difficulty can be overcome. Past experience with
scale-up of large industrial batteries and submarine cells have given the
lead-acid battery companies sufficient experience with large cells so that
state-of-the-art batteries are presently available that can probably meet
the requirements of the BEST facility, with its 1 MW, 10 MWh capacity.

With the 20 MW, 200 MWh facility the situation is somewhat different.
In this case larger modules are proposed than are currently available and
the manufacturers are being asked for designs that they feel would most
economically meet the requirements. One problem that is immediately evident
is that present commercial facilities are inadequate to produce either the
case required for such modules or the separators. Other problems concern
the difficulty of transporting and handling such large modules, intercell
connectors, switching and circuit breakers, but none of these problems
appear to be insurmountable.

There appear to be no problems that cannot be solved in connection with
the use of lead-acid batteries for this project, once the manufacturers are
themselves sufficiently convinced of the commitment of Government and the
electric utilities to the project. On the other hand, Government and the
electric utilities are unwilling to become this committed without a
practical demonstration. Under these circumstances the project could reach
a stalemate, disasterous to all concerned.

This idea of energy storage in batteries, however, suggests a potential
market for lead acid batteries that is enormous. Since in such a market,
almost all parts of the battery would be salvageable, it would seem that
batteries for this use can be made cheap enough to appear attractive as
energy storage facilities. The initial amount of lead required would be
very large, but lead is a metal that is plentifully distributed in the
U.S. and friendly nations, so that it is not in critical supply. In
addition, there is the added advantage that there are not so many competing
uses for lead that supply would become a problem. There is, of course,
the additional factor that after a few years and as batteries began to
fail there would be practically 100% recovery of lead from facilities
installed earlier. There is the final advantage that in this application
the problem of energy density would not be as important as in some other
applications, so that the weight of lead-acid batteries would not be as
great a determent as in the usual case.

Because of already demonstrated reliability and low maintenance
requirements, as well as long life (yet to be demonstrated for this type of
application) it is possible that the lead acid battery could remain
competitive for power storage even with the arrival of the higher energy
density batteries that have been promised in the next decade.

LEAD-ACID BATTERY RESEARCH

The foregoing discussion should serve to illustrate that there has been
an on-going improvement in the lead-acid battery during the past several

decades. This has been accomplished for the most part by engineering advances. Because of the highly cost competitive nature of the lead-acid battery business there has been very little funding available for research of a basic nature. Emphasis has been rather on engineering development, automation of processes, improved machinery and other achievements aimed toward improved efficiency and reduced cost.

Although each of the major battery companies has maintained a research laboratory, much of the effort of such laboratories has in the past been expended on the development of other battery systems or to other products that would help the economic position of the company. Relatively little effort has been directed toward understanding the basic mechanisms of the battery. This may be due in part to the fact that the lead-acid battery has proven sufficiently reliable and satisfactory for past needs without the necessity for research of this nature.

Eventually a point will be reached in engineering development where little or no additional improvement will be possible, and at this point further improvement will require more fundamental research than has been used up to this point.

That such research is needed is indicated by the low efficiency of active material utilization; the continuing problems of grid corrosion; the lack of knowledge as the causes of poor active material cohesion and its lack of adhesion to the grid in the positive plate; the continuing problems with both positive and negative degradation with continued cycling; and by the continuous loss of capacity that takes place throughout life. Very little is known about the effect of impurities; what impurities are harmful, and what impurities are of possible benefit; what constitutes acceptable limits of impurities, etc.

Fortunately, in the past few years a change has taken place and the principal battery manufacturers are now employing more people to study battery problems and are obtaining the necessary equipment to make more fundamental research possible.

A very good sign is a cooperative effort being made by a number of American and European manufacturers who have banded together to support a study of grid corrosion at the Battelle Research Institute in Switzerland.

An example of battery research by other than battery manufacturers is that furnished by the International Lead Zinc Research Organization, Inc. (ILZRO). For over a decade ILZRO has been supporting various projects that pertain directly or indirectly to battery problems. While this organization naturally cannot address itself to problems resulting from the use of proprietary products or procedures, it can and does undertake research on some of the lead acid battery problems that are of general importance to the entire industry, and more specifically to those problems likely to be encountered in the use of lead-acid batteries for electric vehicle propulsion.

Aside from an occasional project supported at the universities by one of the battery manufacturers, there seems to be no university interest in lead-acid batteries, at least in the United States. In England, Europe,

Japan and the eastern European countries there is a great deal more research into lead-acid battery problems than has been evident here, in universities, government supported institutions and the battery companies themselves.

One government laboratory in this country, the Naval Research Laboratory, has supported basic work in lead-acid battery research, but this effort has been small in the past, involving no more than two or three people at any time, except during the period of World War II. This effort, although small, has been given considerable cooperation from the battery companies and NRL has received contributions of battery components, test specimens and cooperation in various experiments from the various manu-facturers, as well as valuable information, advice and criticism and all of this cooperation is hereby gratefully acknowledged.

Work at NRL has been primarily directed toward study of the properties of the positive and negative active material and in the determination of how these properties are affected by changes in such factors as paste composition, discharge or charge rate, acid specific gravity, temperature, etc. We have also directed a great deal of effort toward determining those factors responsible for capacity loss, battery failure and poor utilization of active material.

In making these studies we have confirmed that the physical condition of the active material plays a most important part in battery performance and life. For upon the physical nature of the active material depend such characteristics as surface area, porosity, chemical activity, solubility and other factors upon which the rate and extent of reaction occur.

Although the reactions of the lead-acid battery are theoretically completely reversible, they are less so in practice and each charge-discharge cycle yields slightly less capacity than the preceding. This loss of capacity may be traced in some cases to purely physical factors such as crystal size, cases in which the crystal becomes passivated by a reaction layer that stops reaction before the entire crystal remainder can be utilized; or by much more complex factors, such as changes in lattice spacing due to adsorbed impurities that may deactivate the crystal.

Two recently found properties of the positive active material appear to be of great significance in the search for better active material utilization and more prolonged capacity retention. The first of these was the discovery that part of the PbO_2 formed initially was electrochemically inactive in subsequent charge-discharge cycles and, moreover,'that the inactive portion increased with increasing number of such cycles. This discovery was the result of earlier investigations based on thermal methods of analysis.

These thermal studies showed that the thermal degradation process was different for chemically and electrochemically prepared PbO_2. Differential thermal analysis studies also showed that after long periods of cycling, the PbO_2 in the plates yielded a decomposition curve very similar to that of the chemically prepared samples. Differential scanning calorimeter studies showed that an exothermic peak occurred at 180°C for the electrochemically prepared material but no similar curve appeared when the chemically prepared samples were run. High temperature mass spectroscopy at 10^{-6} mm showed that this peak was associated with water evolution. To determine more about the structure that yielded this water on decomposition, pulsed NMR studies of the hydrogen nuclear relaxation times were made of the chemically and electrochemically prepared PbO_2 and of the product obtained after prolonged cycling. It was evident from the relative magnitudes of the hydrogen signals that the electrochemically prepared form contained more hydrogen than the other two. The nonexponential character of the relaxation time also indicated that the electrochemically prepared samples had hydrogen present in at least two different configurations. The chemically prepared and cycled samples showed uniform hydrogen behavior, as evidenced by a single exponential relaxation curve.

The course of the relaxation curve for the electrochemically prepared PbO_2 can be separated into short term gaussian behavior and long term exponential behavior. The long term exponential portion of the curve is similar to that found for the chemically prepared form. From the magnitude of the short term gaussian portion of the curve one can calculate a hydrogen-hydrogen separation of slightly more than 2Å. Since water molecules have a lower separation of 1.6Å, this means that the fast relaxation component is not due to water but corresponds to some other form of hydrogen bonding, the exact nature of which has not yet been determined.

Another recent discovery concerning the lead-acid positive plate that has been made at NRL was that the originally compact and dense PbO_2 found in the formed plate was transformed by cycling into a much more porous and open structure that appears to be continuous throughout the plate. This new structure has been named coralloid because of its resemblance to coral. The transformation to the coralloid form takes place as a gradual process, beginning at the surface and proceeding to the center of the plate. The length of time for the conversion to the new structure varies, decreasing with increasing depth of discharge and with increasing discharge current density.

The configuration of the PbO_2 is apparently an ideal structure since it has strength, rigidity, good electrical conductivity and a large surface area, while providing adequate porosity for electrolyte flow. Unfortunately, any benefits derived from the occurrance of this type of structure are masked by the simultaneous increase in the inactive form of PbO_2.

Eventually, as the battery approaches the end of its life, this coralloid structure disappears and is replaced by a very loosely connected, nondescript structure that resembles neither the original as-formed nor the coralloid structures. Numerous tests have shown that this metamorphous seems to take place in all plates, regardless of their source or method of manufacture. However, the study was originally made only on battery plates that had lead-antimony grid alloys. We are now repeating these investigations to determine whether the absence of antimony, as is the case with plates with lead-calcium grids, will give different results than those previously obtained.

Auger spectroscopy has also been used to study the distribution of impurities within the electrodes during charge and discharge and it has been found that the mobility of some of these impurities is greater than had previously supposed. Results so far have been too rudimentary to analyze but it would appear that some of these migrations of the elements may correlate with capacity loss and disruption of the coralloid structure. This is also an area where more thorough investigations are planned.

We consider these recent observations to be highly significant in the quest for longer life, higher energy density and higher capacity. In particular the discovery that there is an inactive form of PbO_2 present is important. If this inactive form can by some means be prevented from forming, or can be converted to the active form, significant improvement in the battery is possible, both in performance and life.

The Energy Research and Development Administration has recently sponsored this work at NRL, so that additional people can be assigned to the project. This additional support, plus the cooperation of the battery companies and ILZRO, should enable answers to be found to some of the principal problems that limit energy and power density as well as cycle life of the lead-acid battery.

5

Copyright © 1935 by The Electrochemical Society, Inc.

Reprinted from *Electrochem. Soc. Trans.* **68**:159–165 (1935), by permission of the publisher, The Electrochemical Society, Inc.

QUALITY ADVANCES IN THE DRY BATTERY INDUSTRY.[1]

By C. A. GILLINGHAM.[2]

ABSTRACT.

The dry battery is a complicated chemical article incorporating 32 different materials. These are rigidly controlled by chemical and physical tests. Further improvements in quality may be brought about by the introduction of more expensive ingredients. The service performance of the No. 6 (63.5 mm. diam. x 152.5 mm. height) battery has been more than doubled during the last 35 years, and shelf deterioration has dropped to 20 per cent of the former value. Similar marked improvements are recorded for other standard types of dry cell. 　　　　　　　　　　　　　　　　　　　　　[C. G. F.]

INTRODUCTION.

The dry battery is usually considered a product of the chemical industry. To a greater extent than most chemical products it is a complicated article. It combines many constituents, most of them of a chemical nature, in a very compact form. In a D size flashlight cell, for instance [1¼ in. x 2¼ in. (32 mm. diam. x 57 mm. high) in size], there are combined thirty-two different materials. Their chemical and physical requirements are rigid, and the proportions and methods required in their assembly are critical. The finished product must perform well under a wide variety of severe conditions and in the hands of inexperienced users. Its behavior must be good in the tropics and in the arctic in spite of the fact that its reactions are strongly affected by temperature. The so-called "dry battery," of course, really possesses a large content of water, but must not leak during its useful life in spite of the fact that it generates gas within itself and this must be allowed an exit. Its weight and size and cost are all vitally important and must be maintained at a minimum in spite of the steadily increasing demands made upon it for electrical output.

[1] Manuscript received July 12, 1935.
[2] Works Managers Dept., National Carbon Company, Cleveland, Ohio.

The problem of the dry battery industry, therefore, has been a difficult one, but, none the less, very admirable advances have been made in the improvement of the product. The improvements referred to have been possible largely through the ample appropriations for research and development which have been made annually by most of the battery manufacturers since the earliest days of the industry. The dry battery industry was one of the first to realize the necessity for chemical and physical control of its materials and processes and the wisdom of carrying on a continuous program of research and development to improve the characteristics of the product and reduce its cost. Extensive laboratories, capably manned and directed, are now very important divisions of the larger dry battery manufacturing companies. The work of these laboratories has necessitated the development of a series of testing methods representative of the applications to which dry batteries are put, and this phase of the problem has involved a great deal of study and investigation. To carry out these tests some very extensive and elaborate installations have been made by the larger companies.

The work of the industry has been greatly aided by the activities of a few laboratories outside the industry, notably that of the National Bureau of Standards, and the Bell Telephone Laboratories. Particularly helpful has been the co-operation of these laboratories in establishing the standard testing procedures, the importance of which cannot be overestimated as a factor in the industry's quality advances.

In the early days the dry battery industry naturally had many difficulties due to lack of knowledge as to the requirements of materials and manufacture necessary to produce a usable article, and the first efforts, of course, were directed toward the elimination of these difficulties. The next problem of research was to increase the efficiency in the use and combining of the constituent materials and so reduce costs. Following this came a period of effort to improve performance in general applications, usually confined to two or three grades of product to take care of all demands. The last phase of development and the one in which we now find ourselves is an attempt to provide greatly improved behavior in a wide variety of grades and types of dry batteries, each directed toward one of many specialized uses. This diversification in applications has led to the manufacture by the larger battery companies of a large variety of grades and qualities, thus greatly complicating the problems of control, research and production, and vastly increasing the scope of and need for the first two mentioned. The day is past when further major economies can be introduced into the battery industry. Materials are now purchased and used efficiently, and operations are largely performed mechanically. Prices of material are rising and the cost of supervision and control with the present complex situation is increasing. Further improvements in quality which are constantly being introduced can no longer be made through better proportioning of constituents with little or no rise in cost, but must now be brought about by the introduction of more expensive materials, or through the use of larger quantities of the higher priced standard materials, such as zinc and sal ammoniac. In general, however, the increases in quality now being made are greater in proportion than the increases in cost which follow, so that the public is still benefiting by the industry's progress.

TABLE I.

Quality Progress of the No. 6 (63.5 mm. diam. x 152.5 height)
General Purpose Dry Cell.

Year	Initial Short Circuit Amperage	Shelf Deterioration 6 Mos.	Heavy Intermittent Service Test	Light Intermittent Service Test
1901	16	35%	..	...
1910	24	20	40 hr.	130 days
1916	27	25	35	140
1926	35	10	60	180
1930	40	8	78	230
1934	43	7	77	250

The oldest form of dry battery is the so-called No. 6 size (63.5 x 152.5 mm.) of the grade used by the general trade for home, office, and most industrial applications. This has come to be known as the "general purpose" type. Records of its performance on a comparable basis can be traced back to about 1901. Typical performance figures for the best available brands of this type of cell are shown in Table I and indicate a fine improvement with the years.

It is apparent from the record of improvement that short circuit amperage (which has no significance regarding capacity) is evidently considered important. The retail trade has very generally adopted this readily used test in demonstrating batteries for sale to the public, based on the unfortunate misconception that such a test is significant of output in service. Shelf-life deterioration and performance on heavy and light service are, of course, much more significant, but must be combined in a salable general purpose cell with low cost and good appearance.

The telephone No. 6 cell was the first specialized industrial type. Its development started early and was continuously stressed with very satisfactory results as indicated in the following table:

TABLE II.

Quality Progress of No. 6 Telephone Dry Cells.

Year	Light Intermittent Service Test (20 ohm)
1910	155 days
1916	165
1926	230
1930	360
1934	450

In the No. 6 size the radio demand about 1926 induced the production of another specialized type specially designed for the heavier service demands of radio receiver "A" circuits. This cell later developed into a type now on the market for special industrial uses such as are met with in the railroad trade and certain governmental demands. This type of cell shows remarkable performance on both heavy and light classes of work, but, of necessity, short circuit amperage is largely disregarded in this type. As it is not sold to the general public, this is not detrimental. A comparison of characteristics with the general purpose cell will show the lack of significance of this attribute. Typical performance figures for the radio and industrial types are shown below. Cells of this type are more costly to produce than the general purpose type.

TABLE III.

Quality Progress of Special Purpose No. 6 Dry Cells
(Other Than Telephone).

Year	Initial Short Circuit Amperage	Shelf Deterioration 6 Mos.	Heavy Intermittent Service Test	Light Intermittent Service Test
1926 Radio A type	35	10%	70 hr.	180 days
1930 Radio A type	40	8	82	240
1934 Industrial type	28	10	115	400

A very large demand exists today for the D size flashlight cell, and this is steadily increasing both in this country and abroad, particularly in the Orient. While this type of cell is considered a general purpose type, it is used almost exclusively by the general public in the common variety of tubular flashlight and it is designed, therefore, with this service particularly in mind. During the past twenty-five years an improvement in quality has occurred as indicated in the following table. The figures given are typical of the best brands on the market.

TABLE IV.

Quality Progress of D Size (1¼ in. x 2¼ in. or 32 mm. x 57 mm.) General Purpose Flashlight Cells.

Year	4-ohm Continuous Test	4-ohm Intermittent Test
1910	170 min.	260 min.
1916	325	380
1926	425	550
1930	500	625
1934	510	750

As in the case of the No. 6 cell, many specialized industrial and governmental demands have developed for small dry cells of this D size. The most important among these is in the public utility field where millions of D size cells are used annually for meter reading. A special cell has gradually been evolved for this purpose, superseding the regular flashlight cell originally employed. In the table below, the progress of this cell is shown and for the short time involved is considered remarkable.

TABLE V.

Quality Progress of Industrial Type, D Size Flashlight Cells.

Year	Heavy Industrial Test (to 0.9 volt)
1929 (General Purpose Type—used for industrial purposes)	200 min.
1930 (First Special Industrial Type)	250
1931	335
1932	390
1933	650
1934	725
1935	975

Cells of special grades in the D and other small sizes have been developed for many other applications such as portable radio receivers, telephone exchange work, portable telephones (U. S. Army), hearing-aid devices, industrial inspection, railroad trainmen's lanterns, theatre ushering, electrical toys, moving sign motors, testing devices, electric clocks, etc. Cells of special type for such purposes are not of sufficient importance to justify tabulation of data in this paper, but in many cases great improvements over the general purpose cell have been introduced for the benefit of such special demands.

Radio still calls for a large volume of B batteries, regardless of the general use of power-operated receivers, and improvements are still in progress in the common variety of radio B batteries as indicated below.

TABLE VI.

Quality Progress of Radio B Batteries (D Size Cells).

Year	5,000-ohm Continuous to 17 Volts
1918	377 hr.
1923	680
1926	1,000
1930	1,400
1934	1,500

The table above does not include data on intermittent tests because records of earlier years are not in terms of the intermittent tests now current, so that comparisons are not possible.

There have been so-called "super" grades of B batteries introduced and grades specially designed for auto radio applications as well as specials for the U. S. Army and Navy which show higher performance figures than listed above. Such batteries now may run up to 1,800 or 1,900 hours on the 5,000-ohm continuous test (D size cell).

TABLE VII.

Quality Progress of Hearing-Aid Batteries (CD Size Cells,
1 in. x $3\frac{3}{16}$ in. or 25 mm. x 81 mm.).

Year	Heavy Earphone Service Test
1932	18 hr.
1933	27
1934	38
1935	50

The data in the tables in this paper are selected from records available to the author and are indicative of the behavior of the best brands on the market at the time. There have been, of course, many brands sold which fall far below the values reported, but it is gratifying to find that the proportion of poorer quality brands is continually decreasing and is less today than ever before. Many industrial users are protecting themselves against lower quality batteries by installing testing equipment and adopting the standard tests as a basis for purchase. The general public, however, have only their knowledge of the reputation of the maker as a protection against poor quality in dry batteries.

DISCUSSION.

J. D. Ceader[3]: I am particularly familiar with the improvements which Mr. Gillingham notes in his remarks on the quality of dry cells for flashlight service. Without the splendid work that has been done in the development of the primary battery, flashlight lamps would not be a commercial product today, even though the lamp itself had been greatly improved.

W. G. Waitt[4]: The dry battery industry needs new uses for dry cells in addition to technical improvements. With the substitution of dynamo power for dry cells in radio and with the technical increase in service life of dry cells the industry has had to face a steadily decreasing national volume of sales. This condition has forced many dry battery companies out of business and only a few of the more efficient ones have managed to survive. In addition to new uses, the industry needs radical technical developments to better combat the inroads of the power and secondary battery engineers into the dry battery field.

C. A. Gillingham (Communicated): Mr. Waitt's point that the dry battery industry is facing a diminishing market is very well taken, although the decreasing demand does not apply to all types of batteries. Certainly the dry battery industry could very advantageously assimilate a number of new battery uses and applications. However, in the face of a declining market, the majority of dry battery manufacturers, I feel sure, are striving not only to manufacture more efficiently, but to give the customer greater satisfaction, which necessarily involves quality advances and improvements. Certainly increased satisfaction on the part of the user can be expected to aid materially in prolonging the demand for dry batteries in all present applications.

E. W. Chambers[5] (Communicated): While the subject of dry batteries is being dealt with, it may be of interest to say that if Australian patent No. 24,779/30 is referred to, it will be seen that a rechargeable dry cell has been invented. The patentee subsequently stated that he discovered an even better cell. At the time the former were available, I tested some on continuous discharge and found that, provided the maximum discharge rate was not exceeded, they had a remarkable straight line characteristic tapering off suddenly at the end, as was claimed to be the case. The patentee of the cell has since turned his attention to a non-sulfating accumulator which is referred to in my comments on the paper by Haring and Thomas which appear on page 319 of this volume.

[3] General Electric Company, Cleveland, Ohio.
[4] Vice-President, General Dry Batteries, Inc., Cleveland, Ohio.
[5] Yartmore, 39 Sutherland Rd., Armadale S E 3, Victoria, Australia.

6

Copyright © 1935 by The Electrochemical Society, Inc.

Reprinted from pages 167–171 of *Electrochem. Soc. Trans.* **68**:167–176 (1935),
by permission of the publisher, The Electrochemical Society, Inc.

THE CURRENT-PRODUCING REACTION OF THE LECLANCHÉ CELL.[1]

By M. deKay Thompson.[2]

ABSTRACT.

The literature on the current-producing reaction of the Leclanché cell is reviewed, and the electromotive forces corresponding to air depolarization, and to the reduction of manganese dioxide to Mn_2O_3, Mn_3O_4 and MnO are recalculated.

The purpose of this paper is to review that part of the literature which contains original experimental work on the current-producing reaction in the Leclanché cell, and to calculate thermodynamically the electromotive force for the reaction that seems to be nearest the truth, and compare it with observed values. The most important question to answer is, what change takes place in the manganese dioxide when the cell operates? The MnO_2 might be reduced to Mn_2O_3, Mn_3O_4, or MnO, or the corresponding hydrates. The valence to which manganese is reduced has been determined[3] by depositing manganese dioxide on a platinum electrode electrolytically and noting how much of the titratable oxygen is used up when the current begins to fall rapidly on discharging this electrode. It was found that the manganese is reduced to the trivalent state. At the same time it was determined on discharging a cell that only half as much dioxide is used up as corresponds to the quantity of electricity furnished. From this it is concluded that oxygen of the air contributes to the current, confirming previous observations by others. On pumping the air from a cell it was found that the discharge took place for only a few seconds; consequently it is concluded that air is a necessary part of the mechanism of discharge.

A. Keller[4] finds that the current is furnished principally by the dioxide and very little by adsorbed oxygen, but explains the difference

[1] Manuscript received April 23, 1935.
[2] Professor of Electrochemistry, Massachusetts Institute of Technology, Cambridge, Mass.
[3] K. Arndt, H. Walter and E. Zender, Z. angew. Chem., 39, 1426 (1926).
[4] Z. Elektrochem., 37, 342 (1931).

between his results and those of Arndt, Walter and Zender by the higher current density in his own experiments. However, no attempt is made to explain the absence of discharge when the air is removed from the cell.

The relation between the hydrogen ion concentration and the potential of the dioxide electrode does not give any information as to which oxide is produced, as will be seen from the following deductions. First, suppose the dioxide to be reduced to Mn_2O_3 reversibly in alkaline or neutral solution. Acid solutions will not be considered. The path by which this occurs makes no difference as long as all reactions are reversible. Therefore the following may be assumed:

$$
\begin{array}{lll}
MnO_2 \ \text{Solid} & = MnO_2 \ (\text{Sat. Soln.}) & (1) \\
MnO_2 \ (\text{Sat. Soln.}) + 2H_2O & = Mn^{4+} + 4OH^- & (2) \\
Mn^{4+} & = Mn^{3+} + F & (3) \\
Mn^{3+} + 3OH^- & = \tfrac{1}{2}\,Mn_2O_3 \ (\text{Sat. Soln.}) + 3/2\,H_2O & (4) \\
\tfrac{1}{2}Mn_2O_3 \ (\text{Sat. Soln.}) & = \tfrac{1}{2}Mn_2O_3 \ (\text{Solid}) & (5) \\
\hline
\end{array}
$$

Total:
$$
MnO_2 \ (\text{Solid}) + \tfrac{1}{2}H_2O = \tfrac{1}{2}Mn_2O_3 \ (\text{Solid}) + OH^- + F \quad (6)
$$

where F stands for faraday.

Therefore the potential of this electrode should depend on the hydrogen ion concentration as given by the formula

$$
E = \acute{E}^\circ - \frac{RT}{F}\ln(OH^-) = E^\circ + \frac{RT}{F}\ln(H^+) \quad (7)
$$

If the oxide is reduced to MnO we have (1) and (2) the same as above, but (3), (4) and (5) become:

$$
\begin{array}{lll}
Mn^{4+} & = Mn_2^+ + 2F & (8) \\
Mn^{2+} + 2OH^- & = MnO \ (\text{Sat. Soln.}) + H_2O & (9) \\
MnO \ (\text{Sat. Soln.}) & = MnO \ (\text{Solid}) & (10) \\
\hline
\end{array}
$$

Total incl. (1)&(2):
$$
MnO_2 + H_2O = MnO + 2OH^- + 2F \quad (11)
$$

$$
\text{and } E = \acute{E}^\circ - \frac{RT}{2F}\ln(OH^-)^2 = E^\circ + \frac{RT}{F}\ln(H^+) \quad (12)
$$

the same as (7). Therefore in either case the difference in E for two hydrogen ion concentrations $(H^+)_1$ and $(H^+)_2$ would be given by

$$
E_1 - E_2 = 0.059 \log \frac{(H^+)_1}{(H^+)_2} \text{ at } 25^\circ \text{ C.} \quad (13)
$$

This equation has been tested repeatedly with more or less approximate agreement with theory. Though these results do not show which oxide is formed, it may nevertheless be interesting to summarize these results for alkaline and neutral solutions.

O. F. Tower's[5] results with electrolytic dioxide agree to within about three to six per cent of the theoretical. H. D. Holler and L. M. Ritchie[6] find the coefficient of the logarithm 0.067 for Caucasian ore, 0.080 for domestic ore, and 0.064 for Brazilian ore, in place of 0.059. The potential of the electrolytically prepared oxide was found by Holler and Ritchie to be independent of the hydrogen ion concentration, while F. Daniels[7] finds the coefficient 0.050. B. M. Thompson[8] finds the coefficient 0.07 for dioxide from Montana and from the Caucasus, and 0.10 for artificial dioxide. T. J. Martin and A. J. Helfrecht[9] obtain potentials that are nearly constant between $pH = 2$ and $pH = 7$. A. Keller[10] measured natural dioxide in normal and in $0.1\ N$ sodium hydrate; the coefficient of the logarithm was 0.057.

Arndt, Walter and Zender[11] say their own measurements between pH values of 1 to 3 and 12 to 14, as well as unpublished work by Lange (at the Dresden Technical High School) give a coefficient of 0.23. This we consider the correct value for *acid* solutions, but not for neutral solutions; for neutral solutions it is 0.059 [see equation (13)] The authority quoted for the value 0.23 is F. Foerster, Elektrochemie wässeriger Lösungen, 4th Ed., p. 218, but the derivation there given is for acid solutions only. Therefore, the results of Lange and of Arndt, Walter and Zender do not agree with theory nor with the results of others.

A method of determining the cell reaction is to compute thermo-dyamically the electromotive force which a given reaction would produce and compare it with the measured value. S. Kaneko[12] has done this for a part of the reaction. The reactions and Kaneko's computed electromotive forces are as follows:

Reaction	Volts, Computed	
$Zn + 2MnO_2 + 2NH_4Cl = Zn(NH_3)_2\ Cl_2 + Mn_2O_3 + H_2O$	1.605	(14)
$2Zn + 4NH_4Cl + 3MnO_2 = 2Zn(NH_3)_2\ Cl_2 + Mn_3O_4 + 2H_2O$	1.367	(15)
$Zn + 2NH_4Cl + MnO_2\quad = Zn(NH_3)_2\ Cl_2\ MnO + H_2O$	1.160	(16)
$Zn + \frac{1}{2}O_2 + 2NH_4Cl\quad = Zn(NH_3)_2\ Cl_2 + H_2O$	1.720	(17)

[5] Z. Phys. Chem., **18**, 17 (1895); **32**, 566 (1900).
[6] Trans. Electrochem. Soc., **37**, 607 (1920).
[7] Trans. Electrochem. Soc., **53**, 45 (1928).
[8] J. Ind. Eng. Chem., **20**, 1176 (1928).
[9] Trans. Electrochem. Soc., **53**, 83 (1928).
[10] Z. Elektrochem., **37**, 342 (1931).
[11] See footnote 3.
[12] J. Soc. Chem. Ind. (Japan), **32**, 120 (1929).

The method used for the first reaction is to break it up into the following:

$$Zn + 2NH_4Cl = Zn(NH_3)_2 Cl_2 + H_2 \qquad (18)$$
$$H_2 + \tfrac{1}{2}O_2 \qquad H_2O \qquad\qquad\qquad (19)$$
$$2MnO_2 \qquad Mn_2O_3 + \tfrac{1}{2}O_2 \qquad\qquad (20)$$

Reaction (18) is not calculated thermodynamically, but is taken from potential measurement by D. A. MacInnes[13] which gave 0.514 volt. The free energy change of (19) is taken from Lewis and Randall's "Thermodynamics." The free energy of (20) is calculated from the dissociation measurements of P. Askenasy and S. Klonowsky,[14] from which and the specific heats the following free energy equation is calculated:

$$\Delta F = 18,690 - 1.29T\ln T - 13.736T \qquad (21)$$
$$\Delta F \text{ at } 25° \text{ C.} = 6,202 \text{ cal.}$$

The voltages corresponding to the other equations are calculated in a similar way. Kaneko set up and measured a cell and the electromotive force was found to be 1.63 at 25° C. He therefore concluded that the first reaction is the one taking place in the cell.

C. Drucker[15] has made a similar calculation. He computes 59,500 calories as the heat evolved by the reaction:

$$Zn + 2MnO_2 + 2NH_4Cl = Mn_2O_3 + (ZnO . 2NH_4Cl) \text{ aq.}$$

using C. Drucker and R. Hüttner's[16] determination of the dissociation pressure of manganese dioxide. He also measured the temperature coefficient of the electromotive force and found $\dfrac{dE}{dt} = 0.0006$ volt per degree. From these values the cell voltage is computed as 1.46.

According to Drucker, the cells measure 1.5 volts for natural and 1.7 for artificial dioxide. Calculated and measured values do not agree; therefore Drucker comes to the conclusion that nothing else is left to account for the initial voltage than the oxygen adsorbed on the dioxide. He does not consider that the formation of the compound $Zn(NH_3)_2Cl_2$ contributes any to the voltage, for the same voltage is obtained with magnesium chloride as electrolyte. Furthermore, ammonia and zinc could not come in contact with each other until an appreciable quantity of electricity has been drawn from the cell, unless zinc is added to the electrolyte at the start.

[13] Trans. Electrochem. Soc., 29, 315 (1916).
[14] Z. Elektrochem., 16, 104 (1910).
[15] Z. Physik. Chem., Bodenstein Festb., 912 (1931).
[16] Z. Physik. Chem., 131, 237 (1928).

A calculation will now be made of the free energy of the reactions:

$$Zn + 2MnO_2 + H_2O \quad = Zn^{++} + Mn_2O_3 + 2OH^- \quad (22)$$
$$Zn + O_2(\tfrac{1}{5}\text{ atm.}) + H_2O = Zn^{++} + 2OH^- \quad (23)$$

for depolarization by dioxide and by air. The first of these equations evidently is not exactly correct, for pyrolusite is hydrated, and the unhydrated dioxide, polianite, does not depolarize.[17] Also the lower oxide is probably hydrated. However, the free energy of hydration is probably small in both cases.

Equation (22) may be broken up as follows:

$Zn + 2F$	$= Zn^{++}(\text{act.} = 1)$	$\Delta F = -34{,}980$ cal.	(24)
$Zn^{++}(\text{act.} = 1)$	$= Zn^{++}(\text{act.} = 0.1)$	$\Delta F = - 1{,}370$ cal.	(25)
$2MnO_2$	$= Mn_2O_3 + \tfrac{1}{2}O_2(\text{equil. pres.})$	$\Delta F = 0$	(26)
$\tfrac{1}{2}O_2(\text{equil. pres.})$	$= \tfrac{1}{2}O_2(1\text{ atm.})$	$\Delta F = +13{,}750$ cal.	(27)
$H_2O(\text{soln.})$	$= H_2O(\text{pure})$	$\Delta F = + \quad 80$ cal.	(28)
$H_2O(\text{pure})$	$= H_2(1\text{ atm.}) + \tfrac{1}{2}O_2(1\text{ atm.})$	$\Delta F = +56{,}560$ cal.	(29)
$H_2 + O_2 - 2F$	$= 2OH^-(\text{act.} = 1)$	$\Delta F = -74{,}910$ cal.	(30)
$2OH^-(\text{act.} = 1)$	$= 2OH^-(\text{act.} = 10^{-7})$	$\Delta F = -19{,}200$ cal.	(31)

Total: $Zn + 2MnO_2 + H_2O(\text{soln.}) = Zn^{++}(\text{act.} = 0.1) + 2OH^-(\text{act.} = 10^{-7})$,
$$\Delta F = -60{,}070, \quad E = 1.30$$

The ΔF values for (24), (29) and (30) are taken from Lewis and Randall's "Thermodynamics," and the dissociation pressure of manganese dioxide at 25° C. is calculated from the values determined by Drucker and Hüttner.[16] These pressures cannot be very accurate, for, as Drucker and Hüttner point out, the heat of the reaction

$$4MnO_2 = 2Mn_2O_3 + O_2 \quad (32)$$

does not agree even approximately with that calculated from the dissociation pressures. Drucker and Hüttner calculate

$$\Delta H = 26{,}000 \text{ calories.}$$

The heat of this reaction may be calculated from thermal data as follows:

$4MnO_2$	$= 4Mn + 4O_2$	$\Delta H = +501{,}200$ cal.[18] (33)
$4Mn + 3O_2$	$= 2Mn_2O_3$	$\Delta H = -454{,}440$ cal.[19] (34)

$4MnO_2 = 2Mn_2O_3 + O_2$, ΔH at 298° K.$= \quad 46{,}800$ cal.
which does not agree with 26,000. According to Drucker and Hüttner, however, X-ray analysis shows the product of the dissociation consists of dioxide and manganic oxide.

[17] G. W. Vinal and L. M. Ritchie, Bur. Standards, Circ. No. 79, p. 10 (1923).
[18] R. W. Millar, J. Am. Chem. Soc., 50, 1875 (1928).
[19] Int. Crit. Tables V, 190 (1929).

[Editor's Note: Material has been omitted at this point.]

7

Copyright © 1939 by The Electrochemical Society, Inc.

Reprinted from *Electrochem. Soc. Trans.* **76**:435–451 (1939), by permission of the publisher, The Electrochemical Society, Inc.

THE CADMIUM-NICKEL STORAGE BATTERY.[1]

By Anna P. Hauel.[2]

Abstract.

The cadmium-nickel battery is widely used in Europe but not in America. It is particularly applicable to severe service conditions since the active material does not shed readily and wide temperature fluctuations have but little effect. The active material of the positive plates is nickel hydroxide plus pure crystalline graphite. In place of graphite, metallic nickel flakes may be used. The active material of the negative plates is cadmium oxide or hydroxide. The effect of the presence of iron in the cadmium paste is discussed at length. Similarly the addition of lithium hydroxide to the KOH electrolyte is treated at length. On the basis of the author's experiments, the lithium appears superfluous in positive electrodes containing graphite except that it counteracts the deleterious effects of traces of iron. The adsorption of CO_2 by the KOH electrolyte reduces the capacity of the battery, due to cadmium carbonate formation on the surface of the CdO grains. The new starting battery has very thin plates and very close spacing. It is replacing the lead storage battery in trucks and buses.

INTRODUCTION.

The cadmium-nickel battery, strange to say, is not made in the United States. It is only slightly known in this country and it is not imported. In Europe, however, the cadmium-nickel battery has been made since the beginning of the century, although it has been used industrially to a large extent during the last eight or ten years only. Recently the manufacture has been taken up by a number of factories. Factories which heretofore had made only lead or nickel-iron batteries changed over partially or completely to the manufacture of cadmium-nickel batteries. Nickel-iron batteries are made on but a

[1] Manuscript received January 16, 1939.
[2] Consulting Chemist, 317 West 76th St., New York City.

small scale in Europe nowadays, and then only for special purposes, principally traction.

The cadmium-nickel battery employs an alkaline electrolyte, just as the Edison nickel-iron battery does. It differs, however, from the Edison battery in the use of a cadmium, in place of an iron, negative electrode. With cadmium as the active cathode, a number of applications are made possible for alkaline storage batteries which have heretofore been impossible due to certain specific shortcomings of iron. Iron, as a cathode material, has the disadvantages of self-discharging and becoming passive at a high rate of discharge at low temperatures; furthermore, iron has the disadvantage of requiring a higher charging potential.

As active material surrounding the cadmium anode of the cadmium-nickel battery, nickel oxide is used exactly as with the anodes of the nickel-iron battery. The electrolyte is a caustic potash solution. The average discharge potential amounts to 1.2 volts per cell.

The reactions which take place during charging and discharging can be given only empirically. They represent the changes taking place only so far as the oxidation and reduction processes of the active substances are concerned.

$$Cd + Ni_2O_3 \leftrightarrows CdO + 2NiO \text{ (Main reaction)} \dots\dots\dots\dots (1)$$

$$Cd + NiO_2 \leftrightarrows CdO + NiO \text{ (Secondary reaction)} \dots\dots\dots\dots (2)$$

The major part of the current is due to reaction (1). The reaction represented by the second equation plays only a very small part as far as useful current is concerned. It can be entirely disregarded if the discharge of the battery takes place several hours after charging.

Actually, *anhydrous* nickel and cadmium oxides are not formed at the electrodes but instead we get hydroxides, and these are probably hydrated besides. As the role of the water in the charge and discharge reactions has not yet been entirely elucidated, we prefer to give merely the oxidation-reduction steps of the metals involved. The electrolyte, in contradistinction to that of the lead battery, does not appear to be involved in the current-discharge process. It seems to serve merely as a conducting medium.

The original form of the cadmium-nickel battery may be traced back to the patents of the Swede, E. W. Jungner (1899).[3] At about the same time, Edison applied for his patents on the iron-nickel battery here in the United States. Since then the cadmium-nickel battery has

[3] Ger. Pat. 110,210; 113,726; 114,905.

been developed into its present form not only in Sweden but in many other European countries as well. This development was, in part, quite independent of the Jungner patents which were concerned chiefly with the details of construction of the battery rather than with its active materials.

Fig. 1 and 2 show the electrodes of the cadmium-nickel batteries as they are made today in most of the European factories. Fig. 1

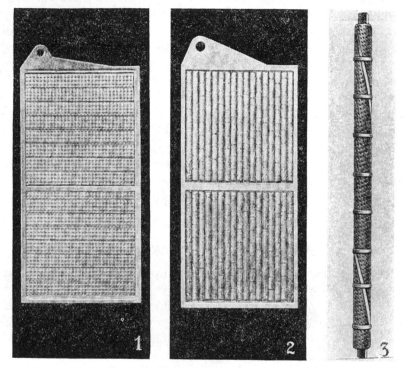

FIG. 1. Positive and negative pocket plate.
FIG. 2. Positive tube plate.
FIG. 3. Positive tube.

shows a typical positive or negative plate of the cadmium-nickel battery. The construction of the positive and negative plates is exactly the same. In the case of the positive plates the pockets are filled with nickel hydroxide; the negative plate pockets are filled with cadmium or cadmium oxide.

Fig. 2 shows a positive "tube" type electrode, which in certain cases is coupled with a negative cadmium electrode of the type shown in Fig. 1. The tube type pocket plate is used, for instance in certain traction batteries which are subject to particularly heavy vibrations. The tube type pocket plate resembles in every particular the Edison plate construction, which is well known in America.

PRACTICAL APPLICATIONS.

The field of commercial application of the cadmium-nickel battery is practically the same as that of the lead and the nickel-iron batteries. On account of certain special advantages, the most important applications of the cadmium-nickel battery are the following: train lighting; flash lights; signal and mine lamps; radio set batteries and all types of military and emergency batteries. In general, the cadmium-nickel battery is selected where special emphasis is placed on absolute dependability even under most severe conditions and where the longest possible life is required.

Cadmium-nickel batteries have been introduced and successfully used as train lighting batteries in all European countries and their colonies. In particular in those colonies where wide atmospheric temperature ranges prevail and where trained servicing personnel is not available, the cadmium-nickel battery has displaced all other types of train lighting batteries. Thus, for example, on railroads in India the batteries are operating successfully where the trains in the mountains are subjected to extreme cold, and in the valleys to extreme heat.

The life of lead storage batteries is very unfavorably affected by overcharging. Cadmium batteries, on the other hand, are entirely resistant to overcharging as well as to deep discharges and occasional short circuits.

For traction purposes, the cadmium-nickel batteries have been successfully introduced into tractors, automobiles, electric locomotives and mine locomotives. Their chief advantage, compared with lead storage batteries, is that even under severe mechanical shocks their active material does not shed, due to the fact that it is enclosed in perforated steel pockets. Furthermore, the steel battery jars are not fragile, as is the case of the hard rubber jars of lead storage batteries.

As a mine lamp battery, radio battery and for military and emergency purposes, the important advantage of the cadmium-nickel battery over other batteries is the very low self-discharge factor. After completing the charging, no gas is evolved. This is particularly important for batteries used in mines. For coal mines, makers of other batteries stress the point that their batteries are sealed in order to minimize every risk of explosion. However, as is well known, the formation of H_2 gas never ceases entirely in the case of the lead battery; this defect is even more serious with the nickel-iron alkaline batteries. The metallic spongy lead of the negative lead electrode, while standing idle as well as during discharge, reacts with the electrolyte, and lead sulfate and H_2 gas are formed due to local action. Likewise, the finely divided metallic iron in the negative electrode of the Edison battery, even on open circuit, oxidizes to iron oxide, whereby the oxygen of the electrolyte is absorbed and a corresponding amount of hydrogen is liberated. In contradistinction to this defect, the metallic cadmium of the cadmium-nickel battery, even in the most finely divided condition, does not react with water or with the caustic potash electrolyte. Furthermore, the hydrogen overvoltage of cadmium in alkaline solution is relatively high, which is an advantage. There is a smaller self-discharge than in the case of all other types of batteries made commercially today. Due to this low self-discharge of the cadmium-nickel battery, about 80 per cent of the original charge is available after the battery has stood charged but idle for six months.

As the result of this property of holding its charge, the cadmium-nickel battery is suitable for all emergency and military purposes. It is very widely used by all European countries for military purposes.

MANUFACTURING DETAILS.

For the most frequently used types, in which the positive and the negative plates are of identical construction, the manufacturing process is very simple. On the other hand, the manufacture of the positive tube type electrodes is very complicated, and requires expensive machinery. Introduced into the tubes under pressure are approximately three hundred alternate layers of nickel oxide and metallic nickel flakes. The latter serve as conductors, as the conductivity of nickel oxide by itself is very poor. Further, the tubes (Fig. 3) are very rigid and their walls do not respond to the pressure developed in the nickel oxide during the first charging and discharging of the battery. In contrast with the tube type electrode, the pocket type electrode (Fig. 4 and 5) is so constructed that it will respond to the pressure generated in the active material. The pocket strips are folded and joined into one another in such a way that, as the pressure increases, the strips push apart as the active mass swells, and the whole plate thereby becomes uniformly thicker.

The pocket type construction was first developed by Jungner, and is today adopted by most manufacturers of cadmium-nickel batteries. The perforated sheets from which the pocket strips are made, are of steel which has been carefully nickel-plated (Fig. 6). In order to homogenize the nickel plating, the strips are annealed in hydrogen. It is important that the nickel plating be impervious, as the contact between nickel oxides and iron is to be avoided. In the positive pocket electrode, graphite is added to the active material as an electrical conductor; it is not applied in layers as in the case of the nickel flakes in the tube type electrode but, instead, the finely divided graphite is introduced as part of a homogeneous nickel oxide-graphite mixture.

It is interesting that T. A. Edison originally used graphite as an electrical conducting medium in his alkaline battery. But he later

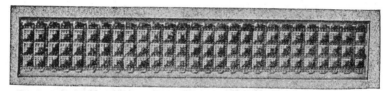

FIG. 4. Positive and negative pocket strip.

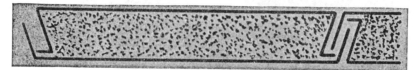

FIG. 5. Connection of two pockets.

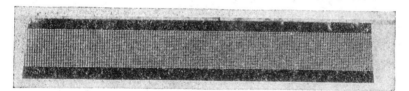

FIG. 6. Perforated steel ribbon.

abandoned graphite when he found that the graphite available was impure. Edison then evolved his tube type positives loaded with nickel flakes. Very pure varieties of graphite are now available to the storage battery industry, with less than 1 per cent ash content. For cadmium-nickel batteries today only natural hexagonally crystallized graphite is used. The flakes are thoroughly purified before use. These flakes have the highest electrical conductivity of any graphite. As small an ash content as possible is desirable for two reasons. Next to having a graphite of good electrical conductivity, it is essential to keep the nickel oxide as iron-free as possible. Graphite ash consists approximately of ⅓ iron oxide. Iron is an impurity which greatly reduces the activity of the nickel oxide as soon as it absorbs any of it.[4] The purest graphite is the artificial or Acheson graphite. However, on account of the comparatively low electrical conductivity, the Acheson graphite is not as well suited as the natural flake graphite.

One of the most difficult problems in the development of the alkaline battery was the production of a suitable nickel oxide material for the positive electrodes. Many patents covering a solution of this problem have been issued. Most manufacturers, we believe, obtain their $Ni(OH)_2$ by interaction of $NiSO_4$ and $NaOH$ solutions. Like many metal hydroxides, $Ni(OH)_2$ precipitates out of solutions as a finely dispersed, very voluminous precipitate, which strongly adsorbs the ions of the solution. The original structure of the nickel hydroxide is of the utmost importance as far as the later functioning of the alkaline battery is concerned, for in the alkaline battery during charging and discharging—just as in the lead battery—reactions take place in which solid chemical compounds change over into other solid insoluble compounds. In the first steps of producing the nickel oxide, a certain degree of dispersion is advantageous which should correspond to a predetermined surface area and to a predetermined structure. These characteristic properties remain intact because subsequently, during manufacture and life of the battery, the material is always in the solid state; the reactions that do take place are on the surface of the particles. Properly prepared oxide will retain its physical properties and ensure a long battery life. The potential, capacity, and length of life depend to a great extent on the colloid structure of the original nickel material. The production of suitable $Ni(OH)_2$ is not so much a chemical problem as it is a colloid chemical problem, just as is the case in the production of a satisfactory lead oxide paste for lead storage batteries.

The structure of the precipitated $Ni(OH)_2$ depends to a great extent on the precipitation and drying conditions. One may, with the two reagents $NiSO_4$ and $NaOH$, obtain the most varied types of $Ni(OH)_2$, depending on the concentration, temperature, or excess of one or the other of the ions. These types differ from one another in color, specific gravity and surface characteristics.

The battery manufacturers treat the precipitated nickel hydroxides at high temperatures in order to obtain a primary material with the desired quality. Factory-produced $Ni(OH)_2$ contains somewhat less than 63.3 per cent nickel, corresponding very closely to the formula $Ni(OH)_2$. Apart from this, individual factories also manufacture

[4] Dvorak, Batterien, Jahrgang 4, 461-3, Berlin (1935).

106

mixtures of nickel hydroxide and nickel oxide, as primary active material. Before the material is filled into the pockets of the electrodes, it is subjected to physical tests.

The negative plates of the cadmium-nickel battery are of exactly the same construction as the positive "pocket" type plates except that the pockets of the negative plates are filled with cadmium, CdO or $Cd(OH)_2$, which during the first charging of the battery is reduced to metallic cadmium. The cadmium oxides used in most factories contain iron. The presence of iron is to be traced back to the basic patent of Jungner and Estelle.[5] According to this patent, finely dispersed cadmium sponge is precipitated electrolytically onto a cathode from a $CdSO_4$ solution. In order to obtain the metal in the desired degree of fineness, the precipitation is carried out so as to co-deposit cadmium with iron from a solution of $CdSO_4$ plus $FeSO_4$. Estelle calls the resulting material an "alloy" of the two metals. We have had the opportunity of making cadmium sponge in the laboratory and on a commercial scale under the most varying conditions. We have definitely established that "alloys" are never precipitated in this way. The presence and co-precipitation of the iron is such that it interferes with or retards the growth of the crystals into more or less coarse metal needles. In the presence of iron we get a very fine cadmium crystal slime. The iron co-precipitated with the cadmium is without doubt to be regarded as an admixture, and not as a constituent of an alloy. For instance, the iron particles are attracted by a magnet. Furthermore, the fine crystalline cadmium precipitate can be obtained in the presence of cations other than iron.

In the meager literature on cadmium-nickel batteries, we find the view expressed that iron participates during the discharge process of the negative cadmium electrode. This simply implies that during the discharge all of the cadmium is oxidized and, after that, part or all of the iron is also oxidized. This viewpoint is based on the charging curve of the iron-containing cadmium plates. We have made cadmium electrodes with and without the addition of iron and established that the charging and discharging curves are identical. We also found that only the amount of cadmium metal present determines the ampere-hour capacity of the plates under otherwise like conditions, an indication that the cadmium alone is the active substance in the plates and not cadmium plus iron.

We also find in the literature the view advanced that the iron in the cadmium electrodes acts as a so-called "expander", similar to

[5] U. S. Pat. 983,430 (1911); Ger. Pat. 229,542 (1910); Brit. Pat. 9,964 (1910).

$BaSO_4$, lampblack and other substances in lead battery electrodes. The CdO is reduced to metal at each charging. Metallic cadmium tends, as does also metallic lead, to contract; that is, its active surface becomes smaller at each following charging cycle. The iron addition is supposed to diminish and counteract the extent of this contraction, lessening the active surface of the cadmium sponge and by this its capacity. On the basis of our own research we could not establish that the iron acts as an expander in the negative plate. Manufacturers today use more effective processes than iron for preventing the contraction of the active surface of the cadmium after repeated cycles of charging and discharging. Furthermore, cadmium electrodes may be made without the addition of any iron, which furnish a battery life of many years, and the batteries are always of the same capacity. The reason we nowadays find a certain percentage of iron in practically all of the products of the battery factories is, according to our view, the fact that cadmium is cathodically co-precipitated with iron; and that iron is merely an aid for the precipitation of a finer crystal of cadmium. However, we also find active cadmium materials in electrode pockets which are made in an entirely different way from the above and which contain 5 to 30 per cent iron. Some manufacturers fill their electrode pockets with CdO or $Cd(OH)_2$ which is originally produced in the pure form and which they then intentionally mix with iron. According to our view, the iron here merely serves to increase the electrical conductivity of the CdO material. CdO conducts electricity noticeably better than FeO; however, the internal resistance of the batteries is decreased by the addition of iron. We may therefore assume that iron in the presence of cadmium is always present in the metallic state, and is not oxidized even during discharge, and therefore acts as a conductor in the discharged plates. We base our opinion on the fact that it makes no difference whether we replace the iron in the negative plate of the cadmium-nickel battery by a corresponding amount of graphite. We have made cadmium-nickel cells which contained only graphite as a conducting addition to the cadmium of the negative electrodes, the graphite functioning just as in the positive electrodes. As CdO itself is a relatively good conductor, we could use an appreciably smaller amount of graphite in the negative electrode than in the positive. The iron-free graphite-containing cadmium electrodes have also proved satisfactory even at the highest rates of discharge, for instance, in automobile starting batteries, where it is important that the negative plates have the smallest internal resistance. These iron-free cadmium batteries were in no way inferior to the iron-containing batteries.

The cadmium-nickel battery contains a caustic potash solution of 1.19-1.25 specific gravity (5 N), which acts as electrolyte, and to which LiOH is ordinarily added. According to Edison, the LiOH serves principally to increase the capacity and the life of the positive tube type electrode. Those cadmium-nickel batteries in which the positive plates consist of tubes made according to the Edison type of construction—therefore principally traction batteries—always have LiOH added to them. We generally use 0.8 to 1.2 g. LiOH per ampere hour rated capacity; this amount cannot be added to the cell at one time, as LiOH is only slightly soluble in KOH solutions. We therefore carry out the first two or three chargings in such a manner that the electrolyte is drained after each cycle, and fresh electrolyte added which is practically saturated with LiOH. The nickel oxide will absorb only a part of the lithium present in solution. Nevertheless, in each cycle an adsorption equilibrium is attained between the LiOH in the nickel oxide on the one hand and the LiOH in the electrolyte on the other. LiOH is commercially obtainable in a very pure form as crystallized $LiOH.H_2O$.

The positive electrode with pockets filled with nickel oxide plus graphite was developed by Jungner originally without the lithium addition. Even today some manufacturers build nickel-cadmium batteries using a KOH electrolyte free from lithium. As a matter of fact, the conditions under which the nickel oxide acts in the batteries today are such that an increase in capacity upon adding lithium no longer takes place. The nickel oxide develops its maximum efficiency even in pure KOH. In such cases where in spite of this fact lithium additions are made to the electrolyte of these batteries, the pocket type positive plates require considerably less, say only a third, the lithium required by the batteries with tube type positives. In the batteries with the pocket type positives the lithium serves merely the purpose of increasing the life of the positives. Lithium may also serve to extend the life of the plates, in particular in cases of severe service such as operating at higher temperatures than the maximum permissible, viz. 45° C. Under normal operating conditions it was established that a favorable action of the lithium on the nickel oxide plus graphite electrodes takes place only after six hundred to a thousand charging and discharging cycles, therefore after two or three years of service life.

What the actual effect of the lithium is in the positive electrodes has not yet been satisfactorily accounted for. Edison maintained that it is adsorbed by the nickel oxide and, as he stated, causes this to expand in the tubes. To this expansion or volume increase, Edison ascribed the favorable effect of lithium. We find that the active nickel material mixed with graphite does expand in the pockets during the first chargings but irrespective of whether or not the electrolyte contains lithium. This expansion is so noticeable that the pocket type plates, after the forming process, increase in thickness by as much as 30 to 35 per cent over the original thickness. Sometimes this increase in thickness occurs after only four to five cycles. This increase in size of the nickel positive plates is not the same as the well known increase in size of the positive lead peroxide plates of the lead battery. Since in the case of the nickel-cadmium batteries no allowance has to be made for an increase in the size of the plates in later life, space for the first increase may be allowed for during the original assembly of the plates into cells. With the lead battery this is not possible.

We have undertaken very exhaustive investigations on the increase in volume of the positive pocket type electrodes in batteries operating both with and without lithium. In one case we charged and discharged plates in KOH solution of 1.2 specific gravity, free of lithium; and in the other case we used KOH solutions of the same concentration as in the first cell, but with various additions of LiOH. We even exceeded the amount of LiOH usually used with tube type electrodes, introducing as much as 2.4 g. of LiOH per ampere-hour capacity. From time to time we determined the thickness of the plates and found that all of the plates were of the same thickness, even after more than 1,000 charges and discharges; an excessive expansion or bursting of the pockets due to "too much" lithium was not observed in any case, even when the largest amounts of lithium were used. In spite of these findings, certain battery manufacturers continue to fear excessive expansion of the active material due to lithium and accordingly specify the omission of lithium from the battery electrolyte.

From actual experiments it seems that lithium hinders or delays in some way the adsorption of deleterious iron impurities by the nickel oxide. Iron is relatively insoluble in KOH of 1.21 specific gravity. Nevertheless, it cannot be entirely avoided, traces being supplied by the various battery constituents. The iron impurity penetrates the nickel active material during the course of a long period of time, and finally causes the nickel oxide to become inactive. We had the opportunity of analyzing nickel oxide which had been in use for a long time in battery plates of various manufacture, and found from 0.1 to 1.4

per cent iron. These large amounts of iron could not have originated from either the original solid battery materials or from the graphite; they must therefore have been furnished by the electrochemical processes. The presence or absence of lithium in the electrolyte has no effect on the negative cadmium electrode.

The alkaline electrolyte in the course of time absorbs CO_2 from the air and from the water added. K_2CO_3 solution is a poorer electrical conductor than KOH solution. On this account, manufacturers of cadmium-nickel batteries specify a change of electrolyte at least once a year. With normal service demands on the battery, for instance, a five hour discharge rate, the K_2CO_3 content of the electrolyte can reach a maximum of 90 g./L. without perceptibly diminishing the potential and capacity of the battery. After this concentration is exceeded, the electrolyte should be changed.

As to the higher electrical resistance of a K_2CO_3 solution, our own potential measurements lead us to conclude that the reduced efficiency of the battery is due to a decrease in capacity of the negative electrode rather than to an increase in resistance of the electrolyte which is in between the plates. We undertook potential studies with cells containing various amounts of CO_2 in the electrolyte and measured the potential against a zinc auxiliary electrode which dipped into the electrolyte. In this way we determined the individual capacity of the positive nickel and negative cadmium electrodes. We could establish that the capacity of the positive nickel electrode remained entirely unaffected by the increasing content of CO_2 in the electrolyte, whereas the capacity of the negative electrode decreased with increasing CO_2 content. This decrease in capacity of the negative electrode takes place even with low CO_2 concentrations but very high current demands. However, at low discharge rates, for instance, at a five hour rate, a serious drop in capacity takes place only when the CO_2 concentration reaches an equivalence of about 90 g./L. K_2CO_3. We believe, therefore, that the deleterious effect of CO_2 is to be attributed to a layer of poorly conducting $CdCO_3$ which builds up on the surface of the CdO particles and thus causes a decrease in capacity of the cells. We could also establish that negative electrodes with fine-grained CdO are much less sensitive to an increase of CO_2 in the electrolyte than are electrodes with coarse-grained CdO. This would indicate that the smaller total active surface supplied by the coarse grains is polarized more rapidly, due to carbonate formation. The $CdCO_3$ is decomposed when the electrolyte is changed; the pure KOH solution converts the $CdCO_3$ back to CdO, thus restoring the original capacity of the battery.

We need not discuss the charging and discharging characteristics of the cadmium-nickel battery (Fig. 7a) because much has already been published on this subject.[6]

A NEW S. L. I. BATTERY.

The newest type of cadmium-nickel battery is a starting battery (S. L. 1.—Starting-Lighting-Ignition battery) which has been developed during the last five years and which has been widely used during the last three years. The older literature states that alkaline bat-

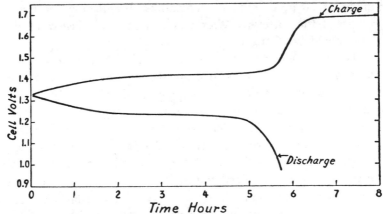

FIG. 7a. Charge and discharge of a low-resistant cadmium-nickel battery at 28° C. at the 5-hour discharge rate.

teries are basically unsuited as starting batteries because the voltage drop of these batteries in the early days was too big when there was a sudden heavy current demand. But commercial development has gone a long way in the modern construction of cadmium-nickel batteries. Batteries that have a very low internal resistance are now made which have proved successful as starting batteries for all types of automobiles. On account of the higher price for these alkaline starting batteries as compared with that for lead batteries, they are not much used in the small passenger automobiles. However, they are used a great deal as starting batteries for trucks and buses. Cadmium-nickel batteries have proved particularly satisfactory in diesel-engined cars where starting conditions are often very severe. They are also used in special cases where absolute reliability and long life are important factors. A

[6] Crenell and Lea, "Alkaline Batteries" (1928); C. Drucker and A. Finckelstein, "Galvanische Elemente und Accumulatoren" (1932).

number of the large European cities are changing their motorbus batteries over to cadmium-nickel batteries. The motorbus companies specify a guaranteed life of six years for the cadmium-nickel batteries.

Alkaline batteries have an inherently higher internal resistance than lead batteries. The problem of decreasing this resistance to a sufficient extent so that a battery of normal size would respond to a sudden high potential demand when starting an automobile, has been solved in the following manner: Batteries were made having very thin plates. Due

FIG. 7b. Group of plates with insulating separators.

FIG. 8. Six-volt cadmium-nickel starting battery for passenger cars.

to the increase in surface area of the plates and the decrease in the distance between plates, these batteries fully complied with the specifications for the starting type.

Factory-produced nickel-cadmium batteries usually have plates that are 2 to 4 mm. thick and are spaced 1 to 2 mm. apart. However, for starting batteries of the cadmium-nickel type the plates are 1 to 2 mm. thick and less than 1 mm. apart. These S. L. I. batteries have a total (positive plus negative) plate surface of about 130 cm.^{2}/amp.-hr.

Some companies resort to schemes other than the above in order to decrease the internal resistance of the cells. For instance, some in-

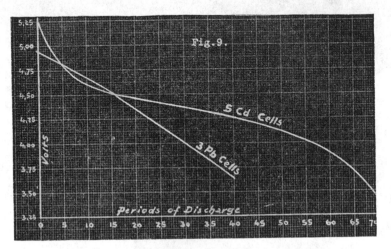

FIG. 9. Potential drop of lead cells compared with that of cadmium-nickel cells at room temperature.

crease the conductivity by making certain additions of conducting material to the mass in the pockets and by inserting a thin corrugated pure nickel band.

Fig. 7b shows a group of plates comprising the typically thin plates closely spaced. Also in Fig. 7b are shown insulating separators made of hard rubber or of alkali-resistant plastic material. Fig. 8 shows a six volt cadmium-nickel starting battery for passenger cars. As a rule three lead cells are replaced by four or five cadmium cells. Fig. 9 shows the potential drop of three 150 amp.-hr. ten-hour capacity lead cells connected in series as against the potential drop of an equivalent battery consisting of five cadmium-nickel cells of 140 amp.-hr. capacity. The discharge consisted of sudden discharges at 500 amp. After each ten-second discharge there was a ten-second idle period.

The curves show that at the beginning the potential of the lead battery drops more slowly. However, for repeated high current discharge requirements the voltage performance is better in the case of the cadmium battery. The lead battery potential dropped to less than 4.0 volts after thirty discharges. In other words, the lead battery discharged to 4.0 volts in a total of 5.0 minutes, whereas the cadmium battery required 9.3 minutes before dropping to 4.0 volts. Therefore, for heavy current demands the capacity of the cadmium battery is almost double that of a lead battery of practically the same capacity at a low discharge rate.

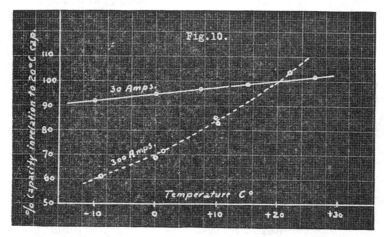

FIG. 10. Capacity of a cadmium-nickel starting battery at various temperatures.

Cadmium-nickel batteries give a better performance also when subjected to extreme temperatures. This ability to withstand high and low temperatures is one of the chief advantages of the cadmium electrode alkaline battery over the iron electrode alkaline battery. With a rise in temperature, the capacity of the cadmium battery increases without a marked increase in the amount of self-discharge, as is the case with the iron electrode alkaline battery. At low temperatures the cadmium electrode, in contradistinction to the iron electrode, does not become passive. The cathode capacity naturally decreases to some extent. The temperature coefficient depends upon the construction of the plates, their thickness and on the current load. With starting batteries with small internal resistance the temperature coefficient is particularly low. The curves in Fig. 10 show the capacity of a cadmium-nickel starting battery at various temperatures as against the capacity at 20°

C. taken as 100. The solid line curve represents the discharge of the five hour current at 30 amp.; the dotted curve the corresponding discharge at 300 amp. From these curves we note that the temperature coefficient is 0.24 per cent for the 30 amp. discharge, and 1.5 per cent for the 300 amp. discharge. For cadmium-nickel batteries with thicker plates the temperature coefficients are essentially higher. The electrolyte of the cadmium-nickel battery freezes at about — 28° C. regardless of the state of charge or discharge.

Cadmium-nickel batteries with nickel oxide electrodes should not be continuously charged at temperatures above 45° C. Otherwise, irreversible changes occur in the active nickel oxide material which result in a lasting loss in capacity of the cell. However, an occasional overstepping of the above temperature limit does not damage the battery. Finally, we should emphasize that the efficiency of the active material in the thin plates of the cadmium-nickel starting batteries is particularly good. Seventy per cent and more of the active oxide functions during the five hour rate discharges when discharged to 1.0 volt per cell. With the thicker plates (2 to 3 mm.) with pocket type construction, 50 to 60 per cent of the active material functions during a five hour discharge. As a matter of comparison, only 33 to 40 per cent of the active material in the lead battery functions. In the Edison iron electrode alkaline battery only 17 per cent of the active iron functions.

[*Editor's Note:* The discussion has been omitted.]

8

Paper presented at *Symp. Workshop Advanced Battery Research and Design*, Chicago Section, The Electrochemical Society and Argonne National Laboratory, 1976, 20pp.

ALKALINE BATTERIES

Jack T. Brown
Westinghouse Research and Development Center
Pittsburgh, Pennsylvania 15235

ABSTRACT

Commercial secondary alkaline batteries include those based on the electrochemistry of the nickel-cadmium, silver-zinc, silver-cadmium, nickel-iron, and manganese-zinc couples. From a market value standpoint, the most important of these is the nickel-cadmium type. However, the total market is less than $100 million per year, and less than 10% of that of the lead-acid battery.

These various cell types are manufactured in several construction technologies for various applications. For example, nickel-cadmium batteries are made as flat-plate pocket or sintered types, with the electrolyte flooding the cell, thus requiring a vent in the case, or retained only in the separators and plate pore volume, known as starved, and enabling the cell case to be sealed. Flat plates are stacked along side each other, resulting in a rectangular shaped cell, or sintered plates can be spirally wound, making a cylindrically shaped cell. Batteries with sintered, flooded, vented, rectangular cells are used for aircraft engine starting, and sintered spiral wound, starved, sealed, cylindrical cells are used in small calculators, etc. Some silver-zinc cells are manufactured in sizes up to 7000 ampere-hours capacity for underseas vehicle propulsion power sources.

In the near term, new secondary alkaline systems are being readied for market introduction. These are a technology modification called the iron-nickel system and the nickel-zinc and nickel-hydrogen batteries. The iron-nickel and nickel-zinc batteries are intended principally as candidates for reasonably-low capital cost, high specific energy content ground transportation or underseas electric vehicle propulsion power sources. Nickel-hydrogen batteries are intended principally as improved satellite batteries.

For the longer term, several battery groups (researchers and technology developers) are intrigued with the combination of potential high performance characteristics and low cost of metal-air type systems. Here the prime candidates use alkaline electrolytes, and the air cathode in combination with either zinc or iron anodes. The technology opportunity is for, perhaps, doubling the specific energy content, and cutting the cost in half, compared to the nickel electrode counterpart systems.

This paper will briefly cover the materials, processes, designs, performance characteristics and applications of state-of-the-art secondary alkaline batteries. It will also preview these same considerations for some of the anticipated advanced technology systems.

I. T. Brown

INTRODUCTION

Battery systems using alkaline electrolytes have uniformly suffered with a similar problem since their commercial introduction which occurred about the beginning of this century. The problem was, and is, that alkaline battery systems have significantly higher initial capital cost than other battery systems which might be used for the same applications. The battery cost problem has little or nothing to do with the cost or use of the alkaline electrolyte per se, but rather with the basic cathode materials costs, i.e., nickel and silver which only have the requisite thermodynamic stability when used with the high pH electrolyte, and the relatively complex and capital intensive processes needed for their manufacture.

Why then has market growth occurred and such a relatively large amount of research and technology advancement been pursued? The answer to this question defines the three eras of technology development of alkaline secondary batteries. The first which lasted about fifty years, centered around the Edison nickel-iron, pocket-type plate battery, which featured rugged construction for long life and high reliability. These were developed for industrial uses for traction power, standby emergency power, signaling devices, and heavy duty starting and lighting. The second era featured the development of the sintered nickel electrode and the silver cell systems (about 1950). This technology provided very high power delivery capability as well as high specific energy content. Applications, utilizing this advancement, were for airplane engine starting, communications, and specialty military propulsion and electronics. The third era of alkaline batteries began in the early 1960's with the perfection of sealed cells. These had an engineered design of over-capacity in the negative plate which permitted controlled recombination on it of the overcharge oxygen from the positive plate. No gassing of H_2 occurred on the negative and the oxygen recombination reaction made sealed cases possible. These batteries were used in emergency equipment, portable tools, instruments, and appliances, and perhaps most importantly in space exploration as satellite power supplies.

Looking ahead to the end of the first 100 years of availability of alkaline batteries, which, interestingly, coincides with the end of the 20th century, we can scenario two additional eras. The first which can begin in the latter part of the 1970's is the introduction of low cost batteries, storing some tens of kilowatt-hours. These will use the readily available anode materials iron and zinc for negative plates, combined with new construction technology nickel plates that have considerably lower process and basic material cost than presently used plates. These batteries with nearly twice the specific energy content and half the cost of presently available nickel-cadmium batteries will find applications as electric vehicle propulsion power sources, both for ground transportation and undersea systems.

Widespread use of alkaline batteries in electric personal commuter vehicles will probably not occur until the second new era when battery costs are again significatnly reduced (perhaps by a factor of two from the first new generation batteries) and the specific energy content again doubled. This will occur with the technological development of long life, low cost, bifunctional air electrodes, which will be the cathodes to replace nickel plates. Low cost but improved technology iron and zinc anodes will be retained in this second generation new battery technology.

118

This personal electric commuter vehicle application era allows us to contemplate conservation of several millions of barrels of critical oil resources per day. This, of course, implies the substitution of coal and nuclear energy sources (when charging batteries) for oil consumed in combustion engines. If one expects a persistent difference of $1 per million BTU between these different fuel costs, the "bottom line" economic picture could easily be a $5 billion per year energy cost savings.

Certainly this, as most, scenarios are fraught with uncertainties, many of which are out of our technological control. As one who has spent a good deal of time and energy over the past ten years developing systems for which no present market exists, I can assure you it is difficult to maintain support for such programs. However, it is certainly one of our purposes at this symposium to raise questions, in thoughtful discussion, about the various alternatives.

STATE-OF-THE-ART SYSTEM MATERIALS, PROCESSES, AND DESIGN

Relative inertness in the electrochemical environment is a prime requirement for the type of structural and conductive materials which are used to support the electrochemically active chemicals. Examples are nickel, silver, iron, copper, or carbon (usually graphitic). They have a low surface area to volume ratio compared with the active materials and are essentially passive in the electrolyte, are able to be formed into a variety of shapes rather inexpensively, and have good electrical and thermal conductivity.

The electrochemically active materials are usually oxide or hydroxide compounds of the metals mentioned above. They are generally rather poor electrical conductors, however, they are rarely stoichiometric or highly crystalline. Off stoichiometry usually is due to an oxygen deficiency and some electronic conductivity is attributable to lattice vacancies. Amorphous compounds are usually more conductive than crystalline ones but mixed bonding in the lattice can give good conductivity in spite of good crystallinity as, e.g., in Ag_2O_2. It is necessary it achieve a high surface area on these materials to be able to use them in cell designs which require high discharge power. This is done by milling, coprecipitating, laminating and other mechanical and chemical dispersing procedures. A tendency to reagglomerate or densify by reducing the surface energy after solid state transformations accompanying the electrochemical reaction often takes place. One often uses an expander incorporated in the material to retard this tendency.

Various alkaline battery electrochemically active materials are listed in Table I. The list is not exhaustive or necessarily accurate in detail, but is shown only as illustrative.

Polyvinylchloride, nylon polyamide, polystyrene, polyethylene and polypropylene are some of the best accepted plastic materials used for cell cases, gaskets, insulators, spacers, and separators. Requirements of these materials are that they be non-conductors, have chemical resistance to the alkaline electrolyte, be moldable into various shapes, have good impact resistance and low-weight and cost.

An understanding of the reaction mechanisms including the kinetics is the basis for the choice of optimizing cell designs, and this also reflects into the choice of processes used. It is not the purpose of this section to describe this methodology, but rather to state what the approximate reaction equations are for the various systems, and describe what is available in cell and batteries in various designs.

Table I - Alkaline Battery Electrochemically Active Compounds

Cell Type	Positive		Negative	
	Charged	Discharged	Charged	Discharged
Nickel-Cadmium	β-NiOOH	$Ni(OH)_2$	Cd	$Cd(OH)_2$
Nickel-Iron	β-NiOOH	$Ni(OH)_2$	Fe	$Fe(OH)_2$ Fe_3O_4
Silver-Zinc	Ag_2O_2 AgO	Ag	Zn	$Zn(OH)_2$ ZnO
Silver-Cadmium	Ag_2O_2 AgO	Ag	Cd	$Cd(OH)_2$
Manganese-Zinc	α-MnO_2 MnOOH	$Mn(OH)_2$ Mn_3O_4	Zn	$Zn(OH)_2$ ZnO

Nickel Cadmium Cells

A simple overall cell reaction really is not able to be accurately written as the actual reaction changes with the preparation of the active materials, their volumetric loading and available surface area, the amount, concentration and analysis of the electrolyte, operating temperature, rate of charge and discharge and perhaps, most insidious, even the cycle history. We know on the nickel electrode, nickel-to-oxygen bond lengths change, water is both absorbed and adsorbed, some potassium hydroxide is absorbed during charge and released during discharge. Lithium can also be incorporated into the lattice. Also, there are several steps in the solid state transformations associated with both charge, overcharge and discharge. Similarly, the cadmium electrode has discharge sequential reactions. Realizing the above, shown below nevertheless are the individual simplified electrode and overall cell reactions:

$$\beta\text{-NiOOH} + H_2O + e^- \underset{\text{Charge}}{\overset{\text{Discharge}}{\rightleftarrows}} Ni(OH)_2 + (OH)^-$$

$$Cd + 2(OH)^- \underset{\text{Charge}}{\overset{\text{Discharge}}{\rightleftarrows}} Cd(OH)_2 + 2e^-$$

$$(\text{Cell}) \quad 2\beta\text{-NiOOH} + Cd + 2 H_2O \underset{\text{Charge}}{\overset{\text{Discharge}}{\rightleftarrows}} 2 Ni(OH)_2 + Cd(OH)_2$$

A more correct equation would identify peroxide formation aspects, water adsorption potassium hydroxide adsorption, other elements incorporated in the lattices, etc. This understanding is needed, for example, to design stable capacity, minimum electrolyte requirements, water consumption and other important application needs, as in starved electrolyte cells compared with flooded electrolyte cells.

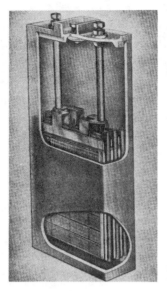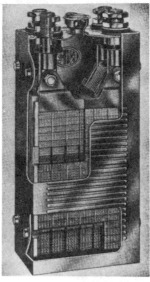

Figure 1.   Cutaway views of typical pocket nickel-cadmium cells.

Included in the types of cells and batteries available are those which
have flat plates into which the active materials have been placed in individual pockets. These plates are stacked together like cards in a deck with
separators between them and placed in a rectangular cross-sectioned case of
either metal or plastic (Figure 1).
One modification uses the so-called tubular positive plates where now
the positive active material has been placed into a cylindrical structure,
Figure 2, and assembled into plates with pocket negatives, Figure 3.
It is possible to obtain pocket plate cells either vented or sealed. Each
has operational advantages and limitations. A third modification of plate
structure is the porous sintered plaque. It is made by sintering metal
powder with a centrally located metal wire screen or perforated metal sheet.
The characteristics of this structure are high porosity (in order to hold or
retain the active material), large surface area, high electrical conductivity
and good mechanical properties. This structure now has the largest production by far of any type. It leads to a cell design with an excellent combination of energy content and power capability, as will be discussed later.
It is used in the construction of prismatic cells with plastic (Figure 4) or
metal cases, either vented and flooded or starved and sealed. The sintered
plaques' mechanical properties have led to the development of the widely
used sealed cell and battery for consumer appliance applications. The
coiled plate, cylindrical, starved and sealed cell looks very much like
the ordinary dry cell, Figure 5.

J. T. Brown

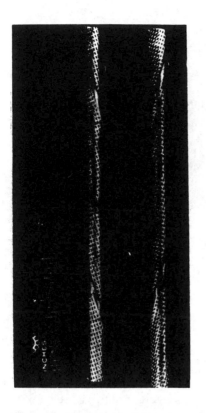

Fig. 2 Positive Plate Tubes

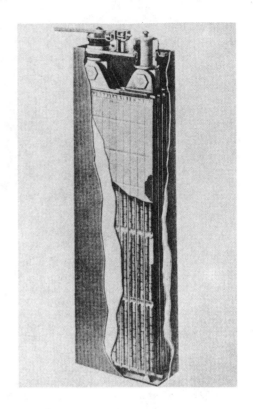

Fig. 3 Tubular Positive-Pocket
Negative Type Cell

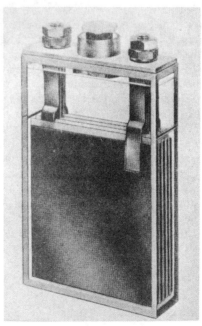

Fig. 4 Sintered Plate Vented Type
Nickel-Cadmium Cell

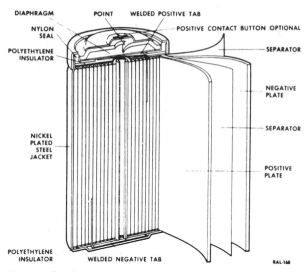

Figure 5. Cutaway drawing of sealed, coiled type, sintered Ni-Cd cell.

Nickel-Iron Cells

The overall cell reaction of the Nickel-Iron cell has all the nickel electrode complications, alluded to earlier, plus now the complex iron electrode reactions. This latter has at least the two-step mechanism shown below:

$$Fe + 2(OH)^- \xrightarrow{Discharge} Fe(OH)_2 + 2e^- \qquad \text{1st step}$$

$$Fe(OH)_2 + 2(OH)^- \xrightarrow{Discharge} Fe_3O_4 + 4 H_2O + 2e^- \qquad \text{2nd step}$$

$$3 Fe + 8(OH)^- \xrightarrow{Discharge} Fe_3O_4 + 4 H_2O + 8e^- \qquad \text{overall}$$

Combined with the nickel electrode this gives:

$$8 NiOOH + 3 Fe + 4 H_2O \underset{Charge}{\overset{Discharge}{\rightleftarrows}} 8 Ni(OH)_2 + Fe_3O_4$$

Edison nickel-iron type cells incorporate tubular positives, shown before, and flat plate pocket negatives, quite similar to the comparable nickel-cadmium technology. These cells only have prismatic configurations and use flooded electrolyte designs which are not sealed. Provision for escape of gases on overcharge must be allowed because of the unusually low hydrogen overvoltage on iron which requires considerable overcharge in order to get maximum iron plate capacity. This, of course, must be accompanied by frequent water additions to the cells. Although this battery's rugged construction and basically stable thermodynamic reaction characteristics enabled it to have very long life, its high manufacturing cost and price in comparison with industrial lead acid batteries caused a declining market over the past twenty years. As a consequence the Edison type battery is no longer manufactured in this country.

J. T. Brown

Silver-Zinc Cells

In general, these cells are used where their high rate capability and
energy content are needed and effect their cost. This is usually in military
and space applications. The silver electrode reaction proceeds in two steps,
with the so-called divalent state progressing to the monovalent state, and
then to reduction to silver. Actually the divalent state is probably a mixture of a crystal structure with a pair of silver atoms with monovalent bonds
and a pair with trivalent bonds. The reaction is written as:

$$Ag_2O_2 + H_2O + 2e^- \underset{Charge}{\overset{Discharge}{\rightleftarrows}} Ag_2O + 2(OH)^-$$

$$Ag_2O + H_2O + 2e^- \underset{Charge}{\overset{Discharge}{\rightleftarrows}} 2\,Ag + 2(OH)^-$$

The voltage plateau for the monovalent oxide reduction to silver occurs
about 300 mv lower than the divalent oxide reduction and consumes, in practice, about three fourths of the discharge time instead of the theoretical
50 percent.

The zinc electrode reaction has been the subject of a great deal of ,
study over the last twenty years. It seems clear that complexities that are
not fully understood as yet exist but that at least three steps are involved.
First, zinc is oxidized to form ZnO or $Zn(OH)_2$ in solid form, followed by
dissolving of these compounds to form zincate ions ($Zn(OH)_4^{-2}$), and at a
saturation point in the electrolyte of these ions a passivating film of ZnO
forms on the electrode. The overall cell reaction is given below, in again
simplified terms.

$$Ag_2O_2 + Zn + H_2O \rightarrow Ag_2O + Zn(OH)_2$$

$$Ag_2O + Zn + H_2O \rightarrow 2\,Ag + Zn(OH)_2$$

(Cell) $\quad Ag_2O_2 + 2\,Zn + 2\,H_2O \underset{Charge}{\overset{Discharge}{\rightleftarrows}} 2\,Ag + 2\,Zn(OH)_2$

Virtually all silver-zinc secondary batteries are built from cells using
flat plates in prismatic cases. They generally use a restricted amount of
electrolyte and have a vent which allows for gases to escape but prevents the
atmosphere from entering the cell continuously. Periodic water additions can
be made through the vent but are not required often due to the care in charging which is necessarily exercised.

Although essentially no mention has been made of separators used in the
other alkaline secondary cells, the silver-zinc cell places the greatest demands on them and consequently requires some comment. When used at high discharge currents, a separator system must have low electrolyte resistance.
This usually requires it to be thin. In addition, both the silver oxide and
zinc oxide electrode materials are soluble to an important extent in the
electrolyte. Therefore the separator must have low permeability to these
materials. Inertness to the electrolyte and any chemical deterioration with
the electrode materials is desirable. This is particularly difficult with
silver oxide which is a strong oxidizing agent. Wetting and wicking of electrolyte is also required because of the restricted amounts of electrolyte.

Unfortunately a totally acceptable separator has not yet been developed for silver-zinc cells but multilayers of unplasticized cellophane have proven an acceptable solution for some applications.

PERFORMANCE CHARACTERISTICS

Battery properties of general importance are the gravimetric and volumetric energy content, power capability both sustained and peak, voltage-time characteristics on both charge and discharge at different rates and temperatures, internal resistance, coulombic and energy efficiency, charge retention, and degradation of capacity and voltage with cycling, i.e., life. For specialty uses there are other characteristics that become important, such as, operating in any position, ability to be overcharged when sealed, etc.

It is a problem of great difficulty to summarize and fairly compare the characteristics of all technology modifications of the alkaline systems and also reference them to the well-known lead-acid system. However, this property summary has been done rather effectively in the John Wiley and Sons Incorporated publication "Alkaline Storage Batteries" by Falk and Salkind, and it is from a section of that book (with their permission) that the figures and tables in this section of the paper have been collated. Care must be exercised in using generalized data for other than a rough comparison. There are detailed considerations that cannot practically be taken into account in such a summary which might in some cases change the preference of one system for that of another. For example, in an indication of power capability, the typical properties as dictated by the technology used may indicate a low value but the electrochemical kinetics may actually be quite high, and a change in design could result in a cell with high power density.

An overall figure of merit rating for various systems relative to each other is shown for different properties in Figure 6, divided into five general categories with five being the best.

Figure 6. COMPARATIVE SURVEY FOR DIFFERENT SYSTEMS BY MEANS OF FIGURES OF MERIT

System	Energy Density	High-Rate Discharge Properties	Low-Temperature Properties	Internal Resistance	Charge Properties	Ah and Wh Efficiency	Charge Retention	Life	Mechanical Properties
Ni-Cd pocket, vented	2	4	5	4	5	1	4	5	5
Ni-Fe	3	2	1	1	4	3	1	5	5
Ni-Cd sintered, vented	3	5	5	5	5	3	3	4	5
Ag-Zn vented	5	5	3	5	2	5	4	1	4
Ag-Cd vented	4	3	5	2	3	5	4	3	4
MnO$_2$-Zn	1	1	3	3	2	3	5	1	4
Ni-Zn	4	4	4	5	3	4	3	2	4
Ni-Cd sintered, sealed	3	5	4	5	4	1	3	2	4
Ni-Cd pocket, sealed	2	2	3	4	2	.	3	3	5
Ag-Zn sealed	5	4	3	4	1	5	4	1	4
Ag-Cd sealed	4	3	4	2	1	5	.	3	4
Pb-acid, pasted	3	2	2	4	3	5	1	3	1
Pb-acid, tubular	3	1	2	2	3	4	1	4	3
Pb-acid, Planté	1	1	2	2	3	5	1	4	1

Theoretical energy density for various couples are shown in Figure 7. This should be understood to be only an upper bound based on the reactions shown and the Emf considerations indicated. Actual cells usually have only 10 to 30 percent of this theoretical value, and the spread indicates they do not necessarily follow in the same order.

Figure 7. THEORETICAL ENERGY DENSITY VALUES FOR VARIOUS ELECTROCHEMICAL SYSTEMS

System	Theoretical Energy Density		Emf V[a]	Simplified Cell Reactions Used for Energy Density Calculations
	Wh/kg	Wh/lb		
Ag-Zn	434	197	1.856[b]	$Ag_2O_2 + 2Zn + 2H_2O \rightleftharpoons 2Ag + 2Zn(OH)_2$
	273	124	1.602	$Ag_2O + Zn + H_2O \rightleftharpoons 2Ag + Zn(OH)_2$
Ni-Zn	326	148	1.735	$2NiOOH + Zn + 2H_2O \rightleftharpoons 2Ni(OH)_2 + Zn(OH)_2$
Ni-Fe	267	121	1.370	$2NiOOH + Fe + 2H_2O \rightleftharpoons 2Ni(OH)_2 + Fe(OH)_2$
Ag-Cd	267	121	1.380[c]	$Ag_2O_2 + 2Cd + 2H_2O \rightleftharpoons 2Ag + 2Cd(OH)_2$
	171	77	1.155	$Ag_2O + Cd + H_2O \rightleftharpoons 2Ag + Cd(OH)_2$
HgO-Zn	240	109	1.344	$HgO + Zn + H_2O \rightleftharpoons Hg + Zn(OH)_2$
Ni-Cd	209	95	1.290	$2NiOOH + Cd + 2H_2O \rightleftharpoons 2Ni(OH)_2 + Cd(OH)_2$
MnO_2-Zn	191	87	1.52	$3MnO_2 + 2Zn + 2H_2O \rightleftharpoons Mn_3O_4 + 2Zn(OH)_2$
Pb-acid	175	79	2.095	$Pb + PbO_2 + 2H_2SO_4 \rightleftharpoons 2PbSO_4 + 2H_2O$

[a]Values based on normal electrolyte concentrations.

[b]1.679 V, the average value of 1.856 and 1.602 V associated with the higher and lower oxidation states of silver, has been used in the calculation of energy density.

[c]1.268 V, the average value of 1.380 and 1.155 V, has been used in the calculation of energy density.

Figures 8 and 9 summarize the energy content of cells and batteries for the various systems. One should particularly note the overlap making the choice on this basis alone quite difficult.

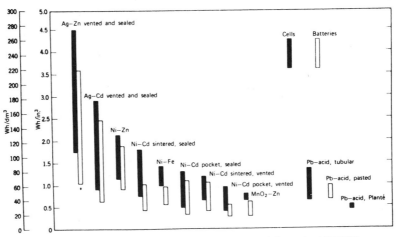

Figure 8. Energy per unit of volume for different systems. Based on nominal energy at room temperature.

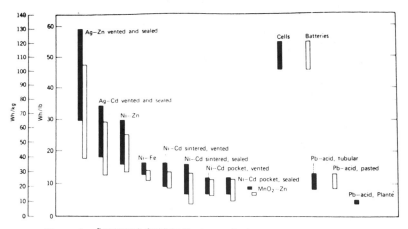

Figure 9. Energy per unit of weight for different systems. Based on nominal energy at room temperature.

Cell voltage as a function of capacity is an important characteristic and is shown in Figure 10. The higher the average voltage, the fewer the number of cells that are required for a battery of specific voltage. Lead-acid leads the list and the silver cells have the best lower plateau voltage regulation.

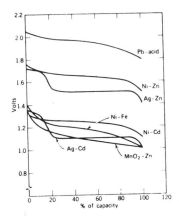

Fig. 10 Typical discharge characteristics of different systems at the 0.2 × C rate Room temperature

The ability of a cell to maintain capability at high discharge rates is shown rationalized to constant capacity in Figure 11. This is an approximation to the sustained power capability of the various systems.

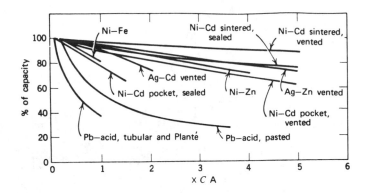

Fig. 11 Percent available capacity versus discharge rate for different systems. Room temperature.

Also important is the effective internal resistance of a cell. Figure 12 shows this relationship rationalized to cell capacity. Sintered Nickel-Cadmium and Silver-Zinc cells have the lowest internal resistance.

Fig. 12 INTERNAL RESISTANCE DATA FOR DIFFERENT ELECTROCHEMICAL SYSTEMS AT ROOM TEMPERATURE

System	Internal Resistance in $R_i \times C$	Notes
Ni-Cd sintered, vented	0.03–0.06	
Ni-Cd sintered, sealed	0.03–0.04	
Ag-Zn vented	0.03–0.05	High rate cells only
Ni-Zn	0.03–0.05	Experimental cells
Ni-Cd pocket, vented	0.05–0.2	
Ni-Cd pocket, sealed	0.05–0.2	
Ag-Cd vented	0.1–0.4	
MnO_2-Zn	~ 0.15	
Ni-Fe	0.4–0.5	
Pb-acid, pasted	0.1–0.15	
Pb-acid, tubular	0.15–0.4	
Pb-acid, Planté	0.2–0.3	

Charge voltage curves at appropriate charge rates are shown in Figure 13, and some generalized data on charging is shown in Figure 14. Overcharge capability without damage to the cells is often important for series connected batteries where slight differences in individual cell characteristics can cause problems. Notice, in general, nickel-cadmium cells are quite good and silver-zinc cells are poor.

Where energy conservation considerations or charging energy cost is important, energy efficiency between discharge and charge can be assessed by consulting Figure 15. Some alkaline systems can achieve 75% efficiency. Lead-acid pasted plate cells are actually the best and can really achieve over 80% turn-around efficiency with appropriate charging control.

Alkaline Batteries

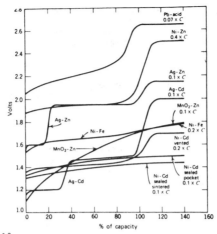

Fig. 13 Typical charge characteristics of different systems at normal constant rates and room temperature.

Figure 14. DATA ON CHARGING OF VARIOUS SYSTEMS

System	Charging Methods		Recommended Constant Current Charge Rate, C·A	Over-chargeability	Temperature Range for Charging, °C	Notes
	Preferred	Not Recommended				
Ni-Cd sintered, vented	cc, cp	—	0.2	Very good	−55 to +75	
Ni-Cd pocket, vented	cc, cp	—	0.2	Very good	−50 to +40	
Ni-Fe	cc, cp	—	0.15–0.2	Very good	0 to +45	
Ni-Cd sintered, sealed	cc	cp	0.1–0.3	Very good	0 to +40	
Ni-Zn	cc, cp	—	0.1–0.4	Fair	−20 to +40	Experimental Cells
Ag-Cd vented	cc	—	0.1–0.2	Fair	−40 to +50	
Ni-Cd pocket, sealed	cc	cp	0.1	Fair	0 to +45	
Ag-Zn vented	cc, cp	—	0.05–0.1	Poor	0 to +50	
MnO₂-Zn	cc, cp*	—	0.1–0.12	Fair	No data available	*Only current limited constant potential recommended
Ag-Cd sealed	cc, cp	—	0.07	Poor	−40 to +50	
Ag-Zn sealed	cc, cp	—	0.07	Poor	0 to +50	
Pb-acid, all kinds	cc, cp	—	0.07	Fair	−40 to +50	

* Constant current (cc) includes two-rate charging, and constant potential (cp) includes modified constant potential charging.

129

Fig. 15   AH AND WH EFFICIENCIES OF VARIOUS SYSTEMS

System	Ah Efficiency %	Wh Efficiency %	Notes
Ag-Zn vented and sealed	90–95	70–75	
Ag-Cd vented and sealed	90–95	~70	
Ni-Zn	~85	~70	Experimental cells
Ni-Fe	80	60	
MnO_2-Zn	~80	~60	
Ni-Cd sintered, vented	71.5–83.5	62–75	
Ni-Cd pocket, sealed	71.5–77	60–65	
Ni-Cd pocket, vented	71.5	60	
Ni-Cd sintered, sealed	67–71.5	60–65	
Pb-acid pasted and Planté	91	75	
Pb-acid tubular	83	68	

All data are related to normal rates of charge and discharge and to room temperature operation.

Information on charge retention is shown in Figure 16. Not surprising is that the best systems are ones that actually are used most often as primary cells, where intermittent service is usually quite important. For most applications of secondary cells this is not a prime consideration.

Fig. 16   TYPICAL CHARGE RETENTION DATA FOR DIFFERENT SYSTEMS AT ROOM TEMPERATURE

System	Available Capacity After Six Months Storage, %	Notes
HgO-Cd	>95	
HgO-Zn	>95	
MnO_2-Zn	95	
Ag-Cd vented	90	
Ag-Cd sealed	85	
Ag-Zn sealed	85	
Ag-Zn vented	75	Considerable spread in values
Ni-Cd pocket, vented	75	
Ni-Cd sintered, vented	70	
Ni-Zn	70	Experimental cells
Ni-Cd pocket, sealed	60	
Ni-Cd sintered, sealed	0	70% available after 1 month
Ni-Fe	0	60% available after 1 month
Pb-acid	0	70% available after 1 month

Last (but not least) is a life assessment comparison shown in Figure 17. Because it is impossible to detail the important restrictions on the interpretation of life characteristics in a paper such as this, caution is advised in using this information for anything but the most cursory consideration. For example, long life is not achieved in manganese dioxide-zinc or silver-zinc cells, but nickel-cadmium cells do pretty well. The important basic life considerations are thermodynamic stability or low corrosion, i.e., stability of the reactant and structural materials in the electrolyte under the complex electrochemical conditions of charge and discharge, and

resistance to the stresses and strains accompanying the solid state transformations as the reactants are charged and discharged.

Fig. 17   DATA ON LIFE OF DIFFERENT SYSTEMS: ROOM
           TEMPERATURE CONDITIONS

System	Life	
	Number of deep discharge cycles. >70%	Total life-time. years
Ni-Fe	2000–4000	7–25
Ni-Cd pocket. vented	500– >2000	8–25
Ni-Cd sintered. vented	300– >2000	3–10
Ni-Cd sintered. sealed	200–2000	2–10
Ni-Cd pocket. sealed	100–250	~5
Ag-Cd sealed	~700	2
Ag-Cd vented	300–500	3
Ni-Zn	100–200	No data available
Ag-Zn sealed	~80	2
Ag-Zn vented	10–150	0.5–1.5
MnO_2-Zn	~50	No data available
Pb-acid, tubular	1400	4–10
Pb-acid, Planté	1000	10–15
Pb-acid, pasted	200–700	3–6

NEW TECHNOLOGIES

Near Term

As mentioned in the introduction, new technology nickel-zinc and iron-
nickel batteries can be introduced in the next few years which have significantly lower materials and process costs than comparable sintered plate
nickel-cadmium batteries or Edison batteries, and twice the specific energy
content. These new batteries may find application in electric vehicles used
in service fleets. Some mention has already been made of nickel-zinc technology, and the new iron-nickel technology will be briefly discussed here.
Pure nickel powder which is sintered is an expensive current collector for
nickel battery plates. Lower cost current collectors can be used and still
achieve high power and long life. Active nickel material, laboriously chemically impregnated into the powder pores and precipitated from a material like
nickel nitrate, has a long process time and twice the nickel metal cost as
electrochemically precipitating, in situ, the active material in electrode
pores from nickel metal plating anodes. Low cost iron plates can be made
from similar technical considerations for current collectors and low cost
iron oxide materials.

The energy content versus current drain is shown in Figure 18 for 180
ampere-hour 16 KWh batteries manufactured in pilot quantities compared to
industrial lead-acid batteries. Typical cycle life tests on similar cells
are shown in Figure 19.

The entire battery system is designed for the intended application and
addresses low maintenance costs, rapid recharge with thermal control and one
point battery electrolyte equalization. The system schematic is shown in
Figure 20. A photograph of the assembled battery with the auxiliary charging
system which is quickely disconnected before discharge is shown in Figure 21.

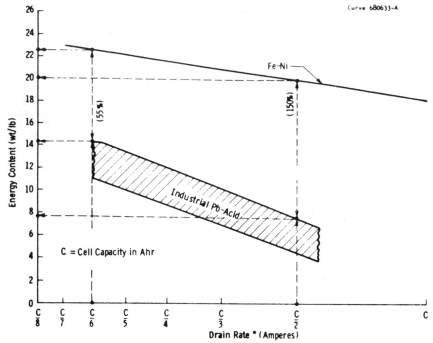

Fig. 18—Energy density vs. drain rate Westinghouse Fe-Ni & industrial Pb-acid

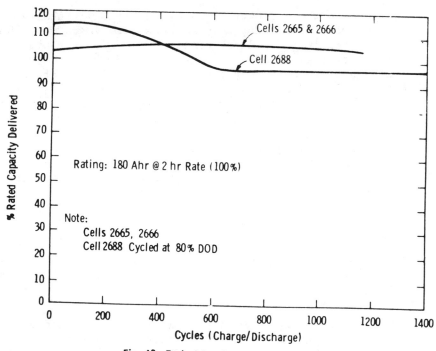

Fig. 19—Typical Fe-Ni life tests type E-180 cells

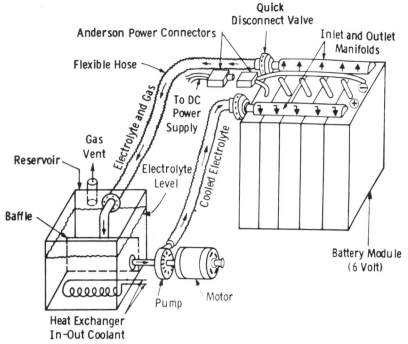

Fig. 20—Fe-Ni Battery system schematic

Fig. 21—Westinghouse Fe-Ni battery for delivery van ready for charge

Longer Term

The type of battery system many technologists still feel offers the most promise for the longer term, perhaps after five to ten years additional development, are the metal-air hybrid type systems. As mentioned earlier, the leading candidates involve zinc and iron negative electrodes. The as yet undescribed plate is the air electrode.

From an energy density standpoint, an air cathode is attractive as a battery component because the oxygen component of the electrochemical reaction is supplied continuously from the air surrounding the battery.

The use of air cathodes is not a recent development. A zinc-air primary battery with air electrodes of wet-proofed carbon has been commercially available for 50 years for use in channel marker bouys, highway flashing systems, railway signals, radio receivers and transmitters, and similar applications. Although reliable, a battery of this type could be discharged only at low rates because of their early technology air electrodes. The power density using such an electrode would be unacceptably low for vehicular applications.

The considerable research effort on fuel cells and metal-air cells in the last 15 years that led to the development of high rate air electrodes has created a renewed interest in metal-air batteries. It has been rationalized the combination of potential high energy content and reasonable power capability can be achieved using new technology air electrodes capable of sustaining high current density.

Oxygen reduction from the air in an alkaline electrolyte proceeds in two steps:

$$O_2(g) + H_2O(soln) + 2e^- \rightarrow HO_2^-(aq) + (OH)^-(aq).$$

This first step involves the formation of the intermediate species the perhydroxide ion which must be decomposed or further reduced with catalysts to achieve the overall four electron reactions:

$$O_2(g) + 2 H_2O(soln) + 4e^- \rightarrow 4(OH)^- (aq).$$

Hypotheses are available to explain the somewhat kinetically sluggish reduction of oxygen on metal surfaces in terms of the high bond-dissociation energy of the oxygen molecule. Other mechanisms have also been discussed which involve availability of favorable surface sites for preferred orientation adsorption.

The physical structure of an air electrode generally consists of several layers. The side facing the air is usually hydrophobic with pores that admit air and allow oxygen molecules to be adsorbed onto the electrolyte side porous surface. Nitrogen molecules are unreactive in the structure. The electrolyte side is usually a high surface area partially conductive material. Various carbons and graphite mixtures are preferred. This surface is usually catalyzed with a small amount of silver and other additives for perhydroxide decomposition and oxygen overvoltage control on charging. A metallic current collector may be sandwiched between the two layers or interspersed with the (hydrophilic) electrolyte side layer.

Overall reactions for both zinc-air and iron-air batteries can be deduced from the above air electrode reaction and the previously mentioned reactions for these anodes. When the upper bound theoretical energy density is deduced from these reactions it is found to be approximately 500 Wh/pound for each system. This is considerably higher than that previously mentioned for any alkaline system, but recall the admonition regarding the actually achievable energy contents of fully developed systems.

134

Various technologies have been developed for zinc-air batteries which have attempted to deal with the basic solubility problems and replating, i.e., recharging morphology difficulties. These have involved, e.g., replaceable zinc anodes and electrolyte change after discharge, zinc conductive slurries being circulated, in situ electrical recharging, usually involving additives to control morphology of the deposits, and rotating anodes to control the zinc deposition. These do get to be rather complex chemical engineering systems, and none of the systems production design have been formulated.

The iron-air system has found several advocates with working groups in Japan, England, Germany, Sweden and the United States. The major drawback vis-a-vis the zinc system is the lower cell voltage, therefore, requiring more cells for a given battery voltage, and low hydrogen overvoltage on the iron electrode, which of course is true of all iron plate systems. The major advantage is the performance stability and long life which can be achieved.

Performance characteristics of an early cell design are shown in Figures 22 and 23.

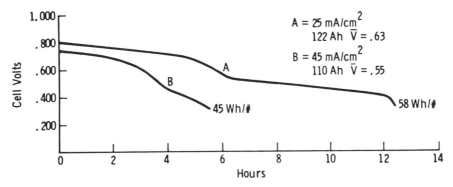

Fig. 22 — Iron-air cell discharge characteristics

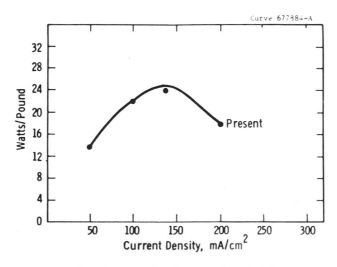

Fig. 23 — Iron-air cell power characteristics

CONCLUSIONS

Most organizations involved with alkaline battery research and development, manufacturing, and marketing expect continued growth in the consumer and special commercial use of nickel-cadmium batteries, and improvements in the performance stability of silver-zinc cells so they can be used to increasing advantage where their high energy content and power capability are critical. Also expected is the major market growth possibilities over the last quarter of this century associated with the introduction of new technologies for ground transportation electric vehicle propulsion power sources, most probably nickel-zinc and iron-nickel first, followed by air hybrid systems.

[*Editor's Note:* Figures 1 through 17 are reprinted with permission from *Alkaline Storage Batteries* by S. U. Falk and A. J. Salkind, Wiley, New York, 1969.]

9

Copyright © 1976 by the Institute of Electrical and Electronics Engineers, Inc.

Reprinted by permission from *IEEE Spectrum* **13**:32–36 (1976)

UNWANTED MEMORY SPOOKS
NICKEL-CADMIUM CELLS

S. F. Pensabene and J. W. Gould, II

[*Editor's Note:* The figure described below appears on the next page.]

A normal sintered-plate nickel cadmium cell discharge curve (blue) [upper] compared with similar data on a cell that has been "memorized" by repeated cycling to 25 percent depth-of-discharge shows the premature output voltage drop associated with cell memory (red) [lower]. Color-keyed to the curves are electron microscope photos of negative-plate active material from both normal and memory cells [red on left, blue on right]. Note the larger crystalline structure associated with the memory effect.

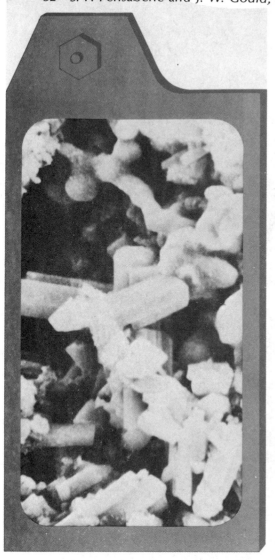
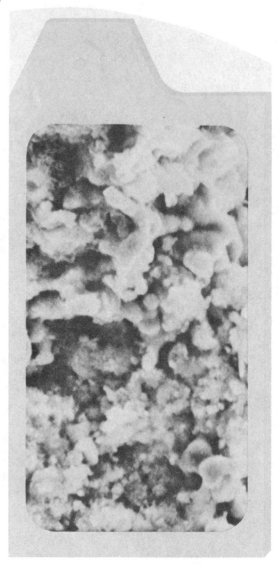

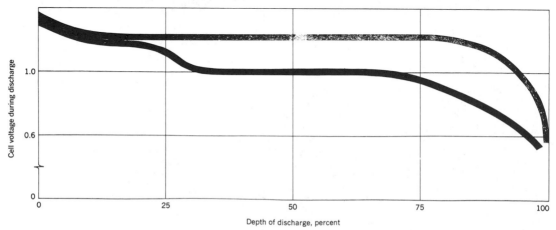

It's been almost 80 years since the first nickel-cadmium cell was developed by Waldemar Jungner in Sweden. Since then, technical refinements and improved manufacturing techniques have made batteries based on this original design welcome in millions of portable electronic products, emergency lighting systems, and standby power applications. What little adverse publicity these batteries have encountered centers on "memory," an effect where otherwise good cells temporarily lose rated capacity in a manner unrelated to normal aging/shelf-life considerations.

Not all varieties of the nickel-cadmium battery are even candidates for this discussion. Actually, two very distinct—and different—nickel-cadmium cell configurations are produced for today's battery market. They are the *sintered-plate* version, commonly available in sealed packages for power tools, calculators, and electric shavers; and the *pocket-plate* design using large, vented cells containing considerable liquid electrolyte. This latter type is best suited for emergency lighting or for providing standby power (see box, p. 35). It is the sintered-plate variety that has been known to exhibit the "memory" effect when subjected to a precise, highly repetitive depth of discharge.

This phenomenon was first noticed in satellite test programs when cells received a carefully controlled charge/discharge regime at constant temperature, over and over again. The effect was defined as "an apparent reduction in capacity to a predetermined discharge voltage cut-off point (usually one volt) resulting from repetitive use patterns." Fortunately, memory is not something most nickel-cadmium battery users must deal with. The suspected cause, and the reasons for its rarity, are explained after a brief discussion identifying the main components within a sintered-plate cell.

Sintered-plate construction

In order to keep the internal electrical resistance low and the capacity high, sintered-plate nickel-cadmium cells are constructed so that the principal three materials needed for the cell to function are kept in balance, both in terms of the relative amounts and how they are placed with respect to each other. These materials are the electron-conducting metal support/substrate (plates), the active materials impregnated into the pores of the plates

(nickel or cadmium hydroxide), and the ion-conducting electrolyte (potassium hydroxide solution). Continuous separator material keeps the closely spaced positive and negative plates physically apart.

These components must be in electrical contact with each other before current can be obtained from the chemical reactions, and the more intimate the contact, the lower the cell's internal resistance. Practical battery design requires the metal substrate to be a highly porous structure with as high a surface area as possible. The active materials must also be distributed relatively uniformly throughout the substrate. The electrolyte must thoroughly penetrate and wet all pores in the structure.

Plates are constructed by first sintering a finely divided nickel powder having a surface area of about one square meter per gram. This produces a "honeycomb" structure that is about 80 percent open pores. Pores are then impregnated by a precise process that brings the active material into the structure, while leaving enough space for electrolyte penetration.

But merely constructing a cell with these intimate component interrelationships is not sufficient to guarantee top performance; these structures must maintain the same (or nearly the same) physical relationship with one another while the chemical reactions involved in both charging and discharging are taking place. This is complicated by the fact that the active materials undergo changes in their own physical structure during charging and discharging.

An uncharged positive plate contains nickelous hydroxide; on charging, this becomes nickelic hydroxide. An uncharged negative plate contains cadmium hydroxide, which is converted to metallic cadmium by charging. When fully charged, continued attempts to charge the positive plate merely produce oxygen gas. At the negative plate, when all the cadmium hydroxide is converted to cadmium metal, overcharging produces hydrogen gas. But since chemical changes in each plate must occur at exactly the same rate, the plate loaded with the least amount of active material will begin to produce gas first.

Designers of sealed, sintered-plate nickel-cadmium batteries have purposely kept the positive plate about 30 percent smaller than the negative plate, favoring the formation of oxygen gas on overcharge—for a very good reason. Oxygen has the ability to diffuse through the insulating separator material (nylon or polypropylene) and react chemically with the cadmium metal to convert it back to its uncharged cadmium hydroxide form. Pressure doesn't build up in the sealed cell to dangerous levels because the oxygen gas is recombined as fast as it forms—providing overcharge rates are kept within reasonable limits. (Hydrogen gas can also diffuse from the negative plate to the positive and recombine, but the rate is unacceptably slow.)

Making a memory

A plot of voltage vs. percent depth of discharge for a cell exhibiting memory is shown on page 32. In this example, the cell was repeatedly discharged to 25 percent, and each time recharged to the initial voltage (but *no* overcharge was allowed) for a great many cycles. The effect became progressively pronounced as the number of cycles increased. But, significantly, the entire characteristic could be wiped out at any time by just *one* complete deep discharge followed by a full charge.

A large number of investigations in various laboratories have attempted to establish the cause of memory. Typical cycling regimes used to test prismatic, hermetically sealed, aerospace cells generally produce the memory effect. This does not always happen. In one particular test program—especially designed to induce memory—no effect was found after more than 700 precisely controlled charge/discharge cycles. In this program, spirally wound one-ampere-hour cells were used. In the follow-up program, 20-ampere-hour aerospace-type cells were used on a similar test regime. Memory effects showed up after only a few hundred cycles.

Analysis of cells exhibiting the memory effect showed that chemically they were no different than cells that did not exhibit memory.

The next step was to study the cells using reference electrodes. When the voltages of the positive and negative plates were plotted against the constant potential of the reference electrode, the voltage profile of the negative

[1] **Certain overcharge conditions can also produce a memory effect in sintered-plate nickel-cadmium cells. Here, the discharge characteristic for a normal cell (A) is compared with similar data for cells with artificially induced memory at various percent discharges (B, C, D, E). This type of memory is induced by long-term continuous overcharge.**

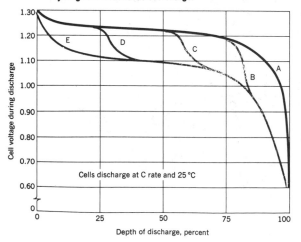

Cells discharge at C rate and 25 °C

Depth of discharge, percent

plate mirrored the voltage profile of the cell, as a whole, much more faithfully than the voltage profile of the positive plate.

At this point, evidence indicated that physical changes at the negative plate produced memory. To confirm this, the investigators subjected both normal and memory cells to scanning electron microscope analysis. Cadmium hydroxide particles on the negative plate from the normal cell were small and needle-like. However, cadmium hydroxide particles from the memory cell were larger and had a fine layered structure.

None of these experiments has proved an exact relationship between the physical condition of the negative plate and the memory voltage profile. But they do tend to support the hypothesis that discharge proceeds uniformly when the active material consists of many small particles. However, when the active material is a mixture of large and small particles, the small particles discharge first. The larger particles then discharge at a greater current density. Associated with this is an increased polarization that appears as a voltage drop in the cell.

Thus, for both the normal and the memory cells the initial discharge curves are similar. But as the memory cell uses up the charge stored in small particles, it must begin to discharge larger particles, leading to a sudden, premature drop in output voltage. Meanwhile, the normal cell is still discharging relatively small particles, and thus continues providing rated terminal voltage until all the active material has truly been depleted.

The "lost contact" theory

As previously mentioned, the chemical reactions of charging or discharging depend on simultaneous, intimate contact between the electron conductor (plates), the impregnated active chemicals, and the electrolyte. For a cell to be completely reversible (all the stored chemical energy produced during charge becomes electric energy on discharge), all the intimate points of contact must remain that way. This must hold true not only while the chemical changes are occurring, but after they have occurred as well.

If, for any reason, some contact points are lost during or after charging, a corresponding loss in recoverable chemical energy will occur. It now appears that memory is associated with such changes; the greater the number of "contact" losses, the larger the "memory" effect. Fortunately, in a well-constructed sintered-plate nickel-cadmium cell, these contacts are not easily lost— particularly when the cells are used in portable power tools and similar applications. Each time the cell is charged from full discharge, a whole new set of contacts are generated that are stable.

Studies made on cells using a reference electrode have shown that there are really two sources of memory effects: one associated with the positive plate and one with the negative. They are both apparently caused by contact losses, but differ in characteristics and magnitude because the two plates differ in the kinds of active material they contain and the fact that the negative plate has a surplus of active material built in. Electron microscope pictures of a negative plate with and without memory are included in the opening illustration (p. 32). The cadmium hydroxide particles in the memory cell are larger and have a fine layered structure.

Exactly how and why contacts are lost is yet to be determined, but it's known that higher temperatures con

The pocket-plate story

Pocket-plate nickel-cadmium batteries are primarily used for standby services such as emergency lighting, switchgear operation and control, diesel and turbine starting, telemetering, microwave communications, uninterruptible power systems, and railway signaling. The battery has wide acceptance in transit cars for emergency lighting and control, controls for electric locomotives, and, to some extent, diesel locomotive starting.

Essentially Waldemar Jungner's original design, the modern pocket-plate nickel-cadmium cell consists of positive active material (nickel hydroxide); negative active material (cadmium hydroxide); a current collector (nickel-plated ribbon steel); and an ion-conducting electrolyte (potassium hydroxide solution). The active materials are automatically loaded into flat pockets of perforated nickel-plated steel. These pockets are pressed, formed into plates, and then assembled into cells. Venting is employed to provide an escape path for gasses generated during charge, since these cells have an ampere-hour efficiency of less than 100 percent. Free-flowing liquid électrolyte helps dissipate heat, making the pocket-plate cell an unlikely victim of thermal runaway during overcharge or cell reversal.

Rugged mechanical design makes pocket-plate cells resistant to vibration, shock, and generally rough handling. They are reasonably forgiving of electrical abuse such as overcharging, short-circuiting, and reversal. Though the stored energy per unit weight and volume is less than found in sintered-plate cells, it is generally greater than for lead-acid cells.

An outstanding pocket-plate performance characteristic is the cell's ability to deliver power at low temperatures. Cells can normally be operated at $-25^{\circ}C$, and with more concentrated electrolyte, will perform down to $-50^{\circ}C$.

In standby service where batteries built from pocket-plate cells are most commonly used, such batteries are continuously connected to a charging source having a voltage high enough to ensure a charged state, but still below the gassing voltage. Usually these chargers have a two-rate feature that allows rapid and complete recharging at a higher voltage level immediately after the battery has been used. Cells subjected to this type of service retain capacity indefinitely, and show none of the "memory" characteristics observed in sintered nickel-cadmium cells.

Typically, failure of sintered-plate cells results from separator breakdown caused by high operating temperatures. Since pocket plates are constructed more rigidly and are not as closely spaced, continuous separator material is not required. Thus, a significant failure mode is eliminated.

The most common failure mode in pocket-plate cells is shorting between adjacent plates (*not* related to separator breakdown) because the proper electrolyte level was not maintained. As the electrolyte volume is reduced by gassing, the potassium hydroxide concentration increases. This results in excessive plate swelling and eventual shorting. However, with proper maintenance, pocket-plate nickel-

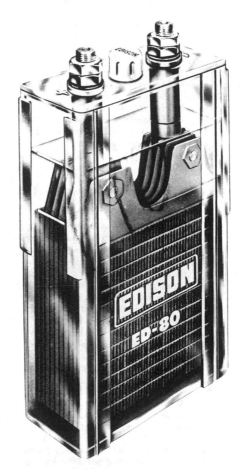

cadmium cells last for many years, even when operated over a wide temperature range.

Though pocket-plate cells are primarily used in heavy commercial and industrial applications, they represent a significant portion of today's total nickel-cadmium battery market. The current U.S. annual production of nickel-cadmium batteries (measured in ampere-hours) is approximately 28 000 000 for pocket-plate, 20 000 000 for vented sintered-plate, and 75 000 000 for sealed sintered-plate cells.—*B. D. Brummet, T. Ulrich, McGraw-Edison, Bloomfield, N.J.*

tribute to negative plate memory, particularly when the cell is "idling" in overcharge. There's also strong evidence that positive plate memory is actually reduced when the cell is overcharged. The latter effect seems to be related to the efficiency of the positive plate, which tends to become reduced when precise cycling (e.g., the satellite test program) is employed. Here "efficiency" means the extent to which charge is accepted by the cell's active chemicals; charge not accepted merely produces gas. During imposed cycling, it is possible that uncharged chemicals and gas can exist simultaneously in a sintered

plate simply because some contacts are lost. However, if the cell is carried on into overcharge, these contacts seem to be largely restored (Fig. 1).

Hard to remember, easy to forget

Though sintered-plate nickel-cadmium batteries can remember, the conditions necessary are almost never encountered in practice. In the vast majority of consumer applications, the variable-use environment created by the customer (a random pattern of partial and complete charge/discharge cycles) insures that memory will not be observed. However, if the situation calls for strictly repetitive cycles or long-term overcharge (greater than four months), allowances should be made for the memory effect.

When the load is sensitive to low voltage, extra cells should be added to the battery to assure the retention of a higher voltage profile. The calculations for memory-induced voltage depression should allow 150 mV per cell (worst case). In other applications that do not include long-term overcharge and are not voltage-sensitive, a periodic deep discharge and subsequent full recharge will return the batteries' performance.

Vented- (flooded-electrolyte) type sintered-plate nickel-cadmium cells can also exhibit the memory effect. But again, the strict charge/discharge conditions necessary to develop memory almost never occur in practical systems. Like their sealed counterparts, vented cells can be restored to normal performance (if memory should appear) by one full discharge/charge cycle.

Cell reversal and extended cycle life

Another subject that often arises when nickel-cadmium memory is under discussion is cell reversal. When several such cells are operated in series (a battery), it is desirable to have cells of matched capacity. If one cell in the string arrives at full discharge before the others, current is forced through that cell in a reverse direction. Although the discharged cell is not contributing to battery output, it is still in the circuit and will pass current proportional to the total voltage supplied by the healthy cells and its own internal resistance added to the total circuit impedance. This is cell reversal. As far as the reversed cell is concerned, it's the same as being hooked up to a charger backwards, and has the potential of damaging or actually bursting the affected cell.

One good way of minimizing cell reversal is to carefully select cells so that all have capacities of the same value. General Electric uses this procedure routinely; each cell is tested individually and only those with closely matched capacities are teamed in the same battery pack.

Cell matching goes a long way toward eliminating the cell reversal problem, but the small differences in cell capacities that remain must be dealt with. This is accomplished by reverse charge "buffering." The idea is to provide some material in both the positive and negative plates that will accept electrons more easily than hydroxyl or hydrogen ions so no gases are formed during charge reversal.

Implementing this idea proves to be disarmingly simple—just add a small amount of the negative plate active material (cadmium hydroxide) to the positive plate and a little of the positive plate active material (nickel hydroxide) to the negative plate. These trace quantities do not affect overall battery performance significantly, but they're there when needed to prevent gas formation while all series cells are reaching full discharge. The amount of buffer material required depends, of course, on how closely the cells are matched in capacity.

With memory and cell reversal eliminated as serious limits to sintered-plate nickel-cadmium battery life, the remaining—and still very important—consideration is the rate of overcharge or discharge during use. This is true because the temperature generated in the battery is the result of the IR drop through the cells. And the higher the temperature, the faster the breakdown of materials used in cell construction, particularly the thin electrode sep-

arators used in sealed cell design.

The separators used in sintered-plate rechargeables are the result of years of research and some performance compromises. They must be wet by the electrolyte, conduct oxygen molecules readily, allow easy passage of electrolyte ions, be true insulators, and be nonbulky in order to keep overall cell dimensions small and the ion-transfer resistance low. The higher the temperature the separator material can withstand, the greater the rate of cell charge/discharge that can occur without cell damage.

As an example of the kind of life available from a nickel-cadmium cell that is charged and discharged at modest rates, GE has some cells that are still in excellent condition after 14 years of use in a "one cycle per day" regime. This represents over 5000 charge/discharge cycles with no significant deterioration in performance. In general, the lower the overcharge and discharge rate, the longer the cell life. Claims by reputable nickel-cadmium-cell manufacturers for cycle life are conservatively made at about 1000 cycles. This represents a fair compromise between useful discharge rates in average applications and cycle life. To maintain cycle life while allowing higher overcharge/discharge rates, GE offers the Gold Top® cell featuring separator construction more tolerant of higher temperatures.

Because temperature of cell operation and cycle life are so intimately related, some care should be exercised in how batteries or cells are positioned with respect to heat-generating components in specific applications. Avoid designs that tightly surround batteries or cells with heat-insulating materials.

Competitive battery systems are frequently evaluated in cost per watthour (simple replacement or "first cost"). When expense is determined in these terms only, nickel-cadmium may be the more costly battery system. However, evaluated over the operating life of the product to be powered, it is often the most economic choice, particularly when the special care and maintenance required by other battery systems are considered. Extensive application information on sintered-plate nickel-cadmium batteries can be found in the General Electric Nickel-Cadmium Battery Application Engineering Handook, available from GE at P.O. Box 861, Gainesville, Fla. 32602.   ◆

Saverio F. Pensabene is an electrochemist at the General Electric Co., Battery Business Department, in Gainesville, Fla., where currently he is conducting morphological studies of nickel-cadmium crystal structures. Since 1962, he has been active in nickel-cadmium batteries' design engineering phases and has served specifically as a scientific and technical contributor in battery development projects. Dr. Pensabene received the B.S. degree in chemistry from LeMoyne College in 1955, and the Ph.D. degree in physical chemistry from the University of Notre Dame in 1960.

James W. Gould, II, is the manager of technical marketing at the General Electric Co., Battery Business Department, in Gainesville, Fla., where presently he is responsible for the application of nickel-cadmium batteries and other battery couples. Mr. Gould, who received the B.E.E. degree from the Georgia Institute of Technology, has been with GE's Battery Business Department since 1975. Prior to that, he spent five years with GE in Huntsville, Ala., working on the Apollo space program. This effort primarily involved developing standby power system ground equipment used with the Saturn 1B and later Saturn 5 launch vehicles.

10

Copyright © 1976 by the Union Carbide Corporation

Reprinted from pages 20–21, 34–36, 287–290, 359, 378–379, 382, 601–603, and 757–758 of *Eveready® Battery Engineering Data*, Union Carbide Corporation, 1976, 830pp.

COMPILATION OF ENGINEERING DATA AND ILLUSTRATIONS OF A VARIETY OF COMMERCIALLY AVAILABLE, SMALL PRIMARY BATTERIES

W. V. Hassenzahl

[*Editor's Note:* In the original, material precedes and follows these excerpts.]

Table A Six principal types of dry batteries and average characteristics.

Usual Name	Carbon-Zinc	Carbon-Zinc (Zinc Chloride)	Alkaline-Manganese Dioxide	Mercuric Oxide	Silver Oxide	Nickel-Cadmium
Electrochemical System	Zinc-Manganese Dioxide (usually called Leclanche or Carbon-Zinc)	Zinc-Manganese Dioxide	Zinc-Alkaline Manganese Dioxide	Zinc-Mercuric Oxide	Zinc-Silver Oxide	Nickel-Cadmium
Voltage Per Cell	1.5	1.5	1.5	1.35	1.5 (monovalent)	1.2
Negative Electrode	Zinc	Zinc	Zinc	Zinc	Zinc	Cadmium
Positive Electrode	Manganese Dioxide	Manganese Dioxide	Manganese Dioxide	Mercuric Oxide	Monovalent Silver Oxide	Nickelic Hydroxide
Electrolyte	Aqueous solution of ammonium chloride and zinc chloride	Aqueous solution of zinc chlorid	Aqueous solution of potassium hydroxide	Aqueous solution of potassium hydroxide or sodium hydroxide	Aqueous solution of potassium hydroxide or sodium hydroxide	Aqueous solution of potassium hydroxide
Type	Primary	Primary	Primary and Rechargeable	Primary	Primary	Rechargeable
Rechargeability (*See pages 4, 27 and 28)	Poor	Poor	Good	No	No	Yes
Number of Cycles	10-20	10-20	50-60 Rechargeable Only			300-2000
Input if Rechargeable			Approximately 100% of energy withdrawn Rechargeable Only			Sealed, minimum of 140% of energy withdrawn
Overall Equations of Reaction	$2MnO_2 + 2NH_4Cl + Zn \rightarrow ZnCl_2 \cdot 2NH_3 + H_2O + Mn_2O_3$	$8MnO_2 + 4Zn + ZnCl_2 + 9H_2O \rightarrow 8MnOOH + ZnCl_2 \cdot 4ZnO \cdot 5H_2O$	$Zn + 2KOH + 3MnO_2 \rightleftharpoons ZnO + Mn_2O_3 + 2KOH$	$Zn + KOH + 2MnO_2 \rightleftharpoons ZnO + Mn_2O_3 + KOH$	$Zn + Ag_2O + KOH \rightarrow ZnO + 2Ag + KOH$	$Cd + 2NiOOH + KOH + 2H_2O \rightleftharpoons Cd(OH)_2 + 2Ni(OH)_2 + KOH$
Typical Commercial Service Capacities	60mAh to 30 Ah	Several hundred mAh to 9Ah	Several hundred mAh to 23Ah	16mAh to 28Ah	35mAh to 210mAh	20mAh to 4 Ah
Energy Density (Commercial): Watt-Hour/Lb.	20	40	Primary: 30-45 Rechargeable: 10	50	50	Sealed: 12-16
Energy Density (Commercial): Watt-Hour/Cubic Inch	2	3	Primary: 2-3 Rechargeable: 1-1.2	8	8	Sealed: 1.2-1.5
Practical Current Drain Rates: Pulse	Yes	Yes	Yes	Yes	Yes	Yes
Practical Current Drain Rates: High (More Than 50mA)	100mA/square inch of zinc area ("D" cell)	150mA/square inch of zinc area ("D" cell)	200mA/square inch of zinc area ("D" cell)	No	No	8-10A
Practical Current Drain Rates: Low (Less Than 50mA)	Yes	Yes	Yes	Yes	Yes	Yes
Discharge Curve (Shape)	Sloping	Sloping	Sloping	Flat	Flat	Flat
Temperature Range: Storage	-40°F to 120°F (-40°C to 48.9°C)	-40°F to 160°F (-40°C to 71.1°C)	-40°F to 120°F (-40°C to 48.9°C)	-40°F to 140°F (-40°C to 60°C)	-40°F to 140°F (-40°C to 60°C)	-40°F to 140°F (-40°C to 60°C)
Temperature Range: Operating	20°F to 130°F (-6.7°C to 54.4°C)	0°F to 160°F (-17.8°C to 71.1°C)	-20°F to 130°F (-28.9°C to 54.4°C)	32°F to 130°F (0°C to 54.4°C)	32°F to 130°F (0°C to 54.4°C)	Discharge; -4°F to 113°F (-20°C to 45°C) Charge: B, BH & CH Types; -20°C to 113°F (0°C to 45°C) 32°F to 113°F (0°C to 45°C) Charge: CF Type; 60°F to 113°F (15.6°C to 45°C)
Effect of Temperature on Service Capacity	Poor low temperature	Good low temperature relative to carbon-zinc	Good low temperature	Good high temperature, poor low temperature—depends upon construction	Poor low temperature—depends upon construction	Very good at low temperature—poor at high temperature
Impedance	Low	Low	Very low	Low	Low	Very low
Leakage	Medium Under Abusive Conditions	Low	Rare	Some salting	Some salting	Very low
Gassing	Medium	Higher than carbon-zinc	Low	Very low	Very low	Low
Reliability (Lack of Duds; 95% Confidence Level)	99% at 2 years	99% at 2 years	99% at 2 years	99% at 2 years	99% at 2 years	99% at 2 years
Shock resistance	Fair to good	Good	Fair to good	Good	Good	Good
Cost: Initial	Low	Low to Medium	Medium Plus	High	High	High
Cost: Operating	Low	Low to Medium	Medium to high at high power requirements	High	High	Low
Features	Low cost; variety of shapes and sizes	Service capacity at moderate to high current drains greater than carbon-zinc; good leakage resistance; low temperature performance better than carbon-zinc	High efficiency under moderate and high continuous conditions, good low temperature performance; low impedance	High service capacity/volume ratio; flat voltage discharge characteristic; good high temperature performance	Moderately flat voltage discharge characteristics	Excellent cycle life; flat voltage discharge characteristic; good high- and low-temperature performance; high resistance to shock and vibration; can be stored indefinitely in any charge state.
Limitations	Efficiency decreases at high current drains; poor low-temperature performance		Primary type expensive for low drains. Rechargeable-limited cycle life, voltage-limited taper current charging	Poor low-temperature performance on some types		High initial cost; only fair charge retention

144

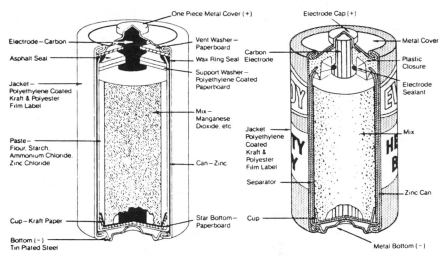

**FIGURE 1 – CROSS SECTION
OF STANDARD ROUND CELL (SIZE "D")**

**FIGURE 2 – CROSS SECTION
OF ZINC CHLORIDE CELL (SIZE "D")**

The rod in the center of the round cell in Figure 1 is carbon and functions
as a current collector.

ZINC CHLORIDE CELLS

The zinc chloride cell is a special modification of the familiar carbon-zinc cell. Zinc chloride cells are very similar to the traditional carbon-zinc Leclanché cells but differ principally in the electrolyte system. The electrolyte in a zinc chloride cell contains only zinc chloride while in a Leclanché cell the electrolyte contains a saturated solution of ammonium chloride in addition to zinc chloride. The ommission of ammonium chloride improves the electrochemistry of the cells but increases the importance of the cell seal. Zinc chloride cells, therefore, have either a new type of seal not previously used in carbon-zinc cells or an improved conventional seal so that their shelf life is equivalent to that of Leclanché cells.

Electrode blocking by reaction products and electrode polarization at high current densities are minimized by the more uniform and higher diffusion rates that exist in an electrolyte that contains only zinc chloride. Because of their ability to operate at high electrode efficiencies, the useful current output of zinc chloride cells is usually higher than that of Leclanché cells and zinc chloride cells will operate at high current drains for a considerably longer time than Leclanché cells of the same size. In addition, the voltage level under load holds up longer.

Because of the electrochemical reactions that occur in the cell during use, water in the cell is consumed (by a reaction product which is an oxide compound) along with the electrochemically active materials, so that the cell is almost dry at the end of its useful life.

A cross section view of a "D" size zinc chloride cell is shown in figure 2.

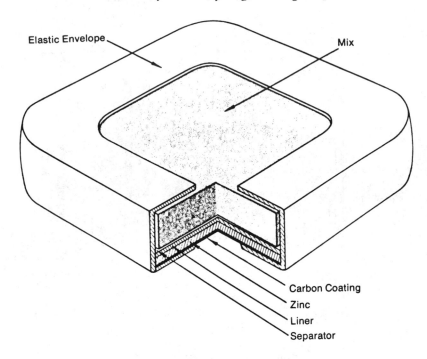

Elastic Envelope

Mix

Carbon Coating
Zinc
Liner
Separator

FIGURE 3 – CROSS SECTION OF "MINI-MAX" FLAT CELL

In flat cells, carbon is coated on a zinc plate to form a duplex electrode — a combination of the zinc of one cell and the carbon of the adjacent one. The "Mini-Max" cell (Figure3), contains no expansion chambers or carbon rod as does the round cell. This increases the amount of depolarizing mix available per unit cell volume and therefore the energy content. In addition the flat cell, because of its rectangular form, reduces waste space in assembled batteries. The energy to volume ratio of a battery utilizing round cells is inherently poor because of the voids occurring between cells. These two factors account for an energy to volume improvement of nearly 100% for "Mini-Max" cells compared to round cell assemblies

PERFORMANCE

The closed-circuit or working voltage of a carbon-zinc cell falls gradually as it is discharged, Fig. 4. The service hours delivered are greater as the cutoff or endpoint voltage is lower.

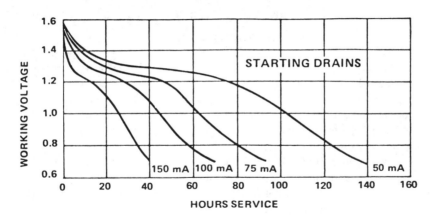

FIGURE 4 — VOLTAGE DISCHARGE CHARACTERISTICS OF CARBON-ZINC SIZE
"D" CELL DISCHARGED 2 HOURS PER DAY AT 70°F (21.1°C).
(FIXED RESISTANCE LOAD)

[*Editor's Note:*Material has been omitted at this point.]

ALKALINE-MANGANESE-ZINC
DRY BATTERIES

General Features: Both primary and rechargeable alkaline cells are comprised of a zinc anode of large surface area, a manganese-dioxide cathode of high density, and a potassium-hydroxide electrolyte.

In addition to the primary type, a rechargeable alkaline battery is also available as a dry-type battery. Discussion of the rechargeable battery follows this section on primary types.

The alkaline-manganese dioxide-zinc battery described on the following pages represents a major advance in portable power sources over the standard carbon-zinc battery. This battery is the result of many years of combined research and development effort.

In answer to a growing need for a high rate source of electrical energy, the Union Carbide Research Laboratories began early in 1955 an intensive study of the caustic-manganese system as a high rate source of electrical energy. Alkaline dry cells, differing in details of construction from the more familiar carbon-zinc dry cells, had not proven commercially practicable up to that time. Alkaline-manganese cells, however, with a zinc anode of high surface area, show amazing depolarizing efficiency. On heavy or continuous drains the alkaline cell is most spectacular and shows an excellent advantage over standard carbon-zinc cells on a performance-per-unit-of-cost basis.

This cell differs from the Leclanché carbon-zinc cell primarily in the highly alkaline electrolyte that is used. The cell is a high rate source of electrical energy. Its outstanding advantages derive from a combination of unique components and construction methods. (See Figure 7).

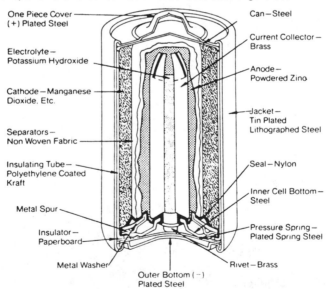

One Piece Cover — (+) Plated Steel
Electrolyte — Potassium Hydroxide
Cathode — Manganese Dioxide, Etc.
Separators — Non Woven Fabric
Insulating Tube — Polyethylene Coated Kraft
Metal Spur
Insulator — Paperboard
Metal Washer
Can — Steel
Current Collector — Brass
Anode — Powdered Zinc
Jacket — Tin Plated Lithographed Steel
Seal — Nylon
Inner Cell Bottom — Steel
Pressure Spring — Plated Spring Steel
Rivet — Brass
Outer Bottom (−) Plated Steel

CUTAWAY OF ALKALINE CELL (PRIMARY TYPE)
FIGURE 7

Two principal features are a manganese dioxide cathode of high density in conjunction with a steel can which serves as a cathode current collector, and a zinc anode of extra high surface area in contact with the electrolyte. These features, coupled with the use of a potassium hydroxide electrolyte of high conductivity give these cells their very low internal resistance and impedance and high service capacity.

The cells are hermetically sealed and encased in steel. The ampere-hour capacity is relatively constant over a range of current drains and discharge schedules.

The voltage of an alkaline-manganese dioxide primary cell is 1.5 volts in standard N, AAA, AA, C and D cell sizes. Batteries are available with voltage up to 9 volts and in a number of different service capacities.
Performance: The closed-circuit voltage of an alkaline primary battery falls gradually as the battery is discharged, Fig. 8.

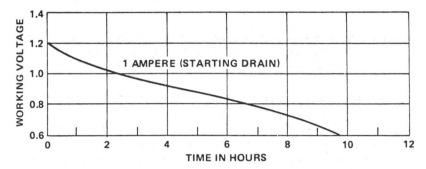

FIG. 8—VOLTAGE DISCHARGE CHARACTERISTIC OF ALKALINE-MANGANESE PRIMARY BATTERY (D CELL) DISCHARGED CONTINUOUSLY.

Similar to carbon-zinc batteries, the service hours delivered by alkaline-manganese primary batteries are greater as the cutoff voltage is lower. The cutoff voltage should be made as low as possible so that the high total energy of the cell can be used.

Service capacity remains relatively constant as the discharge schedule is varied. Capacity does not vary as much with current drain as for the carbon-zinc battery. Service capacity of the entire line ranges from several hundred milliampere-hours to almost 23 ampere-hours depending upon the current drain and cutoff voltage.

The alkaline-manganese primary cell is for applications requiring more power or longer life than can be obtained from Leclanché carbon-zinc batteries. Alkaline cells contain 50 to 100 per cent more total energy than a conventional Leclanché carbon-zinc cell of the same size.

In a conventional carbon-zinc cell, heavy current drains and continuous or heavy-duty usage impair its efficiency to the extent that only a small fraction of the built-in energy can be removed. The chief advantage of the alkaline-manganese battery lies in its ability to work with high efficiency under continuous or heavy duty, high-drain conditions where the standard Leclanché carbon-zinc cell is unsatisfactory. Under some conditions, alkaline cells will provide as much as seven times the service of standard Leclanché carbon-zinc cells. Discharge characteristics of the two battery types are compared in Fig. 9.

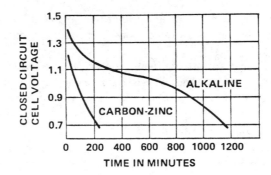

FIG. 9 — COMPARISON OF DISCHARGE CHARACTERISTICS OF ALKALINE— MANGANESE AND CARBON—ZINC "D" SIZE CELLS FOR 500 mA STARTING, DISCHARGED CONTINUOUSLY AT 70° F (21.1° C)

Although alkaline-manganese cells will outperform standard Leclanché batteries in any type of service, they may not show economic advantage over standard cells at light drains, or under intermittent-duty conditions, or both. For example, with intermittent use at current drains below about 300 milli-amperes for the "D" size cell, alkaline cells, while performing very well, will begin to lose their economic advantage over the standard carbon-zinc batteries. "Eveready" alkaline primary batteries are ideally suited for use in many types of battery-operated equipment. Some of these applications are listed on page 290. In addition, the characteristics of these batteries make possible the development of equipment which up to this time has been thought impractical because of the lack of a suitable power source.

Alkaline cells are ideal for motion picture camera cranking and cassettes. They have proved their worth for electronic flash, toys, model boats and automobiles and innumerable heavy drain applications.

In radios, alkaline cells usually last twice as long as standard carbon-zinc cells. In cassettes, in some equipment they last five times as long. In battery-powered toys, alkaline batteries last up to seven times as long as standard carbon-zinc cells.

Alkaline-manganese cells are excellent for photoflash applications, In addition to high amperage, they have more energy than standard carbon-zinc photoflash batteries.

Some electronic flash units use transistor or vibrator circuits in a converter to change low voltage d-c into the high voltage necessary to charge the flash capacitor. The current drains involved strain the capabilities even of high amperage photoflash cells. The alkaline cell provides both a sustained short recycling time and 2-3 times as many flashes as standard Leclanché carbon-zinc photoflash or general purpose cells. This is due to the unusual cell construction which provides a very low internal resistance such that the cell delivers its energy faster than standard carbon-zinc types.

The service maintenance of alkaline primary cells is excellent at both normal and elevated temperatures.

The following information applies to "Eveready" cells stored at 70° F (21.1° C).

Time of Storage at 70°F	Percent Retained of Ah Capacity of Fresh Battery
1 year	95%
2 years	90%
3 years	85%
4 years	80%

"Eveready" alkaline primary batteries are presently available in seventeen types. Physical characteristics for these are listed in Table H.

[*Editor's Note:*Material has been omitted at this point.]

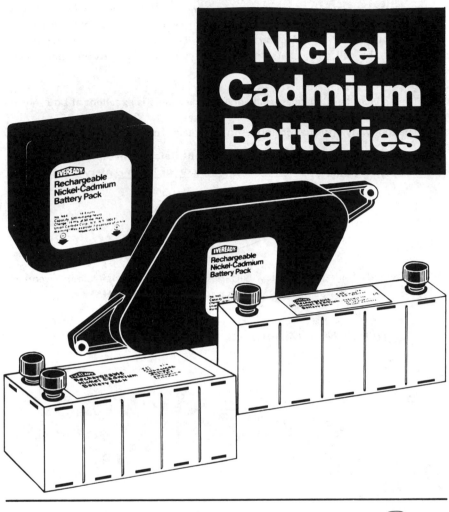

Nickel Cadmium Batteries

Button Cells

Two cutaway views of a standard rate button cell are shown in Figure 15. A cross section of a double plate molded electrode high rate button cell is illustrated in Figure 16. Specifications for all button cells are listed in Tables L and M.

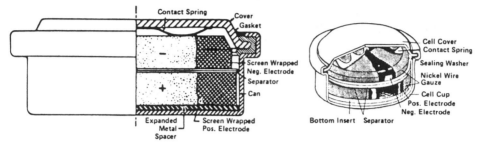

FIGURE 15 — CUTAWAY VIEWS OF STANDARD RATE BUTTON CELL

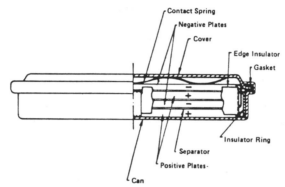

FIGURE 16 — CROSS SECTION VIEW–DOUBLE PLATE MOLDED ELECTRODE HIGH RATE BUTTON CELL

Button type cells utilize a cell cup (postitive pole) and a cell cover (negative pole). The electrodes consist of pressed powder tablets wrapped in nickel wire gauze, separated by a fine pored separator. Sealing is accomplished by crimping the rim of the cell cup over the rim of the cell cover with a plastic washer placed between them. This washer at the same time serves to insulate the cell cup from the cell cover.

Button cells are supplied with and without solder tabs. They may be used in a clip or holder or may be soldered into the circuit by means of the tabs welded on the top and bottom of the cell. *One should never attempt to solder directly to the cell case as the seal may be damaged by heat.*

Since the nickel-cadmium cell is a long-life device, it can be considered an integral electronic component, as any other installed part, and wired directly into the circuit.

Button Cell Stack Assemblies

When required, two to ten button cells may be assembled into a higher voltage series stack by special factory welding techniques. The assembly is jacketed with plastic tubing to provide stack insulation and to improve rigidity. The stacks are usually furnished with solder tabs at both ends. A typical assembly is shown in Figure 17.

It is not recommended that unit button cells be stacked in a pressure held assembly because of possible contact resistance changes in long term use. Cells should be assembled in the factory on a welded cell-to-cell basis.

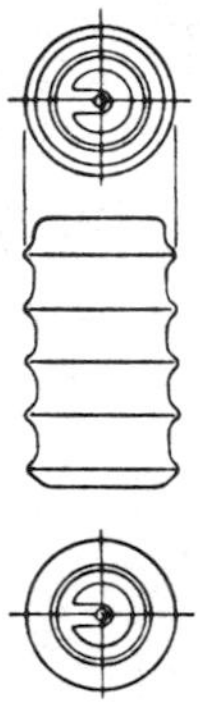

FIGURE 17— BUTTON CELL STACK ASSEMBLY

The table on page 380 lists the part numbers and specifications of button cell stack assemblies made up of 2 to 10 cells of 5 different types. Under special conditions more than 10 cells may be stacked.

NOTE: The stacks of button cells listed in the tables have flat contacts for terminals. The height dimensions shown apply to this type of terminal. Stacks are usually furnished with a solder tab at each end. The solder tab terminals are designated by the suffix letter T. For example:

A stack assembly of 6 B225 cells, with solder tab terminals, would have the following part number:

6/B225T: One solder tab at each end.

The slant line indicates jacketing with plastic tubing.

The height indicated in the table is the maximum height of the assembled stack. The height of the stack is subject, due to internal pressure, to a +0.020" (.51 millimeter) maximum temporary increase in dimension for each cell in the assembly.

Solder Tabs

One at each end.

For assemblies of B150, B225, BH225, BH500, and BH1 cells: No increase in stack height.

"Eveready" nickel-cadmium cells are assembled in a wide variety of battery types. If the standard types listed in the tables do not meet the necessary requirements of an application, contact the nearest Union Carbide Battery Products Division Sales Office for details on special configurations and voltage desired.

[Editor's Note: Material has been omitted at this point.]

A number of "Eveready" cylindrical high rate cells use the "Jelly Roll" construction illustrated below, Figure 18.

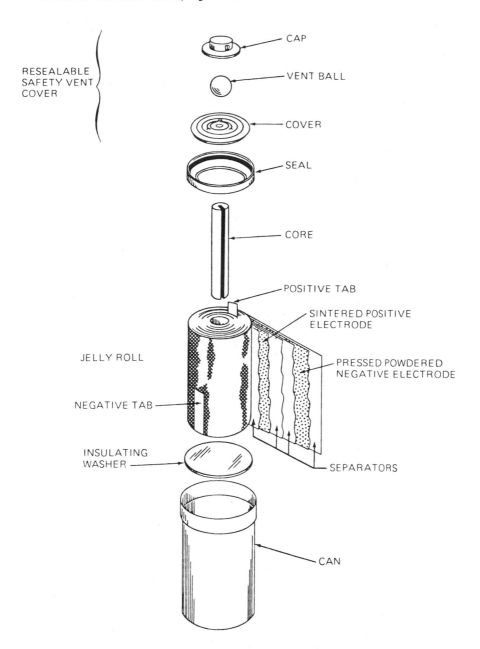

RESEALABLE
SAFETY VENT
COVER

CAP

VENT BALL

COVER

SEAL

CORE

POSITIVE TAB

SINTERED POSITIVE
ELECTRODE

JELLY ROLL

PRESSED POWDERED
NEGATIVE ELECTRODE

NEGATIVE TAB

INSULATING
WASHER

SEPARATORS

CAN

FIGURE 18
HIGH RATE CYLINDRICAL TYPE CELL—EXPANDED VIEW

MERCURIC OXIDE BATTERIES

General Features: The mercuric oxide battery consists essentially of a depolarizing mercuric oxide cathode, an anode of pure amalgamated zinc, and a concentrated aqueous electrolyte of potassium hydroxide or sodium hydroxide. The voltage is 1.35 volts for a mercury cell with a depolarizer of 100 percent mercuric oxide and 1.4 volts for a cell with a mixture of mercuric oxide and manganese dioxide.

Utilization of active materials is 80% to well over 90%. Recommended drains are given for each cell type.

The fundamental components of the mercuric oxide cell are a pressed mercuric oxide cathode (in sleeve or pellet form) and pressed cylinders, or pellets, of powdered zinc or a gelled mixture of electrolyte and zinc. In steel enclosures these provide precise mechanical assemblies having maximum dimensional stability and marked improvements in performance over dry batteries of the Leclanché (carbon-zinc) type.

MECHANICAL CONSTRUCTION

Cells are currently produced in two different designs using either flat or cylindrical electrodes. Electrochemically both cells are the same, differing only in case design and internal electrode arrangements.

1. Depolarizing cathodes of mercuric oxide, to which a small percentage of graphite is added, are shaped as illustrated in Figure 40, and either consolidated to the cell case (for flat electrode types), or pressed into the cases of the cylindrical types.

2. Anodes are formed of amalgamated zinc powder of high purity, in either flat or cylindrical shapes.

3. A permeable barrier of specially selected material prevents migration of any solid particles in the cell, thereby contributing effectively to long shelf and service life.

4. Insulating and sealing gaskets are molded of nylon, polyethylene or neoprene, depending on the application for which the cell or battery will be used.

5. Inner cell tops are plated with materials which provide an internal surface to which zinc will form a zinc amalgam bond.

6. Cell cases and outer tops of nickel or gold plated steel are used to resist corrosion, to provide greatest passivity to internal cell materials and to insure good electrical contact.

7. An outer, nickel-plated, steel jacket is generally used for single cells. This outer jacket is a necessary component for the "self-venting construction" used

on some cells which provides a means of releasing excessive gas in the cell. Venting occurs if operating abnormalities such as reverse currents or short circuits, produce excessive gas in the cell.

At moderately high pressures, the cell top is displaced upwards against the external crimped edge of the outer jacket, tightening that portion of the seal and relieving the portion between the top and the inner steel cell case. Venting will then occur in the space between the internal cell container and the outer steel jacket. Should any cell electrolyte be carried into this space, it will be retained by the safety absorbent ring. Corrosive materials, therefore, are not carried with the escaping gas through the vent hole at the bottom of the outer steel jacket. After venting excessive gas and reducing the internal pressure, the cell stabilizes and reseats the top seal, continuing normal operation in the circuit.

It should be noted that the polarity of some individual mercuric oxide cells is reversed from that of Leclanché types.

Characteristics and features of the mercuric oxide electrochemical mechanical system in the present state of the art include:

- High capacity-to-volume ratio results in several times the capacity of other primary cells (other than silver oxide) in the same volume, or proportionally reduced volume for the same capacity.
- Flat discharge characteristic.
- Higher sustained voltage under load.
- Relatively constant ampere-hour capacity.
- Low and substantially constant internal impedance.
- No recuperation required; therefore, the same capacity is obtained in either intermittent or continuous usage.
- Good high temperature characteristics.
- Good resistance to shock, vibration and acceleration.
- Electrically welded or pressure intercell connections.
- Single or double steel case encapsulation.
- Chemically balanced - all zinc is converted at end of life.
- Automatic vent.

Cutaway views of flat and cylindrical cells are shown in Figure 40.

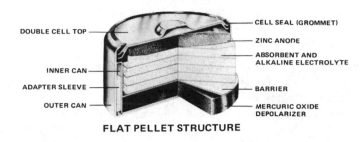

FLAT PELLET STRUCTURE

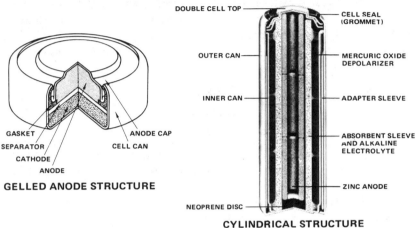

GELLED ANODE STRUCTURE

CYLINDRICAL STRUCTURE

FIGURE 40

CROSS SECTION VIEW FLAT AND CYLINDRICAL TYPE MERCURIC-OXIDE CELLS

Current "Eveready" battery types are available in voltages ranging from 1.35 to 97.2V and in capacities ranging from 16 mAh to 28 Ah.

PERFORMANCE:

The ampere-hour capacity of mercury cells is relatively unchanged with variation of discharge schedule and to some extent with variation of discharge current. They have a relatively flat discharge characteristic, Figure 41.

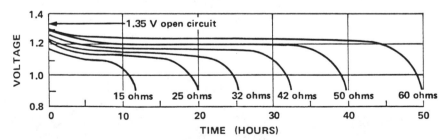

FIGURE 41 — Typical voltage-discharge characteristics of mercuric-oxide cells under continuous load conditions at 70°F (21.1°C). At 1.25 V, equivalent current drains for the resistances are: 15 ohms, 83 mA; 25 ohms, 50 mA; 32 ohms, 40 mA; 42 ohms, 30 mA; 50 ohms, 25 mA; 60 ohms, 20 mA.

[*Editor's Note:*Material has been omitted at this point.]

SILVER OXIDE BATTERIES

The silver oxide-alkaline-zinc (Ag_2O-KOH-Zn) primary battery is a major contribution to miniature power sources. It is the result of a research and development program of many years.

"Eveready" silver oxide batteries provide a higher voltage than mercuric oxide batteries. They offer a flat voltage characteristic. Silver oxide batteries have good low temperature characteristics. Their impedance is low and uniform.

The silver oxide battery operates at 1.5V while mercury batteries operate at about 1.3V. Present types are available in 1.5V cells in capacities of 35 to 210 mAh. A 6 volt battery rated at 190 milliampere-hours is also obtainable. Maximum current output ranges up to 100 mA.

The silver oxide battery consists of a depolarizing silver oxide cathode, a zinc anode of high surface area and a highly alkaline electrolyte. The electrolyte is potassium hydroxide in hearing aid batteries. This is used to obtain maximum power density at hearing aid current drains. The electrolyte in watch batteries may be either sodium hydroxide or potassium hydroxide. Mixtures of silver oxide and manganese dioxide may be tailored to provide a flat discharge curve or increased service hours.

Silver oxide batteries are well suited, for example, for use in hearing aids, instruments, photoelectric exposure devices, electronic watches and as reference voltage sources.

A cutaway of a silver oxide cell is shown in Figure 47.

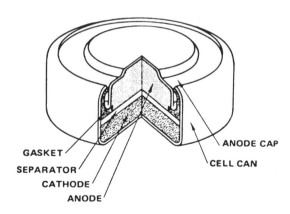

GASKET
SEPARATOR
CATHODE
ANODE
ANODE CAP
CELL CAN

FIGURE 47
CUTAWAY VIEW—SILVER OXIDE CELL

The manner in which the cell is designed results in high volumetric efficiency. An effective radial seal is a unique feature of the construction. This patented radial seal, developed in the late 1950's, results in cells which are excellent in protection against the incidence of salting. Briefly, the radial sealing system incorporates the use of a nylon gasket and a top which is a gold-plated bi-clad, stainless steel anode cup which serves as the negative terminal. The cathode cup is usually a nickel plated steel can which serves as the positive terminal.

The radial seal is formed during the final stages of cell manufacturing. The cell can is subjected to an operation that actually reduces the diameter of the can. This process tightly squeezes the nylon gasket against the bi-clad stainless steel top, creating the initial radial seal.

The above operation is possible because a stainless steel anode top can withstand the extreme pressure during the diameter reducing operation. Secondly, the selection of nylon is important because nylon, after being squeezed, tries to regain its original position.

Once the can diameter is reduced, a secondary seal is effected by crimping the edge of the can over the gasket. Again the use of a nylon gasket is significant since nylon will continue to exert pressure as a result of the second sealing operation.

The radial seal technique is highly effective in providing excellent protection against the incidence of salting, and is found on those silver oxide and mercuric oxide batteries that we produce.

The open circuit voltage of the silver oxide cell is 1.6 volts. The operating voltage at typical current drains is 1.5 volts, compared to 1.3 volts for mercuric oxide cells.

The impedance of silver oxide batteries for hearing aid use is low and consistent. It does not rise appreciably until after the voltage of the battery has fallen below a useful operating level.

Silver oxide cells have excellent service maintenance [generally 90% after one year of storage at 70° F (21.1° C)].

Because of the relatively large surface area of the anode, "Eveready" silver oxide cells exhibit good low temperature performance characteristics.

11

Reprinted from pages A-19, A-21–A-28, A-30, and A-32–A-33 of *Symp. Workshop Advanced Battery Research and Design, 1976, Proc.,* Chicago Section, The Electrochemical Society and Argonne National Laboratory, 1976, 422pp.

SECONDARY BATTERIES FOR ELECTRIC VEHICLES

Albert R. Landgrebe
U.S. Energy Research and Development Administration
Washington, D.C. 20545

An important part of the Energy Storage Program of the U.S. Energy Research and Development Administration is the development of secondary storage batteries for automotive propulsion applications. The successful development of batteries for this purpose will provide significant savings of oil resource and will also have beneficial environmental effects.

Of the battery systems being developed, major efforts are presently or will be concentrated on the following seven systems: advanced lead-acid, nickel-iron, nickel-zinc, zinc-air; iron-air, lithium-metal sulfide and sodium-sulfur batteries. The probability of one or more of these systems meeting the technical requirements for the intended application within the next several years appears good. The ultimate technical/economic feasibility of these systems cannot be assessed at this time because many engineering problems must still be resolved. However, the high temperature systems, Li-MS and Na-S systems appear to be good candidates for electric vehicles in the long term.

Introduction

Major electrochemical storage systems in the form of batteries are used in most facets of everyday life: the starter batteries which provide cranking power for automobiles, flashlight and transistor radio batteries, telephone and emergency standby power batteries, and the small batteries which power consumer products such as toothbrushes and electronic calculators. Also, battery-powered vehicles are used extensively as golf carts and as transportation conveyances in such facilities as airports, large warehouses, and manufacturing facilities. Battery development is being pressed today to power electric and hybrid vehicles through the ERDA's programs.

The common forms of present day electrochemical storage systems such as lead-acid and nickel-iron (Edison) cells were developed during the 19th century. At the turn of the century, electric vehicles were in wider use than heat engine powered vehicles with as many as 10,000 in operation. Batteries also found application in numerous lighting, especially emergency lighting, and signaling applications. With the widespread introduction of low cost, available petroleum products, internal combustion engines overcame the early advantage of electric cars and the major portion of battery use was and still is relegated to engine starting and automobile/truck/bus electric system regulation.

[*Editor's Note:* Material has been omitted at this point.]

<u>Goals</u>

In general, today's batteries have been custom-tailored for specific applications that are frequently designed for either low cost or high performance but not both simultaneously. Table I reviews the performance characteristics and the cost of present secondary batteries.

TABLE I.   Comparison of Today's Secondary Batteries (1)

Battery Type	Cost[a] ($/kWh)	Energy Density By[b] Weight (Wh/kg)	Volume (kWh/ft^3)	Life[c] (Cycles)
Silver-Zinc	900	120	8.8	100/300
Nickel-Cadmium	600	40	3.6	300/2000
Nickel-Iron	400	33	1.4	3000
Lead-Acid				
Motive Power	50	22	2.6	1500/2000
Submarine	80	28	2.0	400
Golf Car	35	35	2.2	300
Elec Vehicle	100	35	2.8	500/800

[a]Cost to the user.

[b]Battery capacity is inversely related to rate of discharge. The values shown are for the 6-hour rate.

[c]Cycle life depends on a number of factors, including depth of discharge, rate of charge and discharge, temperature, and amount of overcharge. Range shown is from most severe to modest duty.

Battery requirements for vehicular applications are represented by a new set of stringent technical and cost goals that will be difficult to meet with the existing commercial batteries. Table 2 shows the interim battery development goals that have been tentatively identified as the battery requirements for viable electric vehicle applications. Of the existing batteries in Table I, the lead-acid battery is the only candidate that comes close to meeting the combined performance and cost goals. Although the lead-acid battery is not expected to meet the interim specific energy goal, its existing technology with one hundred years of effort behind it, argues for a worthwhile pursuit for the near-term vehicle applications. The above discussion points out the need for a concerted development of advanced batteries that have potential of meeting the application goals. Such a concerted effort is expected to be expensive (>100 million dollars) and long

(>5 years) in order to explore and evaluate potential battery systems, but these factors may very well be inconsequential to the impact of the successful implementation and benefits of electric vehicles.

TABLE II.   Interium Battery Requirements for
Electric Vehicle Applications

Duty Cycle	Tentative Goals
	2-4 hour Discharge
	1-6 hour Charge
Energy Efficiency	>50%
Specific Energy	>70 Whr/kg
Specific Power	
Sustaining	>20 W/kg
Peak	>100 W/kg
Cycle Life	>1000 Cycles
	(3-10 years)
Cost	$25-35/KWhr
Environmental Impact	Minimal

The performance requirements for automotive vehicles, Table III, and the effects of battery characteristics on electric car design, Table IV, present some characteristics that may be acceptable for electric vehicles. It is evident that various types of batteries may be acceptable for different types of vehicles depending on the characteristics which are shown as goals. The complex question of trade-off between cost and performance characteristics requires more careful analysis. However, it is evident that there is a need for development of high-energy batteries for electric automobiles.

TABLE III.   Performance Requirements For Automotive Categories (2)

Automotive Categories and Types	STORED ENERGY PROPULSION ONLY			HYBRID PROPULSION* (Heat Engine/ Stored Energy)	
Characteristics and Performance Parameters	General Purpose Family Car	Commuter Car	Utility Car	General Purpose Family Car	Commuter Car
Gross Vehicle Weight (lbs)	4000	2500	1700	4000	2500
Passenger	6	4	2	6+	4
Range Between Recharge (Miles)	150–200	100	50	200+	100+ (Limited By Heat Engine Fuel Capacity)
Top Speed (MPH)	60	60	30	60	60
Maximum Acceleration (MPH/SEC)	4	3	2	4	3
Allowable Storage System Weight (Lbs)	900–1400	600–1000	400–650	300–500	100–200
Storage System Target Performance					
Power (Watts/Lb)	70–100	45–70	30–50	190–300	170–300
Energy (Watt-Hrs/Lb)	85–150	25–45	15–30	<10	<10
Cycle Life	>1000 Deep Discharge Cycles	>1000 Deep Discharge Cycles	>1000 Deep Discharge Cycles	>10^5 Shallow Cycles	>10^5 Shallow Cycles

*Thermal storage/heat engine (Stirling) is not considered a hybrid

Batteries for Electric Vehicles

Seven advanced secondary battery concepts have been identified for their potential for use in electric vehicles and about another six battery concepts have been identified. These are: advanced lead-acid, nickel-iron, nickel-zinc, zinc-air, iron-air, lithium-metal sulfide and sodium-sulphur batteries. Other battery concepts having merit but not covered herein include zinc-chlorine, zinc-bromine, nickel-hydrogen, lead-maganese dioxide, zinc-maganese dioxide and a whole class of batteries containing organic electrolytes. In addition to secondary batteries, at least one primary battery concept is being investigated for its application to electric vehicles, that is Lockheed's Li/water/air battery.

TABLE IV. Effect of Battery Characteristics on Electric Car Design (3)

Type	Battery Specific Energy W-hr/lb	Specific Power, W/lb	Cost, $/kW-hr	Veh, lb	Suggested Vehicle Characteristics Weight Batt, lb	Batt, %	Energy Stored, kW-hr	Power, kW	Battery Cost, $	Range for J-227, mi
Lead-Acid	12.5	12.5	50	3000	1200	40	15	15	750	50
Ni-Alka-line	30	30	100	2500	750	.30	22.5	22.5	2250	80
Near-Term Li/MS	35	35	-	2500	800	32	28	28	-	110
Ad-vanced High-Temp										
High Power	70	100	30	2500	625	25	43.8	62.5	1300	150
Low Power	80	40	25	4000	1200	35	112	56	2050	300

Lead-Acid Battery

Improvements are possible with the lead-acid battery and development of this technology could result in an improved battery in a short period of time. By decreasing the weight of the battery case, increasing the specific gravity of the electrolyte in order to have better utilization of the active materials and decreasing the weight of the plates it should be possible to obtain a battery having a life of 500-700 cycles and a specific energy of 50 W-hr/kg. There is a need to reduce the grid corrosion to increase the cycle life. Major improvements have been made in the past decade and more developmental type improvements will be made in this old technology.

Nickel Systems

Nickel systems such as $Zn//\beta$-NiOOH, $Fe//\beta$-NiOOH and $H_2//\beta$-NiOOH have been developed for other applications and the $Fe//\beta$-NiOOH and $H_2//\beta$-NiOOH have demonstrated very long cycle life, i.e., several thousand cycles. However, the development of such systems into a suitable vehicle battery is contingent upon the successful development of an economical nickel electrode. This must be accomplished without sacrificing the lifetime and performance of the conventional sintered-nickel electrode. A summary of the nickel systems is given in table V.

Because the basic technology for these systems already exist, and substantial effort is still being expended, the likelihood of the successful development of an acceptable battery system (of intermediate specific energy) for the interim electric vehicle market is high and the time frame for development and demonstration would be relatively short.

166

TABLE V.   Present or Estimated Characteristics of Nickel Battery Systems

Battery Type	Energy Density ($KWHR/M^3$)	In-Out WHR Efficiency	Cycle Life (Deep Cycles)	Cost To User ($KWHR) (Est.*)	State Of Development	Critical Technical Problems
H_2/NiOOH	50	75-80	2000	100	Moderate to High	1. Production of an inexpensive high performance NiOOH cathode. 2. Reduce degradation of anode. 3. Improve H_2 storage
Zn/NiOOH	60	65-70%	300	50	Moderate to High	1. Dendrite growth and shape change. 2. Utilization on Ni. 3. Recycle Ni
Fe/NiOOH	30-50	No Data	1000	120	High	1. Reduce self-discharge 2. Improve utilization of Ni

Metal-Air Systems

From an energy density point of view, an air cathode is attractive as a battery component because the electrochemical reactant is supplied continuously from the surrounding battery. Two alternatives that present possibilities for power sources are iron-air and zinc-air battery systems.

The use of air cathodes is not a recent development. A zinc-air battery with air electrodes of wet-proofed carbon has been commercially available for fifty years for use in channel marker bouys, highway flashing systems, railway signals, radio receivers and transmitters, and similar applications. Although reliable, a battery of this type could be discharged only at low rates because of air electrode limitations. Consequently, the power density would be unacceptably low for vehicular applications.

The considerable research efforts on fuel cells and metal air cells in the last 10 years that led to the development of high-rate air electrodes has created a renewed interest in metal-air batteries.[4]

The development and technical problems related to zinc electrode are well known (Table VI).

Edison's work on iron electrodes for nickel-iron cells date back to the early 1900's.[5] At that time, extensive work by Thomas Edison on the iron

167

electrode resulted in his pocket plate electrode that was used in Edison
Ni-Fe cells manufactured up to 1974. This electrode uses a synthesized Fe/
Fe_3O_4 mixture of active materials that is contained within nickel plated
steel perforated compartments. This arrangement represented a durable long-
lived (to 50 years) stable electrode that was very suitable for specialty
applications such as railroad lighting, industrial trucks, and mining
lighting. However, the ruggedness of the iron plate design, coupled with an
equally rugged nickel electrode and case design, led to low stored energy
and power density electrodes (33 Wh/kg and 22 W/kg, respectively, in the
finished cell). Consequently, although iron electrode technology existed
and was available, it was never suitable or considered for use in the construction of a high-energy-density iron-air cell.

Recently, work has been conducted on an iron electrode at Seimens in
Germany[6], Matshushita in Japan[7], SU in Sweden[8], McGraw-Edison[9], GT&E[10], and
Westinghouse in the United States. The development and technical problems
related to the zinc electrodes are well known and will not be discussed
herein. Both the iron-air and zinc-air system have the potential for use in
hybrid vehicles and perhaps in electric vehicles. Table VI lists the
characteristics of these systems.

Table VI. Present or Estimated Characteristics of Metal-Air Battery Systems

Battery Type	Energy Density ($KWHR/M^3$)	In-Out WHR Efficiency	Cycle Life (Deep Cycles)	Cost To User ($KWHR) (Est.*)	State of Development	Critical Technical Problems
Zn/Air Slurry	No Data	40%	3000 Hours	25-40*	Moderate	1. Recycling of slurry 2. Catalyst deactivation 3. Contamination from CO_2
Zn/Air Mechanically Rechargeable	No Data	N.A.	No Data	No Data	Low	1. H_2O loss 2. Leakage of electrolyte 3. Rapid replacement 4. Contamination from CO_2
Zn/O_2	No Data	No Data	200	No Data	Low	1. Zinc dendrite and shape change 2. Control of O_2 pressure
Fe/Air	110	30-40	300	25-40	Moderate	1. Low cost Fe electrode 2. Rechargeable air electrode 3. Low efficiency

[*Editor's Note:* Material has been omitted at this point.]

Conclusion

Secondary batteries for electric vehicles offer several potential advantages including conservation of resources and favorable environmental features such as no thermal and air pollution.

Several advanced batteries under development show promise of meeting the requirements for electric vehicles within the next decade. Potential vehicle batteries include lead-acid, nickel-iron, and nickel-zinc for the near term (one to three years); advanced lead-acid and nickel-zinc for the intermediate term (three-to-five years); and metal-air, lithium-metal sulfide, sodium-sulfur and other advanced systems for the longer term (greater than five years).

[*Editor's Note:* Material has been omitted at this point.]

References

1.   D. L. Douglas, "Batteries For Energy Storage"
     Preprinted papers presented at 168th National Meeting A.C.S.
     September, 1974

2.   George Pezdirtz, ERDA, private communication

3.   Paul Nelson, Argonne National Laboratory, private communication

4.   E. S. Buzzelli "Low Cost Bifunctional Air Electrodes," paper 46,
     ECS Dallas Meeting, October, 1975

5.   V. Falk and F. J. Salkind, Alkaline Storage Batteries, John Whiley
     and Sons, Inc., New York (1969) pp. 16, 21

6.   H. Cnoblock, et al, "Performance of Iron-Air Secondary Cells Under
     Practical Operation Conditions" preprint 17, International Power
     Sources Symposium, Brighton, England (1972)

7.   M. Mori, et al, Japanese Patent Application 50-72-137, June 14, 1975
     (Matshushita).

8.   O. Lindstrom, "Rechargeable Metal-Air Battery Systes,"
     23rd Meeting, International Society of Electrochemistry,
     Stockholm, Sweden (1972)

9.   J. D. Moulton, et. al., U.S. Patent 2, 871, 281
     Alkaline Storage Battery with Negative Iron Electrodes,"
     January 27, 1959

10. E. R. Bowerman, Proceeding, Power Sources Symposium, Atlantic City
     (1968)

11. N. P. Yao and J. R. Birk, "Battery Energy Storage For Utility Load-
     Leveling and Electric Vehicles: A Review of Advanced Secondary
     Batteries," IECEC Newark, Delaware, (August 1975).

12

Reprinted from pages 1–9 and 24–29 of *Report UCRL 51811*, Lawrence Livermore Laboratory, 1975, 36pp.

THE LITHIUM-WATER-AIR BATTERY: A NEW CONCEPT FOR AUTOMOTIVE PROPULSION

L. G. O'Connell, B. Rubin, E. Behirn, I. Y. Borg, J. F. Cooper, and J. Wiesner

Summary

INTRODUCTION

The transportation sector of the economy uses approximately one-fourth of all energy consumed in the United States.[1] From an energy-use standpoint, this sector is dominated by the highway vehicle system. It alone accounted for 73% of the transportation energy used in the United States in 1972.[2] This energy demand is almost entirely supplied by petroleum, and thus transportation is a major contributor to our domestic petroleum imbalance.

Many energy-use sectors other than transportation can change the fuels they use, subject to the constraints of capital equipment inventories. This is not true of transportation at this time. Thus it is imperative that alternative energy sources, and propulsion systems to utilize them, be developed for transportation, in particular for the highway vehicle system. The passenger automobile is the dominant energy user within the highway vehicle system.[3] Clearly the potential for reducing petroleum consumption by developing alternative propulsion systems that use fuels other than petroleum is greatest in the automobile sector.

Electric propulsion is one interesting and challenging alternative to the internal combustion engine. If successfully developed and introduced, the electric vehicle will be energized by electric power plants which can be fueled by coal or uranium. Thus the electric vehicle can be a multi-fuel but nonpetroleum automobile.

Although the concept of the electric vehicle has been with us for as long as the automobile, it has not been acceptable to the driving public because the available batteries had limited energy and power capacity. This resulted in a vehicle of dismal performance and limited range. In addition, secondary-battery recharge times were long, so that the vehicle had to be idled for long periods of time. The key to success for the electric vehicle is an adequate battery system.

This paper discusses the lithium-water-air battery, a new concept for automotive propulsion. It is based on the lithium-water battery developed by the Lockheed Missiles & Space Company, Inc., for marine applications.[4] If successfully developed, it should make the electric vehicle practical by providing good performance and long range between recharges. It is a primary battery and thus cannot be recharged in the normal manner. It must be mechanically recharged by replacing the consumable lithium anodes. With proper design, this can be used to advantage, since the battery can be refueled quickly, which removes the disadvantage of limited range for the electric vehicle.

DISCHARGE CYCLE

The lithium-water battery represents a new class of batteries that employ semi-passivated lithium anodes. A lithium-metal anode and an iron-wire-mesh cathode are brought into direct contact with aqueous electrolytes. The cell is discharged with high energy and power densities at ambient temperature. The cell reactions and standard potentials in volts are as follows:

Anode:

$$Li = Li^+ + e^- \qquad E_a^\circ = -3.04$$

Cathode:

$$H_2O + e^- = OH^- + \tfrac{1}{2}H_2 \qquad E_c^\circ = -0.82$$

Net reaction:

$$Li + H_2O = LiOH(aq) + \tfrac{1}{2}H_2 \qquad E^\circ = E_c^\circ - E_a^\circ$$

$$= 2.22$$

During discharge the iron mesh is pressed against the lithium anode. A thin layer of reaction products (nominally lithium hydroxide) forms at the lithium-electrolyte interface, which exhibits high ionic conductivity and relatively low electronic conductivity. Because of this layer, no separator is needed to protect the lithium. In fact the anode and cathode can be pressed together with pressures on the order of 2 kg-f/cm^2 (30 psi) to reduce internal ohmic losses.

Water serves as the active cathode material in this cell. By discharging an oxidizing agent or by introducing oxygen (air) at a catalytic surface, the electrolysis of water can be eliminated, with an accompanying increase in the battery's energy capacity. Even without the use of an oxidizing agent, and on the basis of demonstrated cell performance data, a lithium-water battery weighing 220 kg could be built to deliver an energy density of 225 W-hr/kg at a power density of 200 W/kg. This battery could power a 900-kg (2000-lb) automobile for 330 km (200 mi). This performance compares very favorably with other available or projected batteries, especially since it is a room-temperature system.

The lithium hydroxide concentration in the electrolyte must be controlled for satisfactory battery operation. In the original design, for marine application, this was done by adding extra water. In an automotive application, the water would impose an excessive weight penalty. Concentration can be controlled more economically by adding carbon dioxide to the electrolyte and precipitating lithium carbonate. This reduces the onboard requirement for water. The overall cell reaction thus becomes

$$2Li + CO_2 + H_2O = Li_2CO_3 + H_2$$

The hydrogen could be released, but this would be an unacceptable loss of energy. It was stated earlier that the electrolysis of water could be eliminated by introducing oxygen (air) into the reaction. This can be accomplished by means of an air cathode. A lithium-water-air cell has been tested and found to give a cell voltage of $E° = 3.4$ V. The energy density of the lithium-water-air battery has been calculated to be 330 W-hr/kg at a power density of 200 W/kg. Furthermore, the hydrogen problem is eliminated. If carbon dioxide is used for concentration control as before, the net cell reaction is

$$Li + \tfrac{1}{4}O_2 + \tfrac{1}{2}CO_2 = \tfrac{1}{2}Li_2CO_3$$

An additional advantage is that water no longer is consumed in the reaction.

Before the lithium-water-air battery can be developed for automotive application, further research is required. An investigation of the protective film must be conducted. It contributes substantially to the internal resistance, and a better understanding of its nature may lead to a reduction of that resistance. The use of an air cathode will greatly increase the energy efficiency of the battery. Improvement of air-cathode technology is required. Alternative electrolyte and cathode materials should be investigated with the intent of simplifying the battery operation and providing a precipitant that would simplify the lithium reprocessing. The sensitivity of the battery to impurities must be understood, and alternative methods of controlling electrolyte concentration should be investigated. The aque-

ous battery suggests the possibility of rechargeable nonaqueous analogs. These require investigation.

BATTERY RECHARGE

From an economic and materials-conservation viewpoint, it will be necessary to recycle the lithium in the battery precipitate. This precipitate will be lithium carbonate or some other insoluble lithium compound. Using present technology, the lithium carbonate would be reacted by one of several methods to produce lithium metal, which would then be fabricated into anodes for reuse. The various methods must be evaluated to determine energy efficiency because the choice of method will affect material conservation and also operational cost.

The best current practices give an electrical efficiency for the recycle process of 50%. An intensive study is required to determine whether this value can be increased. The propulsion efficiency for the discharge cycle is estimated at 52%, so that the overall efficiency of a primary battery system is 26%. Secondary-battery systems are more efficient, although they do not presently supply the needed performance. This suggests that the ultimate solution is a battery/battery (primary/secondary) hybrid.

AUTOMOTIVE SYSTEM FEASIBILITY

In considering the question of the feasibility of the lithium-water-air battery for automotive propulsion, a number of issues should be considered. One must consider the vehicle-design options offered by this battery, the type of per-

formance to be expected, the cost of operation, the servicing requirements, and the
fuel-distribution system that results. In
addition, the nation's lithium resources
must be critically examined. We do not
want to create a future "oil" problem.
Finally, one must be concerned with assessing the effect of introducing a successful
battery-powered electric vehicle. Each of
these issues must be examined in detail,
since each of them affects battery development. The vehicle-design options will be
of prime concern to battery development,
as they will directly affect its output
requirements.

Several battery options are available
for electric vehicle design, including primary, secondary, and battery/battery hybrid. A preliminary study examined a vehicle powered by the lithium-water-air
primary battery aided by a small secondary
battery. The secondary battery acts as a
power buffer and supplies power for starting, lighting, and instrument functions.
A 900-kg vehicle equipped with such a battery system weighing 227 kg would have a
range of 330 km, an acceleration from 0 to
>64 km/hr (40 mph) in 10 sec, and a maximum
sustained speed of 97 km/hr (60 mph).

The electric vehicle could quite rapidly receive a carbon dioxide recharge, re-

moval of lithium carbonate, and a change
of lithium anode. Refueling need not be
tedious if an appropriate modular design
is effected. If an additional 29 kg of
lithium (330-km range requires 7.3 kg of
lithium) is carried, the anode need only be
changed every 1600 km. This could be done
during normal more extensive vehicle servicing. The advent of this vehicle will
require establishing a different fuel-
distribution system.

The operating cost of an electric
vehicle powered by the lithium-water-air
battery is tied inextricably to the cost
of lithium and the battery system chosen
for the vehicle. At present gasoline
prices (which are expected to rise) and
20 mpg (the 1980 goal) the cost of fuel is
3¢ per mile. For comparison, it is expected that the cost of reprocessed lithium in
large quantities will be $1.50 per kilogram.
If the battery system is primary, then the
cost of operation will be 7.4¢ per mile;
if it is hybrid with the secondary battery
sized for 20-mile range, the cost will be
about 3¢ per mile. Other combinations will
give different operating costs.

A more detailed feasibility study must
be conducted, since the type of system will
affect battery design.

Conclusions

• The lithium-water battery can provide
high enough specific energy and specific
power to give an electric vehicle acceptable performance.

• Considerations of cost per mile moti-
vate development of an even higher performance battery: the lithium-water-air
battery.

• Both batteries derive high energy
and power from a combination of high standard potential with a novel cell design in
which anode and cathode are in direct mechanical apposition, separated only by the
anode surface layer.

• There is a good possibility for the
development of high-performance, nonaqueous,

rechargeable analogs to the lithium-water battery.

● Research is needed to improve processes for the regeneration of lithium metal.

● A primary battery with a power buffer makes it possible to design a 900-kg vehicle with a 330-km range at 97-km/hr cruise speed and good acceleration.

● To achieve this superior performance at a reasonable fuel cost, considerable system-design effort will be required.

Delivery of Energy from the Battery

THE LITHIUM-WATER PRIMARY BATTERY CONCEPT

The Lithium-Water Galvanic Cell: A New Battery Concept

A primary battery that uses the reaction between lithium and water has been developed by the Lockheed Missiles & Space Company, Inc., for marine applications.[4] The battery is a representative member of a new class of batteries that employ partially passivated lithium anodes. In these batteries, lithium metal is brought into direct contact with aqueous electrolytes. They give impressively high energy and power densities (compared to other room-temperature batteries) when used in combination with a water cathode or with any of several other cathode materials (including air).

The lithium-water cell can be described by the following reactions (standard potentials are given in volts):

Anode:

$$Li = Li^+ + e^- \qquad E_a^\circ = -3.04$$

Cathode:

$$H_2O = e^- = OH^- + \tfrac{1}{2}H_2 \qquad E_c^\circ = -0.82$$

Net reaction:

$$Li + H_2O = LiOH(aq) + \tfrac{1}{2}H_2 \qquad E^\circ = E_c^\circ - E_a^\circ = 2.22$$

In a simple version of this cell, an iron-wire-mesh cathode is pressed against the lithium anode, and the leads are connected through a load as shown in Fig. 1. The electrode assembly is immersed in lithium hydroxide electrolyte. During cell discharge, a thin layer of reaction products (nominally lithium hydroxide) forms at the lithium-electrolyte interface. The layer exhibits high ionic conductivity and relatively low electronic conductivity. Hydrogen evolution, which requires electron transfer, is severely restricted on the anode surface and occurs almost exclusively (95%) on the iron cathode.

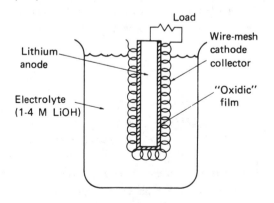

Fig. 1. Schematic of the lithium-water cell.

It is the existence and properties of the anode surface layer that permit the unusual design of the lithium-water cell. No separator is needed to protect the lithium from corrosion in the presence of water. The wire-mesh cathode may be pressed against the anode with considerable pressure [2 kg-f/cm^2 (30 psi)] without shorting the electrodes. The advantages of this configuration are as follows:

1. The ohmic drop through the electrolyte in the interelectode gap is essentially eliminated.

2. The close spacing of the electrodes provides enhanced mass transport by diffusion in the electrolyte.

3. Constant electrode separation produces time-independent discharge voltage and current. These features result in a stable discharge at high power levels.

The film exists in a dynamic steady state, being continuously formed and dissolved at the lithium-electrolyte interface. Maintenance of the steady state requires continuous cell discharge with an electrolyte of sufficiently high lithium hydroxide concentration (greater than 1.5 M LiOH). With the use of appropriate additives (Lewis bases), the parasitic corrosion of lithium can be essentially eliminated over a wide range of discharge-current densities, and coulombic efficiency (i.e., the fraction of lithium utilized in the galvanic reaction) may approach 95%. When the external circuit is broken, the film dissolves and corrosion ensues.

Discharge characteristics are dependent on lithium hydroxide concentration and on electrolyte temperature. At a given temperature, an increase in concentration results in an increase in energy density

but a decrease in power density, as shown in Fig. 2. At a given concentration, increasing the temperature has the opposite effect. Optimum cell performance requires, therefore, control of temperature and of lithium hydroxide concentration.

In the lithium-water cell, water serves as the active cathode material. However, electrolysis of water may be replaced by the reduction of any of a number of dissolved oxidizing agents or by oxygen (air) at a catalytic surface. The existence of the protective film on the lithium is not affected.

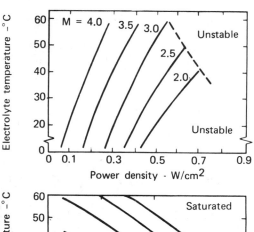

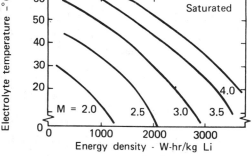

Fig. 2.   Discharge of the lithium-water cell. Dependence on system parameters. (Maximum theoretical energy density, 8500 W-hr/kg Li. Data for cell discharge at 1 V.)

Comparison with Automotive Requirements and Other Batteries

The performance characteristics of various electric batteries can be compared on a graph of specific power against specific energy (Ragone plot[5]), the power and energy being normalized to the weight of the battery system (Fig. 3). In an automotive application, for fixed battery and vehicle weight, top speed and acceleration are determined by the specific power, whereas vehicle range is determined by the specific energy. On the basis of demonstrated cell performance data, a lithium-water battery weighing about 220 kg could be built to deliver an energy density of 225 W-hr/kg at a specific power of 200 W/kg; it would power a 900-kg auto for 330 km.

Shown for comparison are lead-acid, nickel-cadmium, and nickel-zinc batteries, which fall far short of the required specific energy for a 330-km range. The zinc-air battery meets range requirements only at an unacceptably low power. The lithium-iron sulfide battery shows sufficiently high specific power, but the specific energy is too low. Molten-salt batteries promise high performance, but present development efforts are plagued by extremely difficult materials problems.

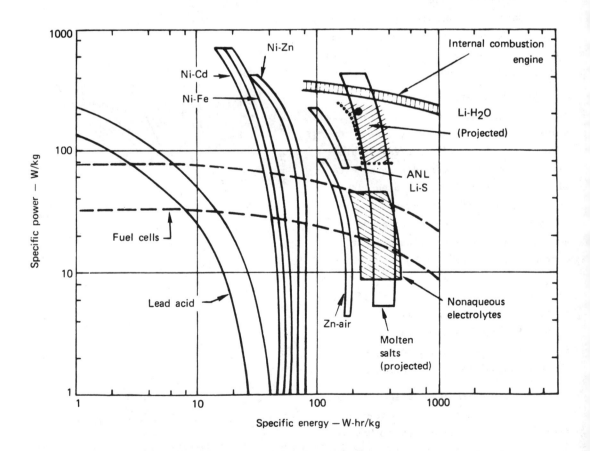

Fig. 3. Performance characteristics of electric batteries. (From ref. 5.)

It should be noted that the performance characteristics of automobile internal combustion engines appear in the specific power range of 200 to 300 W/kg and a specific energy range of 100 W-hr/kg or more. Only the lithium-water and the molten-salt batteries fall in this region of the diagram and thus offer the possibility of meeting the performance requirements for battery-powered automobiles. In particular, the lithium-water battery is the only demonstrated room-temperature system with such characteristics.

Battery Description and Characteristics

Physical Descrpiton and Nature of Operation - Present cell designs provide for control of electrolyte temperature and concentration, as indicated in Fig. 4.

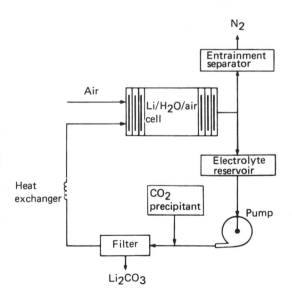

Fig. 4.   The basic lithium-water-air system requires heat rejection and LiOH concentration control.
1. Overall reaction is
$2Li + CO_2 + 1/2O_2 = Li_2CO_3$.
2. Must dissipate heat.
3. Must add 0.5 mole CO_2 per mole Li.

During shutdown periods, the battery is drained and the electrolyte is stored in a reservoir. During discharge, electrolyte is circulated through the battery and through a heat exchanger. The concentration is controlled by precipitating lithium carbonate with carbon dioxide. The overall cell reaction is

$$2Li + CO_2 + H_2O = Li_2CO_3 + H_2$$

Provision can be made for the separation and release of the hydrogen gas. However, operation in this mode would represent an unacceptable waste of energy. Recovery of this energy through the use of an air cathode is discussed in a later section.

Advantages and Disadvantages of the Lithium-Water Battery for Automotive Propulsion - A lithium-water battery is competitive with an internal combustion engine in terms of energy and power characteristics. Nonetheless, it is a primary battery and requires periodic physical replacing of the reactants: lithium, water, and a precipitating agent (e.g., carbon dioxide). The most energetically dense reactant is lithium; an on-vehicle inventory of 65 kg would be sufficient for 1600-km range. The heavier reactants, water and carbon dioxide, limit the range, but these fluids can be easily replaced. The cost of physical replacement of lithium, including the economics of lithium recycling and distribution, will be considered later in this report. For the moment we note that an increase in specific energy for a given vehicle and battery weight will result in a proportionate decrease in the operational cost per mile. Thus there is motivation for increasing the performance of the basic lithium-water cell beyond the minimum level required for acceptable automotive performance.

Necessity for a Lithium-Water-Air Battery (Air Cathode) - Tests have been made on a lithium-water-air cell that uses the following reaction:

$$Li + \tfrac{1}{4}O_2 + \tfrac{1}{2}H_2O = LiOH(aq) \qquad E° = 3.4 \ V$$

An open-circuit potential of 2.9 V and a cell voltage of 2.6 V at 200 mA/cm^2 have been demonstrated in laboratory experimental prototypes. The prototype cell utilizes a porous Teflon cathode with a catalytic coating. The cell configuration is otherwise the same as with the lithium-water battery. If carbon dioxide is used for concentration control, the net cell reaction is

$$Li + \tfrac{1}{4}O_2 + \tfrac{1}{2}CO_2 = \tfrac{1}{2}Li_2CO_3$$

The use of an air cathode is advantageous for several reasons:

1. Calculations indicate that the higher cell voltage increases the specific energy of the battery to 330 W-hr/kg at a specific power of 200 W/kg.

2. Water is no longer consumed as a reactant, and large on-vehicle inventories are unnecessary.

3. The cathode depolarizer, oxygen, is available from the air, and its weight need not be considered in the calculation of specific energy. Hazards associated with the disposal of combustible hydrogen gas are eliminated.

Costs associated with the preparation of lithium metal for replacing the anodes are extremely important. The increased energy per mole of lithium that results from the use of an air electrode is essential for the lithium system to attain reasonable economy of operation. Therefore the development of a suitable air cathode will be essential for the success of a lithium-water automotive propulsion system.

[*Editor's Note:* Material has been omitted at this point.]

178

SYSTEM-DESIGN EFFECTS

Vehicle Design

The effect of the lithium-water power buffered primary system on electric vehicle performance is illustrated in Table 4. As a basis for comparison, we have chosen the 1975 Seneca, manufactured by Linear Alpha, Inc. The specifications for this commercially available electric vehicle are shown in the first column. The specifications are taken (or estimated) from the company's specification sheet. The second column gives specifications for a vehicle with 330-km range, with a lithium-water-air power buffered primary system replacing the

Table 4. Performance capability of the representative battery system and a currently available electric vehicle.

Parameter	1975 Seneca[a] (Pb-acid batteries)	Representative battery system
Range (km)	97	330
Passenger capacity	2	4
Estimated vehicle weight (kg)	1450	1000
Estimated battery weight (kg)	680	220
Battery energy (kW-hr)	23	51
Specific energy (W-hr/kg vehicle weight)	16	51
Battery power (kW)	23	45
Specific power (W/kg vehicle weight)	16	45
Normalized acceleration	1	3
Cruising speed (km/hr)	64	88–96
Cost:		
Unit production (1975)	$7000	--
Mass production (1985)	$4800	$3300[b]

[a]Produced by Linear Alpha, Inc.

[b]Estimated at $3.3 per kilogram of vehicle weight.

lead-acid batteries. Note that this vehicle is about 30% lighter than the Seneca; its increased acceleration and cruising speed are due to the threefold improvement in specific power. The purchase-price estimates are based solely on the vehicle weights and on the simplifying assumption that mass-produced vehicles will sell at $3.3 per kilogram in 1985.

It should be pointed out that the comparison with the Seneca was made for reference purposes only and that the vehicle design we envision is in no way tied to that of the Seneca. Indeed, there are many design options for further improvements that deserve further study.

Drive-motor options should be investigated. Alternating-current motors seem more attractive than dc motors in both specific weight (kg/kW) and specific cost ($/kW). In the past, the ac system has not been frequently used because of its costly, complex, and heavy controller. The recent development of high-current Darlington transistors coupled with integrated-circuit logic elements may represent a very significant technological advance in controller performance for ac systems. This approach, which can also handle regenerative-braking requirements, should be investigated. Chassis and body design, including the reduction of weight, rolling, and drag losses, must also be studied.

Vehicle Refueling and Servicing System

As already indicated, the representative vehicle (900 kg) powered by a lithium-water battery will have a range of 330 km. This is essentially a primary battery system, and refueling would consist of replacing the lithium anodes, refilling the

carbon dioxide reservoir, and removing the lithium carbonate. The present system of service stations is properly distributed for vehicles with a range of 330 km, although not presently equipped to provide service.

The refueling process, assuming that the vehicle had traveled the maximum range, would consist of removing 39 kg of lithium carbonate as slurry and adding 23 kg of carbon dioxide and 7 kg of lithium. Proper design of the vehicle can lessen the task of refueling. Several concepts can be postulated. The system chosen should require minimal time at the service island. Lengthy refueling times will probably be unacceptable to the public.

One concept is to make the battery components modular, particularly those involved in refueling, so that they can be removed easily. The carbon dioxide refueling should be a rather simple matter of recharging tanks. However, this will require equipping the service center to handle high-pressure gas. Since the lithium carbonate will be quite heavy, the package can be divided into a number of small modules; otherwise handling equipment will have to be made available.

Changing the lithium plates will be more tedious. The battery can be designed to have quick disconnects, so that plate modules can be rapidly removed. These modules should not be heavy; at replacement they should weigh about 10 kg, so that no special handling equipment is required. Rebuilding lithium modules at the service center rather than at some central processing center is a possibility. This can eliminate transporting inactive module parts. The decision will most likely be dictated by economics. If transportation costs are low enough, rebuilding at a central plant would be more suitable since better quality control and safety standards can be maintained.

Another refueling approach is to design a system with mechanical equipment (e.g., a slurry pump) available to remove the lithium carbonate from the vehicle. In the process it is transferred to a central collection container at the service center for later removal to the reprocessing center. The carbon dioxide and lithium would be handled as before. The essential difference is that there is less handling of the lithium carbonate and a lower total weight to be transported to the reprocessing center.

The most tedious refueling task in the above systems is lithium replacement. The range of the vehicle, before lithium replacement is required, can be extended by increasing the amount of lithium in the battery. If instead of 7.3 kg the battery carried 36.3 kg of lithium, the vehicle could travel 1600 km between lithium changes. Assuming an average of 16,000 km/yr travel, the lithium need only be refueled every 1.2 months. Lithium replacement would not take place at the "gas" island, but as part of a more lengthy servicing of the vehicle, similar to the 1600-km oil change. Routine refueling would then consist of adding carbon dioxide and removing lithium carbonate.

The refueling of the lithium-water-air battery does not seem to pose any fundamental obstacles. If the range can be extended by increasing the amount of lithium, refueling can approach the ease and speed of that afforded the present-day gasoline vehicle. A more detailed study is required, of course, to determine what is desirable from the standpoints of vehicle design, service-center design, and economics.

The distribution system from the service center to the reprocessing center and return does not exist as such and will have to be established. This will require a large capital expenditure and may require some interim steps at first. Further study is needed.

We have assumed refueling at 330 km, but normally drivers refuel before the maximum range has been reached. The vehicle design must make provision for premature refueling.

Since makeup water and carbon dioxide will be obtained from local sources, the effects of impurities on battery operation must be determined. Water will undoubtedly be the main problem.

Finally, attention must be paid to safeguards against "poisoning" the lithium carbonate so that the reprocessing system is not harmed. Since the carbonate has considerable value, attempts will be made to cheat the system. Considerable thought should be given to finding the best way of providing these safeguards.

ECONOMIC, POLITICAL, AND SOCIAL IMPACT

Economics of Lithium-Water-Air Battery Use

In the nominal system, wet lithium carbonate is removed from spent primary batteries at service stations. This is illustrated in Fig. 15. The lithium carbonate is collected and shipped to regional reprocessing plants, where it is converted into lithium battery anodes. Currently available technology would involve conversion

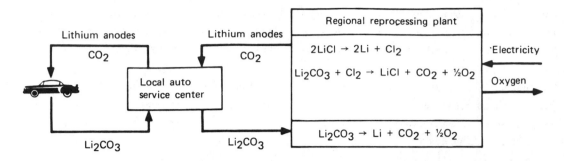

Fig. 15. An automotive system involves recycling of Li_2CO_3 to a reprocessing plant.

of the carbonate to lithium chloride, followed by electrolysis of the chloride to lithium metal and chlorine. Since a number of different chemical routes are available for these steps, a detailed survey of current industrial practices and of possible alternatives will be required to establish the most economical ones.

Initial Inventory Costs – The current cost of new lithium metal produced from ore is $20.64 per kilogram. On the basis of discussions with lithium and sodium producers, it has been stated that this figure could be reduced to about $4.40 per kilogram in quantities of 455,000 kg/yr or more.[16] This figure would thus represent a lithium capital inventory cost of $32 per automobile. In addition, it is assumed that an additional inventory of lithium equal to this amount would be in process through the recycle system.

Charge-Cycle Costs – There is considerable uncertainty about the cost of reprocessing lithium carbonate to lithium metal. Foote Mineral[16] has stated that for quantities greater than 455,000 kg/yr and starting with lithium carbonate, the cost of preparing metal would be in the range of $2.2 per kilogram. Costs of $1 to $2 per kilogram will be required for the lithium-water-air system to be competitive; thus considerable improvement over the Foote estimate will be needed. It should be noted that the current lithium industry is relatively small; U.S. production is only about 4500 metric tons per year. Two other metals produced in large quantities by electrolytic methods are sodium and aluminum. Lithium probably could be produced in currently existing sodium facilities. The price of sodium is about 40¢ per kilogram, and the price of aluminum is 65 to 85¢ per kilogram. On the basis of relative atomic weights, lithium should be about three times as expensive as sodium, or about $1.20 per kilogram. Very large production facilities would be needed to make this price realistic. However, a recycle cost for lithium of about $1.50 per kilogram does not seem unreasonable.

Although lithium is a reactive metal, with proper care and facilities it is not difficult to handle in a dry atmosphere. It is extremely soft and can be easily shaped. Therefore fabrication into anode assemblies should not be difficult or expensive.

Very little can be said at present about the costs of air cathodes. Possible materials are porous carbon or porous Teflon, impregnated with small amounts of activator, such as silver or other metals. Overall materials costs should be low.

Other battery and battery-system components appear to be straightforward in fabrication technology and cost. Such items as tanks, pumps, molded plastic battery containers, and filters are familiar to the auto industry and seem to present no special cost problems.

Carbon dioxide will also have to be supplied at service stations. The current cost is about 44¢ per kilogram, supplied in standard 2-ft^3 pressure cylinders. In large quantities, this cost can be reduced to 22¢ per kilogram or less[17]; in the quantities needed for an automotive industry, this cost probably can be cut in half. The nominal auto starting with a fresh charge of 7.3 kg of lithium will require 23 kg of carbon dioxide.

Discharge-Cycle Costs - The principal costs associated with the discharge cycle are (1) transporting the lithium anode assemblies and carbon dioxide containers from the regional processing plant to the service-station network; (2) installing new assemblies and supplies in automobiles and removing lithium carbonate slurry; and (3) collecting and transporting the slurry

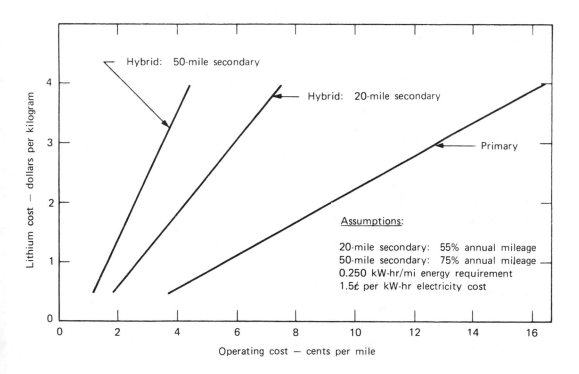

Fig. 16.   Vehicle operating costs are a function of system design.

and spent anode containers to the regional processing plants.

A typical transportation cost for packaged items such as batteries or carbon dioxide cylinders is about 13¢ per ton per kilometer.[18] Assuming an average delivery trip length of 25 km, the average transportation cost for these items would be 0.005¢ per kilogram for lithium, 0.015¢ per kilogram of lithium for carbon dioxide, and 0.025¢ per kilogram of lithium for lithium carbonate.

Typical handling costs at the service station should be analogous to current handling costs of about 2¢ per kilogram of material. For the lithium-metal and lithium carbonate handling, this amounts to about 12¢ per kilogram of lithium. Costs for handling water and carbon dioxide are assumed to be negligible.

The operating cost per mile for an automobile powered by a lithium-water-air battery is highly dependent on the cost of lithium reprocessing. Figure 16 puts these costs in perspective. Assuming a cost of $1.50 per kilogram for lithium, the representative vehicle denoted by the primary curve would have an operating cost of 7.4¢ per mile. Today's vehicle, at 20 mpg, costs 3¢ per mile before tax to operate.* A battery/battery hybrid (e.g., with a 20-mi-range secondary), would have an operating cost of about 3¢ per mile. An appropriately designed lithium-water-air battery automobile thus can have a competitive operating cost. Considering possible future increases in the price of gasoline, comparisons are indeed favorable. Although these cost estimates are very crude, they indicate that the technical possibilities

offered by the lithium-water system must be investigated.

Future Inventory of Lithium

Some estimate of the amount of lithium required can be made from the following:

Lithium consumption: 7.3 kg
Automobile range: 330 km
Number of electric cars[14,19]
    Year 1985   (0.2 to 1.0) $\times 10^6$
    Year 2000   (18 to 25) $\times 10^6$

For every battery in use, an equivalent amount of lithium will be in process. Thus for each electric automobile in use, 14.6 kg of lithium is needed. Acquisition of more exact data may cause this estimate to change in the future, so a range of 13 to 17 kg per electric vehicle is probably a better statement of the requirement.

Taking into account all vehicles powered by the lithium-water battery, we arrive at the following estimate:

Year	Number of electric vehicles	Lithium required (metric tons)
1985	(0.2 to 1) $\times 10^6$	2600 to 17,000
2000	(18 to 25) $\times 10^6$	234,000 to 425,000

[*Editor's Note:* Material has been omitted at this point. Only those references cited in the excerpts follow.]

---

*This accounts for gasoline only.

REFERENCES

1. C. J. Anderson et al., *An Assessment of U.S. Energy Options for Project Independence,* Lawrence Livermore Laboratory, Rept. UCRL-51638 (1974).
2. "Statistics of Energy 1958–1972," *Basic Statistics* (OECD, Paris, 1974).
3. J. J. Mutch, *Transportation Energy Use in the United States. A Statistical History, 1955–1971,* prepared for the National Science Foundation by the Rand Corp., R-1391-NSF (December 1963).
4. H. J. Halberstadt, *The Lockheed Power Cell,* Lockheed Missiles & Space Co., Inc., SAE Paper No. 739008.
5. D. V. Ragone, *Review of Battery Systems for Electrically Powered Vehicles,* SAE Paper No. 680453, SAE midyear meeting, Detroit, Mich. (May 1968).
14. F. Kalhammer, "Energy storage: Incentives and prospect for its development," presented to Amer. Chem. Soc. (Atlantic City, September 12, 1974).
16. Foote Mineral Co., private communication (March 1975).
17. Matheson Chemical Co., private communication (March 1975).
18. Lockheed Missiles & Space Co., Inc., private communication (March 1975).
19. P. A. Nelson, A. A. Chilenskas, and R. K. Steunenberg, *The Need for Development of High Energy Batteries for Electric Automobiles,* Argonne National Laboratory, Rept. ANL-8075 (draft) (January 1974).

13

Copyright © 1976 by the General Motors Corporation

Reprinted from *Research Publication GMR-2065*, General Motors Corporation, 1976, 17pp.

HIGH-TEMPERATURE BATTERIES

Elton J. Cairns and John S. Dunning
Research Laboratories, General Motors Corporation
Warren, Michigan 48090

ABSTRACT

The state of the art for high-temperature batteries will be presented and discussed. Emphasis will be given to the lithium alloy/metal sulfide and sodium/sulfur cells. Other systems to be considered include lithium/chlorine and sodium/metal halide. Cell chemistry and performance and life-limiting factors will be reviewed for all of the systems, and the status of investigations in critical problem areas will be given. Recent advances in the demonstration of high specific energy and expectations for future improvement will be presented.

INTRODUCTION

The current awareness of the developing shortage of inexpensive sources of energy has given new impetus to the search for and development of means for making more effective and more efficient use of the energy sources and energy conversion systems that we possess. The most rapidly-growing sector of our energy economy is that of electrical energy generation. Fortunately, we possess the capability of generating electrical energy from a wide variety of primary fuels, including coal and nuclear fuel, which are in much larger supply in the United States than petroleum. In order to more effectively utilize our electrical energy system, it is important to have an efficient, flexible, economical means of storing off-peak electrical energy for later use during peak demand periods. In addition, the demand for petroleum could be reduced by the use of rechargeable batteries as a power source for automobiles.

The performance, lifetime, and cost goals for the battery applications mentioned above tend to exclude all of the presently-available batteries, and many proposed batteries. The class of batteries which is projected to have the best combination of performance, life, and cost for large-volume application in multikilowatt sizes is that of high-temperature batteries, which are being developed to meet the following general goals:

	Peak Specific Power (W/kg)	Specific Energy* (W·h/kg)	Minimum Cycle Life	Minimum Lifetime (yr)	Cost ($/kW·h)
Off-peak energy storage	15-50[†]	100-200[†]	1000-2000	5	20
Automobiles	200	200	500-1000	3	20

[†] not very important for this application   * at 50 W/kg

Of course, only those systems using abundant materials can be considered for widespread use.

In the field of high-temperature cells, there are two types of electrolytes in use: molten salts (almost exclusively alkali halides), and solids (almost exclusively sodium-ion conductors). Nearly all of the cells with molten-salt electrolytes use lithium as the reactant at the negative electrode (usually as an alloy) because other candidate reactants are relatively soluble in their molten salts. All of the cells with solid electrolytes use sodium as the negative electrode reactant because the only solid electrolytes of adequate conductance for a low electronegativity metal conduct only sodium ions. The positive electrode reactants are elements of high electronegativity and low equivalent weight, or compounds containing them. These conditions result in the focusing of effort on the following systems: lithium/alkali halide/metal sulfide (or sulfur), sodium/solid electrolyte/sulfur, sodium/solid electrolyte/metal halide, lithium-aluminum/alkali halide/carbon-TeCl$_4$, and lithium/alkali halide/chlorine.

In the sections that follow, the systems just listed will be discussed, with emphasis on current status (based on the latest publicly available information) and problems remaining to be solved.

LITHIUM/METAL SULFIDE CELLS

The current efforts on lithium/metal sulfide cells evolved from earlier work on lithium/sulfur cells,[1-3] which experienced gradual loss of sulfur from the positive electrode and a corresponding decline in capacity. The use of a sulfur compound such as FeS_2, FeS, or Cu_2S reduces the solubility of sulfur-bearing species in the electrolyte, and provides for greatly improved stability of operation without capacity loss,[4-6] at the expense of cell voltage (tenths of a volt) and lower specific energy. (Compare the values in Table I to 2600 W·h/kg for Li/S.)

Table I. PERFORMANCE SUMMARY : HIGH TEMPERATURE BATTERIES

System	Theoretical Specific Energy W·h/kg	Operating Temperature °C	Small Cell Tests (<20 A·h)					Large Cell or Battery Tests (>20 A·h)			
			Capacity Density & Current Density A·h/cm²	A/cm²	Peak Power Density W/cm²	Cycle Life	Lifetime h	Specific Energy W·h/kg†	Specific Power W/kg†	Cycle Life	Lifetime h
Li/LiCl-KCl/FeS$_2$	1321	400-450	0.4	0.4	1.4	92	800	-	-	115	617
Li/LiCl-KCl/FeS	869	400-450	-	-	-	-	-	-	-	-	-
Li-Al/LiCl-KCl/FeS$_2$	650	400-450	0.65	0.64	(1.0)	300	6400	80-150	8-80	300	6400
Li-Al/LiCl-KCl/FeS	458	400-450	0.75	0.064	0.8	(300)	(5000)	70-85	8-150	121	3700
Li$_4$Si/LiCl-KCl/FeS$_2$	944	400-450	-	-	-	-	-	-	-	-	-
Li$_4$Si/LiCl-KCl/FeS	637	400-450	-	-	-	-	-	-	-	-	-
Na/β-Al$_2$O$_3$/S *	758	300-400	1.7	0.16	0.3	-	-	77	154	N.A.	N.A.
Na/β-Al$_2$O$_3$/S **	521	300-400	-	-	-	-	-	-	-	-	-
Na/β-Al$_2$O$_3$/Na$_2$S$_{5.2}$ ***	308	300-400	-	-	-	8500	10000 (1000 A·h/cm²)	-	-	-	-
Na/β-Al$_2$O$_3$/NaAlCl$_4$-M$_x$Cl$_y$	792-1034	210	0.4	0.015	0.375	>200	5000 (>60 A·h/cm²)	-	-	-	-
Li-Al/LiCl-KCl/C-TeCl$_4$	N.A.	400	-	-	-	200	-	60-80	468	100	288
Li/LiCl-KCl-LiF/Cl$_2$	2167	450	0.33	1.05	>2.8	325	650	277††	230††	210	668

* Reaction to Na$_2$S$_3$ Excluding weight of thermal insulation

** Reaction to Na$_2$S$_{5.2}$ †† Excluding weight of case, insulation, and Cl$_2$ storage

*** Reaction in the single phase region Na$_2$S$_{5.2}$ → Na$_2$S$_3$

Introduction of a metal sulfide in place of sulfur modifies the chemistry of the cell reactions. For the Li/LiCl–KCl/FeS_2 cell, the predominant discharge reactions and corresponding potentials (*vs.* Li) are:

$$4Li + 3FeS_2 \rightarrow Li_4Fe_2S_5 + FeS \quad 2.1\ V \qquad (1)$$

$$2Li + Li_4Fe_2S_5 + FeS \rightarrow 3Li_2FeS_2 \quad 1.9\ V \qquad (2)$$

$$6Li + 3Li_2FeS_2 \rightarrow 6Li_2S + 3Fe. \quad 1.6\ V \qquad (3)$$

The theoretical specific energy for these cell reactions is 1321 W·h/kg, about half that for the Li/S cell, largely because of the weight of the Fe.

Each of the above reactions is associated with a plateau in the voltage-capacity curve for the cell, as shown in Fig. 1. Reaction 1 is associated with the upper plateau, Reaction 2 with the small second plateau, and Reaction 3 with the longer plateau near 1.6 V. All of the compounds in Reactions 1–3 have been identified in the electrodes of cells that had been cycled and then examined at the appropriate state of charge.[7,8] It is difficult to return to the fully charged state in which only FeS_2 is present, but little loss of capacity is evident. Even though the reactions above appear to be rather complex, FeS_2 electrodes have demonstrated stable operation for extended time periods.[9]

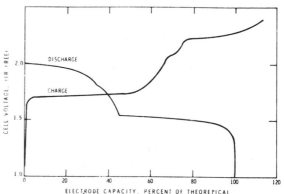

Fig. 1.   Voltage-capacity curve for a Li/FeS_2 cell.

Another cell under investigation makes use of FeS in the positive electrode. The most commonly used electrolyte with FeS is the LiCl–KCl eutectic. For the Li/LiCl–KCl/FeS cell, the reactions appear to be:

1st Discharge $\qquad\qquad 2Li + FeS \rightarrow Li_2S + Fe \qquad\qquad\qquad (4)$

Recharge $\left\{ \begin{array}{l} 8Li_2S + 8Fe + 2KCl \rightarrow K_2Fe_7S_8 + Fe(?) + 2LiCl + 14Li \quad (5) \\ K_2Fe_7S_8 + Fe(?) + 2LiCl \rightarrow 8FeS + 2KCl + 2Li \qquad\qquad (6) \end{array} \right.$

The leftover iron in Reaction 5 has not been found conclusively, but has been included in the reactions to preserve the 1:1 ratio of iron and sulfur. The compound $K_2Fe_7S_8$* has been identified in large amount in electrodes from cycled cells.[8,9] Reaction 6 only proceeds with some difficulty, and $K_2Fe_7S_8$ probably is the major reactant at the positive electrode of the Li/LiCl–KCl/ FeS cell after the initial discharge. The voltage-capacity curve for this cell is characterized by a single, flat plateau at 1.5–1.6 V. It is interesting that $K_2Fe_7S_8$ is not an important phase in the FeS_2 electrode, and that Li_2FeS_2 is not important in the FeS electrode (with LiCl–KCl

---

* The exact formula for this compound is in doubt, but this is the best available.

electrolyte). The theoretical specific energy of the Li/FeS cell is
869 W·h/kg. Improvements in stability of operation can also be obtained with
other metal sulfides, such as Cu_2S,[6] NiS,[8] CoS,[8] and $CuFeS_2$,[8] but the theoretical specific energies for these Li/MS cells are lower than for Li/FeS_2
cells, and these materials are more expensive than FeS_2 or FeS.

Liquid lithium electrodes have not been as stable as desired, because of
capacity loss related to both physical and chemical losses of lithium from the
electrode.[10] The physical losses are caused by a lack of sufficient wetting
of the current collector during recharge, and inadequate wicking of the deposited lithium into the current collector. Various additives to the lithium
have been evaluated in an attempt to improve the wetting and wicking properties of the lithium electrode. Chemical losses of lithium from open cells
with LiCl-KCl electrolyte have been experienced because of the displacement
reaction

$$Li + KCl \rightarrow K + LiCl \qquad\qquad (7)$$

and the evaporation of potassium. The rates of both of the above losses have
been reduced significantly by the proper choices of current collector materials (nickel, stainless steel, low carbon steel), additives (e.g., copper,
zinc), and the use of sealed cells or potassium-free electrolytes (to avoid
the potassium-loss mechanism). Even so, further work is necessary before
liquid lithium electrodes are acceptable for stable, long-lived cells.

As an alternative to the liquid lithium electrode, the solid lithium-
aluminum alloy electrode has been investigated, and shows good stability, at
the cost of a lower cell voltage (by about 0.3 V over a large composition
range, ∿7 a/o to 45 a/o Li) and a greater weight (about 80 w/o of the fully
charged electrode is aluminum).[11,12] Compare the theoretical specific energies of corresponding cells in Table I with lithium and lithium-aluminum
electrodes, which show a loss of 50%. In addition, some of the lithium in
the lithium-aluminum alloy is not available at reasonable current densities
(20-30% unavailable at 0.1 A/cm^2) in contrast to the liquid lithium electrode
which exhibits essentially 100% utilization. The best operation of these
electrodes is found at an electrolyte volume fraction of 0.2 in the electrode.

Solid lithium-silicon is also under investigation as a negative electrode,[13] and at a composition of Li_4Si, shows a significant weight advantage
over LiAl, as indicated by the theoretical specific energy values in Table I.
Since silicon is not as good an electronic conductor as aluminum, a more
elaborate current collector probably will be required for Li_4Si. Voltage
plateaus were found at 48, 158, 280, and 336 mV $vs.$ Li at 400°C, in locations
consistent with the existence of compounds of the following stoichiometries:
Li_5Si, $Li_{4.1}Si$, $Li_{2.8}Si$, and Li_2Si. No information was presented concerning
the ability of the electrode to support current. Lithium-boron alloys have
also been investigated,[9,14] and are able to support high discharge current
densities (up to 8 A/cm^2) at 500°C in LiCl-KCl. Performance is much poorer at
lower temperatures. Little information is available on recharge characteristics or cycle life,[9] but it appears that it is difficult to remove lithium
from the composition LiB_2.

The scaleup and engineering efforts on lithium-aluminum/iron sulfide
cells have progressed to the point of performance and cycle-life measurements
on lightweight cells of about 100 A·h capacity,[8,9,11] of the design shown in

Fig. 2. These cells have typically been operated at current densities in the range 0.04 to 0.25 A/cm^2, corresponding to complete discharge in 4 to 24 h, which is in the range of interest for off-peak energy storage. The cell weights have been near 1 kg, corresponding to a specific energy of 100 to 150 W·h/kg (see Table I and Fig. 2). Repeated cycling of these cells has shown the capacity retention to be good. Lifetimes of more than 3000 h have been demonstrated for these lightweight Li-Al/LiCl-KCl/FeS$_2$ cells.[9] In tests involving heavier cells with larger amounts of electrolyte, lifetimes as long as 6400 h have been achieved.[12] Prismatic cells, of vertical orientation with similar capacities (∿100 A·h) are also being developed.

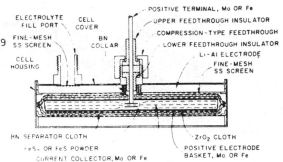

Fig. 2. Lightweight Li-Al/MS cell design. Cell diameter ∿13 cm, weight 1-1.7 kg.[9]

The factors which are serving to limit the performance of Li-Al/FeS and Li-Al/FeS$_2$ cells include swelling of the positive electrode (especially of the FeS cells) which accompanies discharge of the cell. The swelling of the FeS electrode has been associated with the formation of K$_2$Fe$_7$S$_8$. Prevention of the formation of this phase by elimination of potassium from the electrolyte should reduce the amount of swelling.

Incomplete utilization of the FeS or FeS$_2$ has been a problem, and is at least partially caused by inadequate current collection within the positive electrode. The use of CoS$_2$ (or CoS) improves the electronic conductivity of the active material,[15] reducing the current collector requirement so that 40-60% utilization can be obtained at 0.1 A/cm^2, and 415°C. It has also been found that CuS added to FeS (in amounts near 40 w/o) improves the utilization of the active material,[9] and probably reduces the extent of formation of K$_2$Fe$_7$S$_8$, (and perhaps the extent of swelling as well). Further improvement in utilization of the active material is necessary especially at the higher current densities required for automobile propulsion. Values near 70% utilization at 0.4 A/cm^2 and 450°C are desired for this application.

Inexpensive, corrosion-resistant current collectors for FeS and FeS$_2$ electrodes are needed. Various forms of carbon have been used with some success, but carbon has only a marginally-acceptable electronic conductivity, poor strength, and is not easy to join to other materials. Tungsten and molybdenum have also been used, but they are heavy and expensive. A large number of candidate materials have been evaluated for possible use in Li/S and Li/MS cells, with only a few showing reasonable corrosion resistance, as indicated in Table II.[8,16]

The development of a lightweight, corrosion-resistant feedthrough for sealed cell operation requires an electronic insulator which will remain stable at potentials near that of lithium. Many ceramics have been rejected because they are readily attacked by lithium. The most promising materials currently being investigated are high-purity boron nitride and high-purity aluminum nitride. High-purity BeO is also good but poses a health question. Yttria and some double metal oxides may prove to be acceptable (e.g., CaZrO$_3$,

Table II.   COMPATIBILITY OF PROSPECTIVE POSITIVE ELECTRODE MATERIALS
IN LITHIUM/IRON SULFIDE CELLS

Material	Positive Electrode	Results of Corrosion Tests or Cell Tests
Mo	FeS_2, FeS	Little or no attack of properly prepared material
W	FeS_2	
C, Graphite	FeS_2, FeS	
TiN } Coating on Fe	FeS_2	
FeB }		
Fe	FeS	Moderate attack or dissolution
Ni	FeS	
Fe	FeS_2	Severe attack or dissolution
Ni	FeS_2	
Cu	FeS	
Nichrome	FeS_2	
Nb	FeS_2	

$MgAl_2O_4$), but more work remains in this area. Bonding techniques are required
for the most compact feedthroughs. The conductor of the feedthrough which operates at positive electrode potential must resist oxidation at these high
potentials, and must be joined with low resistance to the electrode. The most
popular materials for this use are molybdenum and tungsten. Molybdenum is attacked slowly, and both metals are heavy. At present, mechanical feedthroughs
are commonly used, but they leave something to be desired in terms of size,
weight, and leak rate.

Next in importance to the improvements in the positive electrodes indicated above are better separators. Boron nitride cloth, about 2 mm thick, together with zirconia cloth 1 mm thick has been used most successfully as a
separator, preventing contact between the positive and negative electrodes,
and helping to retain particles of active material. Thinner, less expensive,
and highly corrosion resistant materials are needed for this purpose. Thinner,
nonwoven, high purity boron nitride may serve well, if developed. The separator should have very small pores (a few micrometers or less) in order to prevent movement of fine particles of lithium-aluminum, iron, and other solids
from the electrodes. It must be thinner (perhaps 1 mm thick) in order to reduce the internal resistance of the cell, and should not be sensitive to air
or moisture (for ease in handling and cell assembly). Yttria may be another
good candidate. It has shown good stability in preliminary tests, is thermodynamically stable, but is not yet available as a strong, flexible, high-
purity cloth or mat.

As a convenience in cell assembly (and as a cost-saving measure), it
would be advantageous to develop a simple means for assembling cells in the
discharged state (using Fe and Li_2S in the positive electrode) avoiding the
need for metallic lithium and the handling of it. Some experiments have been
performed on the assembly and startup of discharged cells, with promising
results.

The outlook for the continued development of cells with lithium alloy negative electrodes and iron sulfide positive electrodes is good. It is likely that within the next few years, cells can be developed having specific energies approaching 200 W·h/kg at a specific power of 100 W/kg with a life of over 5000 h and 1000 cycles using FeS_2 electrodes, and 140 W·h/kg at a specific power of 60 W/kg and a life of over 5000 h and 1000 cycles using FeS electrodes. The cost will probably continue to be high (compared to $20/kW h) until inexpensive feedthroughs, current collectors, and separators are available. Figure 3 shows a conceptual design for a 43 kW·h, 270 kg Li-Al/FeS_2 battery for electric vehicle propulsion.

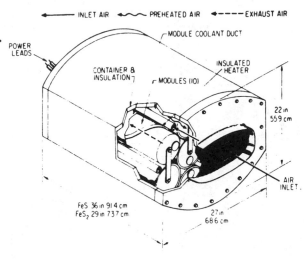

Fig. 3. Conceptual design of an electric vehicle battery using Li-Al/MS cells.[17]

SODIUM/SULFUR CELLS

In 1967, the Ford Motor Company[18] revealed their work on the development of a sodium/sulfur cell with a ceramic, sodium-ion-conducting electrolyte. This cell is appealing because it is very simple in concept: molten sodium separated from molten sulfur by a ceramic electrolyte, penetrable only by sodium ions, as shown in Fig. 4.[19] The cell operates at 300–400°C, sodium serves as its own current collector and a carbon felt serves as the current collector for the sulfur electrode.

The overall electrode and cell reactions are rather simple:

$$2Na \rightarrow 2Na^+ + 2e^- \quad anode \quad (8)$$
$$2Na^+ + 3S + 2e^- \rightarrow Na_2S_3 \quad cathode \quad (9)$$

$$2Na + 3S \rightarrow Na_2S_3 \quad overall \quad (10)$$

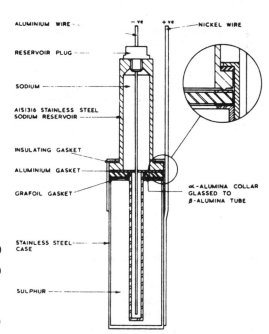

Fig. 4. Conceptual design of a tubular Na/S cell.[19]

The Na_2S_3 is not a compound, but merely the stoichiometry at which the compound Na_2S_2 begins to precipitate from the sodium polysulfide melt at operating temperature. The phase diagram of the Na-S system shows that at 350°C, as the sulfur electrode receives sodium during the discharge reaction, a separate liquid phase, $Na_2S_{5.2}$ forms, causing a voltage plateau at 2.07-2.08 V vs. Na, extending from essentially pure sulfur to $Na_2S_{5.2}$. As more Na is added to $Na_2S_{5.2}$, the voltage declines through a single-phase region, until the composition Na_2S_3 is reached, at which point Na_2S_2 begins to precipitate. This is the end of discharge.

The details of the sulfur electrode reactions are rather complex because two sulfur-containing phases and many sulfur-containing species are involved in both chemical and electrochemical reactions. The present state of knowledge of the reactions is as follows.[20] The major species in the polysulfide melt are believed to be $S_4^=$, $S_5^=$, and $S_2^=$ (and not $S_3^=$ or any singly-charged sulfur species), and of course Na^+. The electroactive species are primarily $S_4^=$, $S_5^=$, and $S_2^=$. The overall discharge of a Na/S cell starts with essentially pure sulfur, but very soon a separate polysulfide phase forms, which is believed to be the seat of the electrochemical reactions, and a number of chemical equilibria. The sulfur-rich phase is believed to be involved via reaction with the polysulfide phase. The discharge reactions in the two-phase region ($Na_2S_{5.2}$ and sulfur) are:

Electrochemical reduction of Na_2S_5,[20] via a two-step reaction

$$^-S_3 - S_2^- + e^- \rightarrow {}^-S_3 \cdots S_2^= \tag{11}$$

$$^-S_3 \cdots S_2^= + e^- \rightarrow S_3^= + S_2^= \tag{12}$$

Equilibria among the polysulfides and disproportionation of $S_3^=$ (nonelectrochemical)

$$S_2^= + 2S_5^= \rightleftarrows 3S_4^= \tag{13}$$

$$2S_3^= \rightarrow S_4^= + S_2^= \tag{14}$$

The sulfur phase reacts with the polysulfide phase, and is consumed in the process by such reactions as:

$$2S_2^= + S_4 \rightarrow 2S_4^= \tag{15}$$

$$2S_2^= + S_6 \rightarrow 2S_5^= \tag{16}$$

$$2S_4^= + S_2 \rightarrow 2S_5^= \tag{17}$$

The potential of the sulfur electrode remains constant at about 2.075 V vs. sodium across this two-phase region of the phase diagram.

After all of the sulfur phase is consumed, the electrochemical reduction of polysulfides continues (with declining potential to about 1.75 V) through the single-phase region extending from $Na_2S_{5.2}$ to Na_2S_3 at about 350°C, according to Reactions 11 and 12 above, and the analogous two-step reduction of $S_4^=$:[20]

$$^-S_2 - S_2^- + e^- \rightarrow {}^-S_2 \cdots S_2^= \tag{18}$$

$$^-S_2 \cdots S_2^= + e^- \rightarrow 2S_2^= \tag{19}$$

Reactions 11, 12, 18, and 19 are very fast, with exchange current densities in the 1 A/cm^2 range. When the overall stoichiometry of the polysulfide melt reaches Na$_2$S$_3$, Na$_2$S$_2$ precipitates, blocking further access of melt to the electrolyte-current collector interface. Normally the discharge process is halted before this occurs.

The recharge process is believed to take place as follows, with the two-electron oxidation of S$_4^=$:

$$S_4^= \rightarrow S_4 + 2e^- \tag{20}$$

followed by the reaction of sulfur with the polysulfide melt:

$$S_4 + 2S_4^= \rightarrow 2S_6^= \tag{21}$$

$$\text{or} \quad S_4 + 4S_4^= \rightarrow 4S_5^= \tag{22}$$

When the solubility limit for sulfur in polysulfide is exceeded, a separate sulfur phase is formed. At high current densities, this sulfur phase, being a poor electronic and ionic conductor, blocks further reaction as it is formed on the electrolyte and current collector. One problem in the design and operation of sulfur electrodes is to prevent this blockage by sulfur until the cell is nearly fully recharged. The voltage-capacity curves of Fig. 5 exhibit the behavior just discussed for the end of charge and end of discharge.[21]

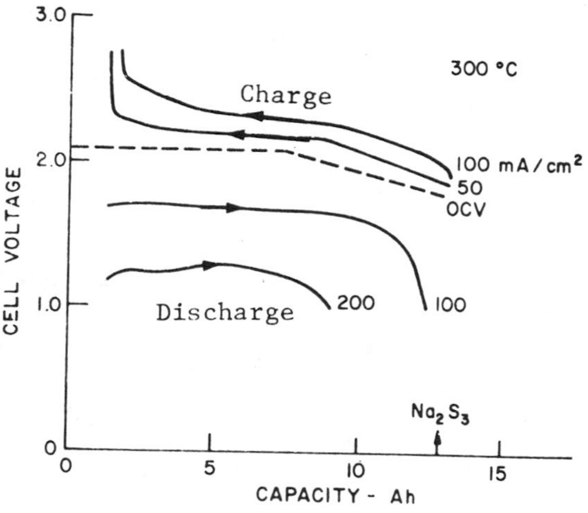

Fig. 5. Voltage-capacity curve for a Na/S cell at 300°C.[21]

The electrolytes which are used in sodium/sulfur cells are ceramics[18-22] or glasses[23,24] which conduct sodium ions. The glass electrolytes have rather low conductivity ($\sim 5 \times 10^{-4} \Omega^{-1} cm^{-1}$), and must therefore be very thin ($\sim 10 \mu m$) in order to yield a cell with an acceptably low internal resistance. In order to have an electrolyte geometry compatible with a 10 μm thick electrolyte, hollow fibers ($\sim 50 \mu m$, OD) are used (Fig. 6). Thousands of borate glass fibers are bundled together with one end bonded to a low-melting B$_2$O$_3$-Na$_2$O glass "tube sheet," and the other end sealed. Sodium is fed to the insides of the hollow fibers from one side of the tube sheet. Among the fibers on their outsides are sulfur and a metal foil current collector.[23] The glass fibers are operated at a current density of 2 mA/cm^2, and have experienced failure by cracking after a number of charge-discharge cycles, and failure by cracking near the joint with the tube sheet. Thicker fiber walls favor longer cell lives. Recently,[23] glass fiber cells of the 1000-fiber, 0.4·A h size have exhibited lifetimes of up to 3300 h, and cycle lives of 1600 cycles at 10-25% depth of discharge (Na$_2$S$_3$ is 100%). Shorter lives are experienced at greater depths of discharge.

The ceramic electrolytes for sodium/ sulfur are composed of Na_2O and Al_2O_3 in varying ratios in the range $Na_2O \cdot 5Al_2O_3$ to $Na_2O \cdot 11Al_2O_3$, usually with small amounts of other oxides such as Li_2O or MgO to stabilize the β'' structure[19,21] which is more conductive (3-5 $\Omega \cdot$ cm at 300°C) than the β structure[22,25] (20-30 $\Omega \cdot$ cm).

The β- and β''-alumina electrolytes (tubular, 1-3 cm OD, 1-2 mm wall, 10-30 cm long) can be prepared by a number of combinations of processing steps, as indicated by Table III. The overall process can be divided into three main steps: powder preparation, green body formation, and sintering. There are a number of options for each of the three main steps, however

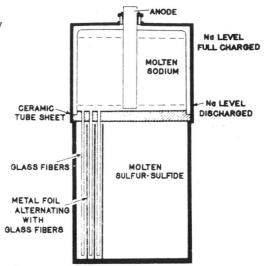

Fig. 6.   Design of a hollow-glass-fiber Na/S cell.[23]

Table III. β- AND β''-Al_2O_3 FABRICATION METHODS

Powder Preparation	I	Direct mixing of compounds
	II	Decomposition of salt(s)
	III	Gel
	IV	Spray drying
	V	Complete reaction to β- or β''-Al_2O_3
Green Body Formation	A	Isostatic pressing
	B	Extrusion
	C	Electrophoretic deposition
Sintering Method	1	Encapsulated (in Pt or inert material)
	2	Enclosed with powder (high-temperature sintering)
	3	Zone pass-through

certain combinations of steps have been more popular than others:

IA1   has been used at Ford,[21] British Rail,[19] and others for preparation of β''-Al_2O_3

VC2   has been used by Compagnie Generale d´Electricite[25] and by General Electric[26] for β-Al_2O_3

IA3   has been used by Chloride Silent Power, and is being investigated in the Ford program[21] for β''-Al_2O_3.

Powder preparation by methods II and III is being evaluated in the Ford program, as well as green body formation by method B, extrusion. Each major step

195

in electrolyte preparation has associated with it a number of important variables that have an influence on the chemical, microstructural and mechanical properties of the final product.

It is beyond the scope of this paper to review the details of β-Al_2O_3 electrolyte preparation, but some information on the major points is appropriate.

Powder Preparation: It is important to have fine-structured, uniformly mixed reactants. This is the objective of methods I-IV of Table III. The compositions favored[21] are in the range 9% Na_2O, 0.8% Li_2O, balance Al_2O_3, from α-Al_2O_3, Na_2CO_3 and $LiNO_3$ starting materials.

Green Body Formation: The emphasis here is to prepare green bodies of optimum density with well-controlled tolerances and good uniformity for subsequent sintering to high density (>96% of theoretical).

Sintering: This may be the most difficult step in the process requiring minimum loss of Na_2O by volatilization at reaction-sintering temperatures in the range 1520-1600°C for a precise time (in the range 1 to 10 min), followed by annealing at a lower temperature ($\sim$1400°C for $\sim$8 h) to complete the formation of the β''-Al_2O_3 phase with good conduction '($\sim$5 $\Omega \cdot$cm), but grain growth (beyond $\sim$25 μm) is to be avoided for good strength ($\sim$1500 kg/cm^2).

Beta-alumina electrolytes have been operated in cells of various shapes and sizes, the most popular being tubular, with electrolytes about 1 cm diameter, and 1-2 mm thick. Various modes of failure of the electrolyte have been observed, the most common being sodium penetration from the sodium side toward the sulfur side, forming "fingers" of sodium which penetrate the electrolyte, eventually causing cracking. At the sulfur side, contamination of the beta alumina by potassium has been troublesome, causing cracking and flaking, as has been the formation of a thin coating of silica on the electrolyte, blocking the electrode reaction. Improvement of the purity and corrosion resistance of the sulfur electrode and housing extends the cell life considerably, as does improvement of the density and strength of the electrolyte.

Cells with two sodium electrodes have demonstrated lifetimes of more than 1000 A·h of charge passed per cm^2 of electrolyte.[21] Carefully prepared Na/S laboratory cells (1-2 A·h size) containing no metals (avoiding iron, manganese, chromium, and silicon contamination) have sustained 600-1000 A·h of charge passed per cm^2 corresponding to 4000-8000 cycles without any loss of capacity, as shown in Fig. 7.[21] However, these cells have been operating only in the single-phase (Na_2S_5 to Na_2S_3) region. This is impractical because it corresponds to a theoretical specific energy of only about 300 W·h/kg, which would probably result in a well-engineered cell of only 60-70 W·h/kg. Some cells with stainless steel containers have shown long lifetimes, but have suffered capacity loss because of silica coating of the electrolyte, and blockage of mass transport in the sulfur electrode because of the formation of solid lumps of iron, manganese, and chromium sulfides. Operation over the composition range S to Na_2S_3 (or nearly so) corresponding to a more attractive theoretical specific energy of 756 W·h/kg has been difficult, especially during recharge, because of problems with segregation of polysulfide and formation of sulfur films on the electrolyte, resulting in capacity loss. Even with these problems, hundreds of cycles have been obtained (Table I), but with declining capacity. Recent work with specially-shaped graphite current collectors for the sulfur electrode[21] has resulted in higher capacity densities at reasonably high current densities because of the promotion of favorable mass-transport conditions, avoiding the buildup of the products of electrochemical reaction at the electrolyte surface, and providing for the transport of reactant to the surface. Other interesting results include the use of metallic current collectors which are well wetted by polysulfide and not well wetted by sulfur, to promote more complete recharge.[21]

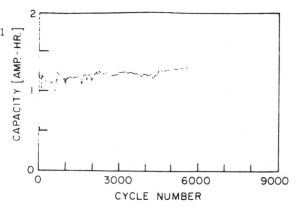

Fig. 7. Capacity-cycle number relationship for a long-lived, metal-free positive electrode $Na/Na_2S_{5.2}$ cell.[21] Discharge current density 250 mA/cm^2, charge current density 125 mA/cm^2.

A few batteries of Na/S cells have been built and operated for relatively short periods of time, demonstrating in small sizes (15 A·h, 11 V, 165 W·h) a specific energy of 77 W·h/kg and a specific power of 154 W/kg exclusive of insulation, etc. A larger battery of 1000 cells[27] yielded 50 kW·h and a power of 20 kW, but with a high total system weight, so that the specific energy and specific power were not very high.[28]

The problems receiving emphasis for the Na/S cell are rather similar to those for the Li/MS cell. Corrosion-resistant metallic materials for use in contact with the sulfur electrode are needed for use as containers and possibly current collectors. A large number of materials have been tested, as shown in Table IV, with only graphite surviving well so far. A number of

Table IV. POSITIVE ELECTRODE MATERIALS FOR Na/S CELLS

Material	Failure Mode	Reference
Tested in Cells		
Stainless Steels	Mn Penetrates electrolyte	21
	Cr \} Blockage of ceramic and/or Fe / current collector	
	Si Oxide coats electrolyte	
Carbon and Graphite	Stable - some interaction with corrosion products.	21
Aluminum (refractory metal coated)	Erratic results	29
Aluminum-Magnesium Alloy (carbon coated)	N.A. (2-year life)	23
Corrosion Resistant in Static Tests		
Cr_2O_3, oxide coated AISI 446 stainless steel, MoS_2, ZrO_2, $La_{0.84}Sr_{0.16}CrO_3$, TiO_2 (single crystal) + Ta_2O_5, $SrTiO_3$, $CaTiO_3$ + 3% Fe_2O_3, polyphenylene thermosetting resins.		21
Unstable		
TaB_2, ZrC, VN, NbB_2, $ZrSi_2$, $TiSi_2$, CrB_2, ZrN, CrC, $CaTiO_3$ + 0.3% Fe_2O_3, TiC		21
ZrB_2 $CrSi_2$ \} with oxide film		
AISI 446 stainless steel		

other candidate materials have been identified in static corrosion tests, and await testing in cells. Chromium oxide is an interesting electronic conductor which might be useful as a coating on a stainless steel cell case. Other problems include the need for corrosion-resistant seals and feedthroughs, and improved joining techniques to seal beta alumina to alpha alumina. More work is needed on the operation of cells across the full composition range S to Na_2S_3 at reasonable rates (>0.1 A/cm^2) without capacity loss for at least 1000 cycles. Of course, inexpensive fabrication procedures for the electrolyte are a necessity. As more work is carried out on battery design and operation, cell charge balancing techniques must be worked out, and thermal control methods must be perfected. These latter two areas are important for all high temperature cells.

SODIUM/METAL CHLORIDE CELLS

Another cell which has recently been investigated[30,31] that makes use of a beta alumina electrolyte is the $Na/\beta-Al_2O_3/M_xCl_y$ in $NaCl-AlCl_3$ cell, which operates at temperatures near 200°C. During discharge, sodium is transferred to the M_xCl_y compartment which contains $SbCl_3$, $CuCl_2$, $FeCl_3$ or $NiCl_3$ as the reactant. Cells have been assembled using either disk- or tube-shaped electrolyte. Early tests have been confined to relatively small (<10 A·h) laboratory cells of low specific energy. Operating current densities of 20 mA/cm^2 are typical.

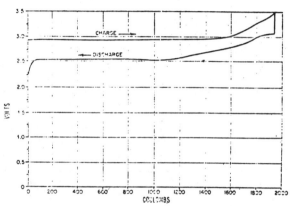

Little information is available on cell performance, other than curves such as those of Fig. 8, which shows a voltage-capacity curve for operation at a constant current of 30 mA (the 18-h rate) for a 0.56 A·h cell.[30] Disk electrolyte

Fig. 8. Voltage-capacity curve for a $Na/\beta-Al_2O_3/M_xCl_y$ cell.[30]

cells have achieved 5000 h and 200 cycles of operation whereas tubular electrolyte cells, failing by electrolyte penetration by sodium or by cracking have not yet demonstrated long life.[31] Information on voltage *vs.* current density has not been publicly available. At this workshop, additional information is expected to be provided.

LITHIUM-ALUMINUM/C-TeCl₄ CELL

Investigation and development of the $Li-Al/LiCl-KCl/C-TeCl_4$ cell operating at 475°C have been performed over about the last decade by Standard Oil (Ohio)[32] and more recently by ESB.[33] The negative electrode is 45 a/o Li-Al alloy, as discussed in the Lithium/Metal Sulfide Cells section of this paper. The electrolyte is the LiCl-KCl eutectic in a boron nitride cloth separator, and the positive electrode is a special porous carbon of high specific area, containing $TeCl_4$. The high-area carbon provides for the adsorption of chloride and alkali metal ions:

$$-\overset{|}{\underset{|}{C}} - Cl + e^- \rightarrow -\overset{|}{\underset{|}{C}}\cdot + Cl^- \tag{23}$$

$$-\overset{|}{\underset{|}{C}}\cdot + M^+ + e^- \rightarrow -\overset{|}{\underset{|}{C}} - M \tag{24}$$

The $TeCl_4$ adds to the capacity of the positive electrode by means of the following reactions:

$$TeCl_x + xLi^+ + xe^- \rightarrow xLiCl + Te \tag{25}$$

$$Te + 2Li^+ + 2e^- \rightarrow Li_2Te \tag{26}$$

The $TeCl_4$ addition yields a flatter voltage plateau at about 2.7 V than is obtained from Reactions 23 and 24. Lightweight, sealed cells having two square, 58 cm negative electrodes between which is a two-sided positive electrode of equal area are now being constructed and tested. A typical voltage vs. capacity curve is shown in Fig. 9.

Single cells have yielded specific energies as high as 79 W·h/kg. A 12-cell experimental battery weighing 5.4 kg delivered 264 W·h (49 W·h/kg) at the 2-h rate, and operated for 100 cycles over a period of about 300 h.[32] Currently, single cells are being constructed and tested after a pause in the program.

The problem areas for this cell are similar to those for other high-temperature cells: seals, feed-throughs, and materials, especially for the current collector of the positive electrode (where tungsten and graphite are now being used). An inherent difficulty for this system, because of the low capacity per unit weight of the positive electrode, is the low specific energy (45-60 W·h/kg); however, the high specific power capability is adequate (450 W/kg). It is expected that batteries of these cells will be constructed for testing in fork lift trucks in the future.[33]

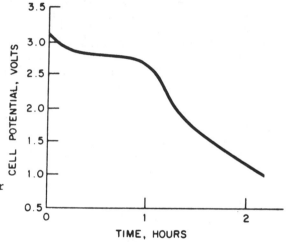

Fig. 9.   Voltage-capacity curve for a Li-Al/C-TeCl$_4$ cell.[32] Discharge current: 6 A, area $\sim$150 cm^2.

LITHIUM/CHLORINE CELLS

Lithium/chlorine cells have been under investigation for over a decade, both as primary and as rechargeable cells. They make use of pure lithium held by capillary forces in a metallic current collector as the negative electrode, a molten LiCl or alkali halide mixture electrolyte, and a porous graphite chlorine electrode. Operating temperatures have been near 650°C, but have been reduced recently to 450°C.[34] The mixed alkali halide electrolyte which has permitted the temperature reduction is 19 m/o LiF-66 m/o LiCl-15 m/o KCl.

The electrode reactions are very simple:

$$Li(\ell) \rightarrow Li^+ + e^-$$
$$\frac{1}{2}Cl_2(g) + e^- \rightarrow Cl^-$$

$$Li(\ell) + \frac{1}{2}Cl_2(g) \rightarrow LiCl\textit{(soln)}.$$

The cell potential is constant at about 3.6 V for the full capacity, in contrast to the situation for all other cells of this paper. The use of chlorine gas necessitates chlorine storage, which may be accomplished by adsorption on carbon.

A recent cell is shown in Fig. 10. Cells of this type have operated for periods up to 668 h, and 210 cycles at a capacity density of 0.34 A·h/cm^2, corresponding to an impressive 277 W·h/kg, counting only the weight of the cell itself, without chlorine storage, insulation, etc.

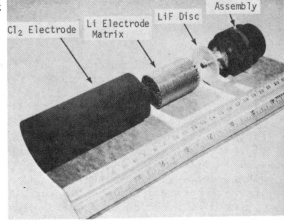

Fig. 10. Stackable Li/Cl$_2$ cell parts including a hot-pressed LiF seal.[34]

The problem areas for the lithium/chlorine cell are largely materials-related, because of the corrosive nature of the chlorine. Reliable seals and conductors are needed, as well as a simple, compact, lightweight means for chlorine storage and handling. The chlorine electrode gradually floods with electrolyte, a life-limiting factor. Control of this problem is essential.

SUMMARY AND PROJECTIONS

The status of high-temperature cell performance and lifetime is summarized in the right-hand portion of Table I. The systems which have shown the most rapid progress are those receiving the most effort: sodium/sulfur, and lithium alloy/metal sulfide. Both cells show promise of achieving 200 W·h/kg in the not-too-distant future. Some difficulty will probably be experienced in attempting to achieve that specific energy at a specific power approaching 200 W/kg. Single cells are now showing lifetimes of several thousand hours, and cycle lives of hundreds, over practical composition ranges. Figure 11 shows the specific power *vs.* specific energy curves for a number of electro-chemical cells and heat engines. The difficulties with regard to the simultaneous achievement of high specific energy and high specific power for Li-Al/FeS and Na/S cells are reflected by the leftward curvature of the solid lines in Fig. 11. The progress in the last seven or eight years has been very good. The projected curves (dashed) represent expectations for the next several years. The prospects for continued progress look bright, but during this period, the difficult materials and seals problems must be solved, and the cost problems must be squarely faced.

200

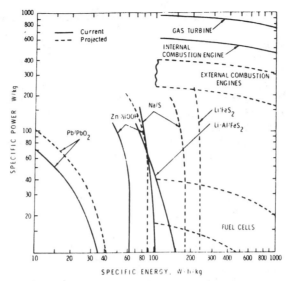

Fig. 11.   Specific energy *vs.* specific power for various systems.

REFERENCES

1. E. J. Cairns and H. Shimotake, Science, 164, 1347 (1969.
2. H. Shimotake and E. J. Cairns, Abstract 206, Electrochemical Society Extended Abstracts, Spring Meeting, New York, N. Y., May, 1969; also Extended Abstracts of the Battery Division, 5, 520 (1969).
3. E. J. Cairns, *et al.*, "Lithium/Sulfur Secondary Cells," presented at the 23rd Meeting of the Inter. Soc. of Electrochem., Stockholm, Sweden, Aug. 27-Sept. 2, 1972; also Extended Abstracts, ISE, p. 432 (1972).
4. D. R. Vissers, Z. Tomczuk, and R. K. Steunenberg, J. Electrochem. Soc., 121, 665 (1974).
5. P. A. Nelson, E. C. Gay, and W. J. Walsh, in Proc. 26th Power Sources Symposium, PSC Publications Committee, Red Bank, N. J. (1974).
6. L. R. McCoy, *et al.*, in Proc. 26th Power Sources Symposium, PSC Publications Committee, Red Bank, N. J. (1974).
7. R. K. Steunenberg, *et al.*, in P. A. Nelson, *et al.*, "High Performance Batteries for Off-Peak Energy Storage, ANL-8057, Argonne National Laboratory (November 1974), p. 21 ff.
8. R. K. Steunenberg, *et al.*, in P. A. Nelson, *et al.*, "High Performance Batteries for Off-Peak Energy Storage, Jan.-June, 1974," ANL 8109, Argonne National Laboratory (January 1975), p. 77 ff.
9. P. A. Nelson, *et al.*, "High Performance Batteries for Off-Peak Energy Storage and Electric Vehicle Propulsion, July-Dec., 1974, ANL-75-1 (July 1975), p. 103 ff.
10. R. N. Seefurth and R. A. Sharma, J. Electrochem. Soc., 122, 1049 (1975)
11. W. J. Walsh, *et al.*, in Proc. 9th IECEC, Amer. Soc. of Mech. Eng'rs, New York, N. Y. (1974), p. 911.
12. E. C. Gay, F. J. Martino, and Z. Tomczuk, in Proc. 10th IECEC, Institute of Electrical and Electronics Engineers, New York, N. Y. (1975), p. 627.
13. S. Lai and L. R. McCoy, Abstract 21, Electrochemical Society Extended Abstracts, Fall Meeting, Dallas, Texas, October 1975.

14. S. D. James and L. E. DeVries, Abstract 22, Electrochemical Society Extended Abstracts, Fall Meeting, Dallas, Texas, October 1975.

15. C. C. McPheeters, W. W. Schertz, and N. P. Yao, Abstract 25, Electrochemical Society Extended Abstracts, Fall Meeting, Dallas, Texas, October 1975.

16. N. Koura, J. Kincinas, and N. P. Yao, Abstract 30, Electrochemical Society Extended Abstracts, Fall Meeting, Dallas, Texas, October 1975.

17. W. W. Schertz, *et al.*, in Proc. 10th IECEC, Institute of Electrical and Electronics Engineers, New York, N. Y. (1975), p. 634.

18. J. T. Kummer and N. Weber, Paper No. 670179, presented at the SAE Automotive Engineering Congress, Detroit, Mich., Jan. 9-13, 1967.

19. J. L. Sudworth, in Proc. 10th IECEC, Institute of Electrical and Electronics Engineers, New York, N. Y. (1975), p. 616.

20. R. P. Tischer and F. A. Ludwig, in "Advances in Electrochemistry and Electrochemical Engineering," V. 10, C. W. Tobias and H. Gerischer, editors, in press.

21. S. A. Weiner, *et al.*, "Research on Electrodes and Electrolyte for the Ford Sodium-Sulfur Battery," Ford Motor Co. Report to NSF, Annual Report for June, 1974 to June, 1975 (July, 1975).

22. J. Fally, *et al.*, J. Electrochem. Soc., 120, 1292 (1973).

23. C. A. Levine, in Proc. 10th IECEC, Institute of Electrical and Electronics Engineers, New York, N. Y. (1975), p. 621.

24. C. Levine, Proc. 25th Power Sources Symposium, PSC Publications Committee, Red Bank, N. J. (1972).

25. J. Fally, *et al.*, J. Electrochem. Soc., 120, 1296 (1973).

26. R. W. Powers, Report No. 73CRD289, General Electric Corp., Corporate R & D, October 1973.

27. R. W. Minck, Proc. 7th IECEC, Paper No. 729009, American Chemical Society, Washington, D.C. (September 1972).

28. "New Battery Will Double Range of Electrics," Commercial Motor, Nov. 10, 1972.

29. J. B. Bush, Jr., *et al.*, "Sodium-Sulfur Battery Development for Bulk Power Storage," Electric Power Research Institute, Research Report 128-2, September, 1975.

30. J. Werth, "Alkali Metal-Metal Chloride Battery," U.S. Patent 3,877,984, April 15, 1975.

31. J. J. Werth, "Sodium Chloride Battery Development Program for Load Leveling," Electric Power Research Institute, Research Report 109-2-1, June, 1975.

32. J. E. Metcalfe, E. J. Chaney, and R. A. Rightmire, in Proc. 1971 IECEC, Society of Automotive Engineers, New York, N. Y. (1971), p. 685.

33. J. C. Schaefer, *et al.*, in Proc. 10th IECEC, Institute of Electrical and Electronics Engineers, New York, N. Y. (1975), p. 649.

34. T. G. Bradley and R. A. Sharma, in Proc. 26th Annual Power Sources Symposium, PSC Publication Committee, Red Bank, N. J. (1974), p. 60.

14

Reprinted from *Am. Electrochem. Soc. Trans.* **2**:113–121 (1902)

VOLTAIC CELLS WITH FUSED ELECTROLYTES.

By Eugene A. Byrnes.

This paper is largely based on experiments which were made three years ago, with a view to determining the source of electrical energy in voltaic cells of the type in which the soluble electrode or anode is of carbon, the electrolyte a fused oxide or salt and the cathode of iron. This question was then the subject of considerable discussion between Ostwald, Anthony, Julius Thomsen, Liebenow and Strasser, Reed and Langley. Some held that the seat of electromotive force is at the surface of contact between the carbon anode and the fused electrolyte and that the current generated is primarily due to, and in greater or less measure proportional to, oxidation of the carbon at this surface. Others considered that the action is merely thermoelectric and due to differences in temperature at the junctions of the electrodes and electrolyte, occasioned by the externally applied heat.

The results of these experiments were not published, and thinking that they might be of interest to the Electrochemical Society, the earlier work has been repeated and a more general investigation of the subject of voltaic cells with fused electrolytes entered upon.

The electrolyte used throughout was primarily sodium hydroxide, contained in an iron or other vessel and maintained in fusion by a gas furnace. The electrodes were of carbon, especially graphite, and of various metals, generally in the form of thin strips. It became desirable to provide a diaphragm or porous cup by which the cell could be divided into two compartments to receive the two electrodes.

The preparation of a diaphragm which would fulfil the necessary conditions of porosity and non-conductivity and which would at the same time resist the fluxing or solvent action of fused basic oxides, was found to be a difficult problem. The ordinary porous clay cup is rapidly dissolved in fused sodium hydroxide.

The first diaphragm tested consisted of a perforated iron cup having a lining of asbestos paper, held in place by iron wire gauze. In use, however, the asbestos was rapidly attacked. A porous cup was moulded from a composition of heavy magnesia and sodium silicate, shaped under considerable pressure in suitable metallic dies and fired at a high temperature while surrounded and filled with sand. This diaphragm was, however, converted into a pasty mass by subjecting it to the action of fused sodium hydroxide for a period of thirty minutes. Similar results followed the use of a porous cup moulded from a mixture of magnesia and coal-tar. A cup was cut out of a block of dense homogeneous quicklime, but it was unable to resist the fluxing and solvent action of fused sodium hydroxide. A diaphragm was now manufactured by taking two concentric cups of perforated sheet-iron, separated by a space of 1 cm., and filling this space with finely pulverized magnesia fire-brick which had been calcined in a steel furnace. This filling, however, became pasty in a short time. Finally an oxide was found which would resist the action of the fused caustic and it was used in granular form as a filling between the perforated iron cups, no binder which would enable a moulded cup to retain its shape being available.

The temperature of the electrolyte was at first determined by the use of mercury thermometers, reading to 360° C., and a resistance wire pyrometer. Afterwards, mercury thermometers reading to 550° C. were employed. The differences of potential at the terminals of the various cells were determined by Weston voltmeters, reading to 1/50 and 1/30 volt, a Weston millivoltmeter and a potentiometer having a resistance of 20,000 ohms, with a D'Arsonval galvanometer.

A normal electrode for the determination of single potential differences, of the type devised by Liebenow and Strasser, was also prepared. This consists of the well-known Ostwald normal electrode, in which mercury covered with mercurous chloride is in contact with a potassium chloride solution, but the tubular extension dipping into the cell in question terminates in a porous clay tube, the extension and tube being filled with the potassium chloride solution. The results of tests with this normal electrode were not satisfactory and it was not used.

In one set of experiments, both electrodes were introduced into an open vessel containing the molten sodium hydroxide ; in the other set, the molten caustic was divided into separate portions by a porous cup and various oxidizing agents or depolarizers were supplied to the portion which contained the cathode.

The electromotive force of various couples in molten caustic held at a temperature of about 360° C. were found to be as follows :

1. Soft Swedish iron—high carbon steel = 0.00 volt.
2. Bright iron—iron coated with Fe_3O_4.

The electromotive force of this couple was found to be a function of the degree of oxidation of the negative strip. The voltage, starting at a low figure, could be progressively raised to a maximum of 0.32 by lifting the oxidized strip out of the caustic several times. Lifting out the bright iron strip simultaneously with the oxidized one made little or no difference in the reading. On removing and washing both strips, sandpapering the bright one and returning both, the reading was 0.30, again rising to 0.32 on lifting out either the oxidized strip or both. While the negative strip seemed to be coated with a continuous layer of the black oxide before its first introduction into the caustic, the experiment indicates that further oxidation or higher oxidation results from its contact with air while hot.

3. Copper—copper oxide = 0.20–0.25 volt.
4. Iron—graphite = 0.44.
5. Nickel—graphite = 0.20.
6. Copper—graphite. Low, but uncertain on account of rapid oxidation of copper.
7. Aluminum—graphite = 0.80. Falling quickly and regularly to about 0.10.
8. Magnesium—graphite = 1.25. Vigorous reaction and boiling.
9. Silicon—graphite = 0.86. Foamed so as to soon render reading uncertain.
10. Chromium—graphite = 0.03.
11. Gold—graphite = 0.12.
12. Silver—graphite = nearly 0.00.
13. Platinum—graphite = 0.19.
14. Titanium—graphite = 0.32.

In the experiments which follow, a porous cup was employed, the electrolyte in both compartments of the cell being fused caustic.

15. Graphite in cup—graphite outside cup = 0.00.

16. Iron   "   " —iron   "   " = 0.00.

17. "   "   " —graphite   "   " = 0.44.

18. Various other metals, *e. g.*, nickel, chromium, silicon, magnesium, titanium, and silver, placed in the porous cup against the graphite cathode outside this cup, gave the same readings as when the cup was omitted.

19. The polarization value of sodium against graphite, due to the passage of two amperes between graphite electrodes inside and outside the porous cup and maintained at a potential difference of 4 volts for one minute, was about 2.30 volts.

20. Graphite cathode in cup—iron anode outside cup = —0.44.

21. With conditions as in preceding test, a few grams of potassium nitrate were introduced outside the porous cup. In one-half minute the voltage fell to 0.00 and began to reverse, rising in one minute to 0.35 and then to 0.385, where it held for ten minutes.

22. A few grams of manganese dioxide were then added outside the porous cup, quickly dissolving to give the green manganate. The reading then rose to 0.46, where it held. The cell was then short-circuited through an ammeter, giving a reading of several tenths, falling until the voltage was 0.30. On opening the circuit the voltage recovered in five minutes to 0.44 and in ten minutes to 0.46, the original value.

23. A strip of bright iron substituted for the iron cathode was almost immediately oxidized black and gave the same reading.

Other metals substituted for iron in the same depolarizer gave readings as follows :

24. Graphite, in cup—nickel, outside cup = 0.46.   Nickel oxidized black when removed.

25. Graphite, in cup—gold, outside cup = 0.46.

26.   "   "   " —graphite, outside cup = 0.46.

27.   "   "   " —aluminum   "   "   Reversed to —0.70, falling quickly to —0.2.

28. Graphite, in cup—iron, outside cup = 0.46.

29.   "   "   " —magnesium outside cup.   Reversed to —1.00.

Various metals in the porous cup against graphite in the nitrate and manganate depolarizer outside gave values as follows :

30. Iron—graphite = 0.80, falling in one-half minute to 0.40 ;

in one minute to 0.35 ; in five minutes to 0.32; in ten minutes to 0.30.   The iron was oxidized black when removed.

31. Nickel—graphite = 0.42, falling in one-half minute to 0.16.   Nickel oxidized when removed.

32. Aluminum—graphite = 1.30, falling in five minutes to 0.24.

33. Magnesium—graphite = 1.94, holding steadily at this figure for fifteen minutes, one-half inch being eaten off end of strip when removed.

In some experiments, anodes of amorphous carbon instead of graphite were employed, giving rather higher readings.

Various other oxidizing agents were tried as depolarizers, for example, lead dioxide, sodium dioxide, potassium bichromate, potassium chlorate, and nitric acid.   Each of these agents, when dissolved in the caustic in one compartment, caused an iron electrode in that compartment to act as cathode toward a carbon electrode placed in fused caustic in the other compartment.   Such different depolarizers, in general, gave different readings, but the tests made with them were not sufficient in number to establish any definite values.

Tests were made to determine the amount of carbon oxidized by the passage of a known current through the cell comprising a graphite anode in caustic and an iron cathode in a depolarizer. The current given by the cell itself being too small and inconstant to permit the loss of carbon by oxidation to be accurately determined, current from an outside source was passed through the cell from the graphite as anode, giving results as follows :

	Current.	Time.				Grams.
1.	9 amperes	183 minutes :	Original weight of graphite,			37.706
			Final	"	"	"	32.128
			Loss in	"	"	"	5.578
2.	1 ampere	600 minutes :	Original weight of graphite,			47.3476
			Final	"	"	"	45.3185
			Loss in	"	"	"	2.0291
3.	2 amperes	210 minutes :	Original weight of graphite,			45.3185
			Final	"	"	"	43.9518
			Loss in	"	"	"	1.3667

No. 1 corresponds to a loss for 96,540 coulombs of 5.445 grams.
No. 2   "   " " " " 96,540   "   " 5.440 "
No. 3   "   " " " " 96,540   "   " 5.242 "

118 EUGENE A. BYRNES.

The reactions which occur within the cell employing a carbon anode, porous cup, and depolarizer, may be expressed by the following equation :

$$Q = (x\ CO + y\ CO_2) + (1/2\ Na_2O + 1/2\ H_2O) - NaOH - X.$$

The second and third terms in the second number of this equation cancel, that is, the sodium hydroxide is reformed by deoxidation of the depolarizer, and the total heat energy of chemical action apparent as electricity is that due to the formation of oxides of carbon minus that absorbed by reduction of the depolarizer. Since we are concerned in the above equation with the reaction of a gram-equivalent of each ion, the amount of carbon oxidized by the passage of 96,540 coulombs, which we may assume from the values given above to be 5.4 grams, must correspond to a gram-equivalent. A gram-equivalent of carbon as a dyad is 6 grams; of carbon as a tetrad, 3 grams. If x represents the fraction of a gram-equivalent of carbon oxidized to CO_2 by the passage of 96,540 coulombs, and y the corresponding fractional portion oxidized to CO, the following relations are obvious :

$$x + y = 1$$
$$3\ x + 6\ y = 5.4$$
$$x = 0.2$$
$$4 = 0.8$$
$$3\ x = 0.60\ gram$$
$$6\ y = 4.80\ grams$$

Thus 0.60 gram of carbon is oxidized to. CO_2 and 4.80 grams to CO, if the proportions of CO and CO_2 produced in the voltaic cell are the same as those due to application of current from an external source.

This result is similar to that obtained by Coehn,[1] who found that carbon employed as an anode in dilute sulphuric acid was oxidized in part to CO and in part to CO_2, though the relative percentages found by him are different.

According to Julius Thomsen, the oxidation of a gram-atom of carbon to CO evolves 29,coo calories and its oxidation to CO_2 evolves 96,960 calories. If these values be applied to this cell, the oxidation of 4.8 grams of carbon to CO gives 11,600 calories and that of 0.60 gram of carbon to CO_2 gives 4,848 calories, a total of 16,448 calories. The voltage corresponding to the heat

[1] Ztschr. Elektrochem., **2**, 541.

208

evolved by such oxidation is 0.72, from which must be subtracted the voltage corresponding to the heat absorbed in removing a gram-equivalent of oxygen from the particular depolarizer employed.

The temperature curve of the same cell was plotted from readings taken at intervals of 5° from 290° C. to 540° C. This curve was found to be irregular and similar in form to those obtained by Liebenow and Strasser for a cell employing carbon and iron electrodes in a bath of fused sodium hydroxide.[1] The temperature coefficient of the cell is therefore not known with sufficient accuracy to enable the electromotive force of the cell to be calculated from the Helmholtz formula representing the total transformation of energy within a voltaic cell.

DISCUSSION.

MR. REED: Mr. President, I would like to ask Dr. Byrnes to define exactly what he means by a voltaic cell, and to state what evidence there is in any of these cases to show in which cases the cell is voltaic and in which cases it is thermoelectric. I think the conclusions we arrive at will depend very largely on the definition of a voltaic cell.

MR. BYRNES: By a voltaic cell I mean one in which electrical energy appears as a result of chemical action. Of course, in a general way, any battery gives electrical energy as a result of thermal action, but the term voltaic cell is commonly used to indicate a cell in which chemical energy, for example, that of oxidation at the anode or reduction at the cathode, is transformed into electrical energy.

MR. REED: In other words, the voltaic cell is one in which the electrical energy originates in chemical change or chemical energy, and not in heat?

MR. BYRNES: Yes.

MR. REED: Now, in the calculation of electromotive force, which has been given here, a large part of that electromotive force was due to the temperature coefficient, was it not?

MR. BYRNES: Yes.

[1] Ztschr. für Elektrochem., **3**, 353.

MR. REED : That part of the electrical energy must then be thermoelectric in origin, must it not?

MR. BYRNES : Yes.

MR. REED : It has been quite a number of years since I have looked into these figures, but some four or five years ago I spent a great deal of time in investigating the energy of the chemical reactions in these cells, and I found that with all the different changes from oxides of iron to metallic iron, and from one oxide to another, with the oxidation of the carbon electrodes, there was no case in which any energy could be evolved as electrical energy by the oxidation of the carbon and the simultaneous reduction of the iron to the metallic state, or from a higher to a lower oxide. I published those results in various publications and have not known of their having been controverted. But it is impossible to go through these figures off hand, and it may be that I was mistaken.

In connection with this subject, some years ago I experimented

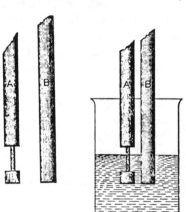

with electrodes of iron of different forms in fused caustic alkali. I found that by taking a round rod of iron and cutting it down to a small stem in one place, then taking another rod of the same size but not cut in that way, immersing them in a bath of fused alkali (see accompanying illustration) I found it very easy to get an electromotive force of 1.2 volts. This was described before the American Institute of Electrical Engineers in 1898. In that case, it seems to me, there is no possibility of any electrical energy being evolved by chemical action. In other words, that it could not be a galvanic cell in the sense in which Dr. Byrnes has defined it. The only source of any electrical energy must be heat. Consequently, it is a thermoelectric cell. It seems to me, therefore, that we cannot settle any of these questions as to whether the cell is thermoelectric or voltaic by merely measuring the electromotive force. In regard to the work of Liebenow and Strasser, I also gave (in the paper above referred to) considerable attention to their researches, and I found that by inverting the sign of their normal electrode, the electromotive force was in all their experiments proportional to the temperature.

DR. BYRNES : Mr. President, it is to be noted that various oxidizing agents give different values.   Not only do they reverse the cell from one in which the iron acts as a soluble electrode or anode, to one in which the carbon acts as such electrode, but that they give different values according, apparently, to their efficiency as oxidizing agents.   As to the thermal reaction which may express a portion of the electromotive force, I feel that it is only partial, and it may not even be the principal one.   Whether the oxidation of the iron is intermediate between the oxidation of the sodium ion and the reduction of the depolarizer, I do not know. The depolarizer may act directly, as in the case of gold.   It is a question whether the gold cathode would be to any extent oxidized.

15

Copyright © 1974 by Academic Press Inc. (London) Ltd

Reprinted from pages 651–659 of *Int. Symp. Power Sources, 9th, Proc.*, Academic Press, 1974, 740pp.

PERFORMANCE CHARACTERISTICS OF A LONG LIFE PACEMAKER CELL

A. A. Schneider
Catalyst Research Corporation, Baltimore, Md. U.S.A.

W. Greatbatch and R. Mead
Wilson Greatbatch, Ltd., Clarence, N.Y., U.S.A.

ABSTRACT

The chemistry of the CRC lithium anode–solid electrolyte cell is presented along with multi-year performance data. The cells have seen clinical use in implanted pacemakers since March 1972. The cathode is a charge–transfer complex of iodine and poly-2-vinylpyridine with excess iodine. The electrolyte, lithium iodide, is formed *in situ* as the cell reaction proceeds. The unit is characterized by a high energy density and low power density and is thus well suited to long life applications. Four A h production versions of the cell, designed to power cardiac pacemakers for 10 years, have been discharged to a 3 A h depth and are still functioning. Discharge data at several temperatures are presented for periods of 2 years or more for the pacemaker cell and for smaller versions designed to power electric watches. Tafel plots using fresh cells show two linear regions suggesting consecutive limiting reactions. Reliability of packaging and containment materials is also discussed.

INTRODUCTION

The lithium iodide solid electrolyte primary cell, first reported by Schneider, Moser, Webb and Desmond (1970), was successfully adapted for medical use in 1971 (Greatbatch, Lee, Mathias, Eldrige and Schneider, 1971). This electrochemical system is characterized by a high energy density and a relatively low power density and is thus well suited to low-power, long-life applications such as implantable cardiac pacemakers. Since the electrolyte is solid, it is possible to avoid many of the difficulties associated with conventional liquid electrolyte cells such as electrolyte leakage, separator failure, etc. No gases are evolved during discharge so that the cell may be sealed hermetically. The system also shows a low self discharge rate.

It would seem, then, that the high reliability, essential for a pacemaker power source, should be available with this solid electrolyte cell. Indeed, high reliability has been the experience with this system.

CELL CHEMISTRY AND CONSTRUCTION

The cell anode is metallic lithium in which is embedded a metal current collector. The surface of the lithium is mechanically cleaned just before cell fabrication to eliminate contaminants, particularly lithium nitride.

TABLE I

Nominal size	1.37 cm × 5.18 cm × 4.41 cm high (31 cm³) 0.54 in × 2.04 in. × 1.74 in. high (1.9 in.³)
Electrode area	24.7 cm² (3.83 in.²)
Weight	80 g (3 oz)
Density	2.5 g/cc (1.6 oz/in.³)
Voltage	2.8 V open circuit
Rated current	30 uA
Energy	9.45 W h
Energy density	*vs* Volume 4.9 W h/in³. (0.30 W h/cc) *vs* Weight 50 W h/lb (0.12 W h/g)
Capacity	Over 3.5 A h
Self discharge loss	Estimated at less than 10% in 10 years
Seal	Plasma needle-arc weld tested to 30 × 10⁹ cc/sec max helium leak after 5 min bomb at 60 psi
Storage temperature	−40°F (−40°C) to 125°F (52°C); with short excursions to 150°F (65°C)

The cathode is a charge-transfer complex of iodine and poly (2-vinyl-pyridine) formed by combining 7 to 24 parts of iodine with one part of the polymer. The average molecular weight of the polymer is 13,000. The cell is constructed entirely in a dry room with moisture content below 1000 ppm.

An open circuit voltage of 2.809 ± 0.005 volts can be measured immediately after anode/cathode contact. No electrolyte is added; rather, solid lithium iodide is formed *in situ*. It is the conductivity of this salt which limits the power output of the cell.

A cutaway view of the E pacemaker cell is shown in Fig. 1. The lithium anode is centrally positioned with the charge-transfer complex cathode on either side. The cell is encased in four layers of iodine resistant plastic and then sealed in an hermetic enclosure. The positive leads from either side of the cell can be welded together internally (as is shown in Fig. 1), or connected independently to two positive feed throughs. The latter configuration allows external diode coupling of the positive terminals.

Pertinent physical and electrical data for the E cell are shown in Table I.

DISCHARGE AT BODY TEMPERATURE

The average load seen by the pacemaker cell is 100 k Ω at 37°C. Cells have been under test at this load since October, 1970. Some of these test cells are of a design in which a single face of the lithium anode is exposed to a single layer of cathode. These cell designs, designated A, B, C and P, use the same volume of active components as the E design of Fig. 1 but one-half the effective electrode area. Discharge data for the clinical E and P versions and the pre-clinical A, B and C versions are shown in Table II.

TABLE II. PERFORMANCE OF PACEMAKER CELLS AT 37°C UNDER 100 k Ω

Cell Type	No. of Cells	Test Duration (Months)	Average Voltage (Volts)	Slope (mV/mo)	Projected Voltage at 6 Years (Volts)
Double Anode	1	20	2.754	2.3	2.634
E Cell	3	15	2.774	1.7	2.677
	42	14	2.762	2.7	2.603
	2	14	2.756	3.1	2.574
	25	13	2.762	2.9	2.592
	189	11	2.784	1.6	2.684
	144	7	2.782	2.6	2.614
	80	7	2.776	3.0	2.584
	47	6	2.783	2.8	2.596
	143	5	2.786	2.3	2.632
	24	5	2.783	3.5	2.548
	115	4	2.790	2.6	2.620
Single Anode	1	20	2.665	6.0	2.353
P Cell	119	16	2.683	5.8	2.339
	192	13	2.693	7.0	2.281
	84	10	2.726	4.7	2.446
	132	9	2.756	5.1	2.417
	105	8	2.767	4.2	2.483
Single Anode	6	40	2.570	5.3	2.398
A, B, C Cells	7	32	2.557	7.1	2.272
	24	29	2.461	11	2.036
	52	14	2.673	7.4	2.252

Based on these data, the graph in Fig. 2 has been drawn, showing the present extrapolated performance of the current production model E cell. The 40 month voltages from the A, B and C single anode cells (Table II) have been used to calculate the 3 year data point in Fig. 2 for the double anode E cell.

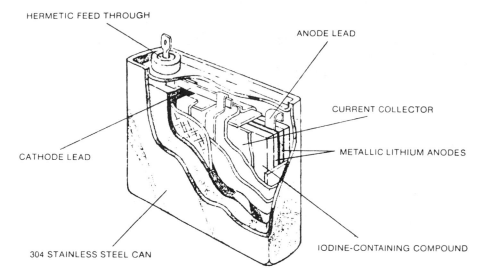

FIG. 1. Cutaway view of double anode E cell.

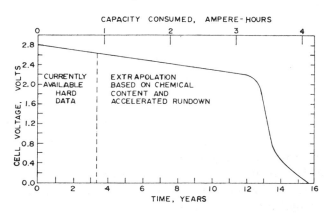

FIG. 2. Performance of pacemaker E cell at 37°C under a 30 microampere load.

Extrapolation of the discharge slope to six years seems reasonable, based on the electrical data above and other multiyear data on packaging materials reliability. The extrapolation beyond ten years is admittedly somewhat tenuous. It is based on the chemical content of the cell and a 10 month accelerated discharge during which more than 4.0 ampere hours was drawn from the cell.

Approximately 9000 clinical (P and E) cells have been manufactured of which 1000 have been implanted as of the end of 1973. To date there have been no failures in P or E cells, either in clinical use, or in test programs conducted in our laboratories and the laboratories of several pacemaker manufacturers.

DISCHARGE OVER A WIDE TEMPERATURE RANGE

Figure 3 shows the effect of temperature on the performance of a fresh E cell. The behaviour of fresh cells is a function of the thickness and area of the cathode. An activation energy of 15.5 K cal/mole has been calculated for such cells,

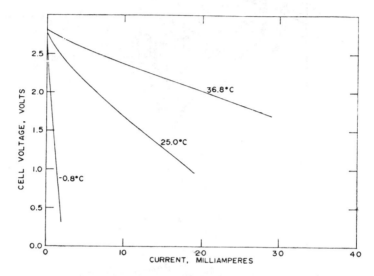

FIG. 3. Typical E *vs.* I plot for a double anode E cell. Total electrode is 24.7 cm².

and may be associated with (1) diffusion of iodine from the bulk of the cathode to the cathode/electrolyte interface or (2) with electron migration from this interface to the cathode current collector.

Tafel plots of IR free polarization *vs* log I show evidence that two consecutive limiting reactions are involved, with a change in slope occurring at 250 mV at room temperature. Interpretation of such plots is difficult, however, because of the large ohmic component of polarization which must be eliminated mathematically before the contribution of transition polarization can be seen.

As the cell is discharged, it is the impedance of the electrolyte which becomes the limiting factor. An activation energy of 8.65 kcal/mole has been found for older cells where appreciable lithium iodide has been generated through electrochemical reaction. This value agrees with that of Haven (1950) for the activation energy of ionic conduction in lithium iodide.

At high discharge rates, the voltage decay is linear with time. Such behaviour is shown in Fig. 4. Generally, such discharge curves obey the equation

$$\eta(t) = \eta_0 + C \cdot j^2 \cdot t \cdot \exp(8650/RT) \tag{1}$$

where η is the polarization, η_0 is an initial polarization, C is a constant, j is the current density and t is time. The dependence on the square of the current

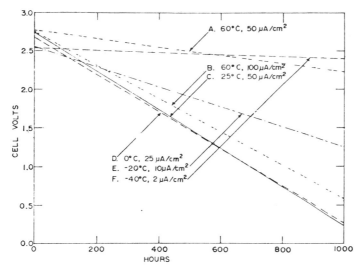

Fig. 4. Discharge at constant current showing linear increase of polarization with time. Cell thickness in cm: A-0.13, B-0.25, C-0.13, D-0.07, E-0.03, F-0.03.

density can be explained in this way: at any time t, polarization is proportional to current density times the thickness of the LiI electrolyte layer ($\eta \alpha j \cdot x$), while the thickness of this layer is itself proportional to the number of coulombs expended per unit area ($x \alpha j \cdot t$).

Although the E package will not tolerate large temperature extremes, the system itself has been investigated at temperatures between $-55°C$ and $+75°C$. At the lower temperatures, cell impedance is extremely high with short circuit currents in the 0.1 $\mu A/cm^2$ range; at $+75°C$ self discharge is very high, although one year tests have proved that the system will survive this temperature.

SELF DISCHARGE

The self discharge reaction for this system involves the diffusion of cathodic iodine through the lithium iodide electrolyte to the anode where additional electrolyte is generated. Impedance increases from this accumulation have been found to be linear with the square root of time. Self discharge losses for the E cell are estimated to be less than 10% in 10 years at 37°C.

217

PACKAGING AND CONTAINMENT MATERIALS

A major effort has been spent on determining the compatibility of containment materials with the active cell components, lithium and iodine. Multiyear studies have eliminated all but a few plastics and cements. Most common packaging plastics react with iodine, forming charge-transfer complexes which are electronically conductive. Only polystyrene and a few fluoropolymers will remain insulating during extended exposure to the iodine-bearing cathode material.

It was failure of packaging materials which caused failures in some of the pre-clinical A, B and C pacemaker cells. More packaging problems will probably be encountered in the next five years of testing, problems which will have to be solved before a ten year reliability is a reality.

CONCLUSIONS

Experience with thousands of cells has shown this solid electrolyte cell to be extremely reliable as a cardiac pacemaker power source. There have been no failures in clinical versions of the cell, either in implanted units or in test programs which range up to 20 months. 40 month tests of pre-clinical cell models allows extrapolation of cell life to at least six years. On the basis of initial capacity and accelerated discharge, lives as long as ten years can be postulated.

The system represents a significant improvement over conventional mercury cells, whose life is typically 30 months in pacemaker applications. It appears that, with a conventional electrochemical power source, it may indeed be possible to power a cardiac pacemaker for the lifetime of most patients.

REFERENCES

GREATBATCH, W., LEE, J., MATHIAS, W., ELDRIDGE, W. and SCHNEIDER, A. (1971). *IEEE Trans. Bio-Med. Eng.*, BME-18 317.
HAVEN, Y. (1950). *Rec. trav. chim.* **69**, 1471.
SCHNEIDER, A., MOSER, J., WEBB, T. and DESMOND, J. (1970). Proc. US Army Power Sources Conf., Atlantic City.

DISCUSSION

G. D. WEST (*Devices Ltd., U.K.*): Figure 2 shows extrapolated data based on accelerated discharge. It is known that conventional zinc-mercuric oxide cells discharged at normal pacemaker drains show much reduced capacity over that expected from accelerated discharges. Do you foresee any mechanism which could result in similar behaviour for your cell?

A. A. SCHNEIDER: We have seen some effect on cell impedance as a function of current density. Cells discharged at a relatively high current density, near 50–$100 \, \mu A/cm^2$, show significantly lower electrolyte impedances than those discharged at the nominal pacemaker drain. This is one of the reasons why we went from a single anode P cell to a double anode E cell—to give us that safety margin. There does seem to be some effect of current density on discharge characteristics. I don't think there is any effect on capacity.

G. D. West: What is the physical appearance of the lithium/lithium iodide interface and how does this progress through the bulk lithium during discharge? Also, what are the changes in volume when going from the charged to discharged state and are these significant?

A. A. Schneider: The growth of lithium iodide via electrochemical reaction occurs at the cathode electrolyte interface. We have seen evidence of dendritic growth of the lithium iodide into the depolarizer layer. As far as volume changes go, we have been measuring over the last three years thicknesses of some of our cells and we have seen no volume change—no thickness increase. We have no evidence of negative changes either. We built a cell with a similar depolarizer which used a silver anode and we saw with this cell, after about a third of the capacity was exhausted, an abrupt rupture of the silver/silver iodide interface. We have seen nothing like that with the lithium anode. I suspect that since both the anode and cathode are somewhat "plastic" that this helps in taking up any small local volume inconsistencies.

G. D. West: The theoretical life of your cell in a pacemaker application can be estimated at greater than 20 years. New developments will tend to increase this predicted life. Are you confident that the iodine resistant encapsulation of the cell internals will remain iodine resistant over these extended periods?

A. A. Schneider: This is the area in which we are working right now. It was the area which gave us problems initially and we are watching it very closely now.

G. D. West: Doctors are looking more and more toward smaller *total* pacemaker packages. Can you foresee any developments which would lead to an increase in the energy density of your system?

A. A. Schneider: We have, as I indicated, made some laboratory models of improved energy density cells with about 30% more active material in the E cell package; these have been under test discharge for about a year and a half and they look very good. We have also gone to very much smaller cells, cells which might be used in electronic watches. The cell is about 25 mm in diameter and maybe 2.5 mm thick and it is fairly efficient in energy density.

B. B. Owens (*Gould Inc., U.S.A.*): The effect of volumetric changes is one of the problems in a completely solid state cell system. In this case you have the advantage that at least two of your components are relatively plastic in nature. In the case of silver/rubidium silver iodide cells of high capacity, when they are subjected to a discharge over several years, we would observe in some cases a fracturing of the electrolyte layer due to swelling in the cathode. You fabricate the cell in a dry room and then you clean the lithium surface off just before the final operation. Is there a particular length of time that you restrict lithium exposure in this atmosphere?

A. A. Schneider: We have been restricting it to perhaps 20 min and will continue to go lower on that restriction. It seems that the lower the exposure time the better the performance later on. We have noticed some effect as far as moisture content goes. Moisture gives us a slightly higher open circuit voltage, perhaps as high as 2.812–2.815, and there appears to be some small effect on discharge slope too.

B. B. Owens: With regards to the coulombic efficiency of your lithium and your iodine, what per cent of them are utilized electrochemically during operation?

A. A. Schneider: There is enough iodine for about 6 A h in this cell; perhaps five of them would be usable. We have problems near end of life because the resistance of the depolarizer increases and I would say that the last A h was not usable limiting us to about 5 A h. We are very near achieving that, about four and a half A h now on accelerated discharge. We still have a way to go at the nominal pacemaker drain.

Lithium has always been in excess; it has not been the limiting electrode. We built it this way simply because lithium is volumetrically very efficient and it is easy to build in an excess here. We have never taken one of these pacemaker cells to depletion as far as the lithium electrode is concerned. We are designing experiments to do that now.

N. Marincic (*GTE Labs. Inc., U.S.A.*): With a cell weight of 80 g there seems to be no room in the pacemaker for a second cell, desired for redundance. This would make your cell the first power source with a lifetime guarantee. Can you comment on this?

A. A. Schneider: Reliability is the criterion here obviously, and one of the nice things about this system is that should it ever fail, it never fails abruptly. I have been working with this for about five years and I have seen many of the very early versions fail and they fail very slowly. We feel that with the reliability we now have, we would probably be building a more unreliable system if we were to use two cells because we would be doubling the chances of single cell failure. So right now the reliability of this cell is better than the electronics and we think that we are going to wait for them to catch up with us.

N. MARINCIC: Are you working on smaller cells so you can combine two of them? And what about little people who cannot take a pacemaker with an 80 g battery?

A. A. SCHNEIDER: We are now trying to qualify cells from 1 A h up to 3 A h; these would be suitable for children.

P. H. HITCHCOCK (*Ever Ready Co. Ltd., U.K.*): Would you describe the factors which lead to the use of zirconium as an anode current collector? What was the cathode current collector?

A. A. SCHNEIDER: The initial search was for a cathode current collector. The one in contact with the iodine bearing depolarizer. We went through a large variety of metals. Only zirconium, in perhaps 50 metals, survived corrosion tests with the depolarizer material. It is not zirconium metal itself, but rather an oxide coated zirconium which is used, and normally the thickness of oxide that one gets on zirconium as received is sufficient for protection. The material in contact with the anode really does not have as many constraints. We chose zirconium simply because it was convenient and because we knew it didn't react with iodine vapour. That was probably the main concern.

B. B. OWENS: You mentioned a small cell designed to power electronic watches. What characteristics are exhibited by this cell?

A. A. SCHNEIDER: The mA h capacity of the small cells is equivalent to the standard mercury battery, but because the voltage is double we have about double the energy density. Because the watch cells see a larger temperature variation than the pacemaker cells we must constrain the drain to a value down in the $1 \mu A/cm^2$ range. We are looking toward interfacing with the next generation of electronic watch circuitry which might use liquid crystals for display. These are very low power devices.

16

Copyright © 1969 by the American Institute of Chemical Engineers

Reprinted from pages 699–704 of *Intersoc. Energy Convers. Eng. Conf., 4th, Proc.*, American Institute of Chemical Engineers, 1969, 1080pp.

SEAWATER POWER SUPPLY
USING A MAGNESIUM-STEEL CELL

C. L. Opitz
Lockheed Electronics Company
Plainfield, New Jersey

INTRODUCTION

A seawater power supply development project was started in 1964 for the purpose of developing a general-purpose-type undersea source of electrical power that would be superior in many ways to then available batteries and radioactive isotope supplies. The primary design objectives were low cost, long life, high reliability, and relative independence of operating depth and salinity. Unit output power levels in the 0. 01- to 1. 5-watt (continuous) range with a 6 months to 4 years plus life bracket were considered. Using these objectives, the feasibility of using the great potential electrochemical energy of metallic magnesium in a free-flowing seawater galvanic cell was investigated.

The use of on-site seawater for an electrolyte in any undersea cell has the distinct advantage of the electrolyte being ever fresh, free, and readily available when required. The stored and transportable power unit could be a "dry" device and thus need not be made as heavy and strong as conventional batteries. Furthermore, the shelf life of a "dry charge" type cell should be superior to cells that contain electrolyte in storage. Being open to the seawater, the cell would require no waterproofing pressure case, oil bath, or auxiliary pressure compensation devices.

The choice of electrode materials in a galvanic cell is influenced by life, cost, and energy per unit weight. In view of these factors, an excellent selection (theoretical) for the anode material is magnesium, which when used in billet form is relatively inexpensive. Although cathode metals other than iron have been shown to provide a greater cell voltage output, the choice of steel would result in a good watt-hours per dollar value with the necessary long-lasting characteristics.

This line of cell design thinking had of course been taken by other workers in the field but (by 1964) none of them had developed a practical seawater magnesium cell with at least 6 months of useful life.

The first limiting problems encountered by other investigators, using metallic cathodes, were polarization and mineral scaling. When conventional parallel-plate electrode cells were operated at relatively high cathode current densities (in order to realize a reasonable watt-hour per pound figure), the cathodes polarized so that most of the theoretical output voltage was lost with a loaded cell. Furthermore, even when this initial low efficiency operational condition was tolerated, the internal voltage loss greatly increased after a few months operation, due to cathode scaling. The required high cathode current densities in these cells caused calcium and magnesium salts in the seawater to collect on the cathode, thereby forming an insulating scale.

A long series of laboratory tests were undertaken to investigate a "critical" cathode current density factor for scaling to the point where the corrosion vulnerable steel was just cathodically protected (Fig. 1). When these data were available, it led to further experimentation, which resulted in the unique use of a steel wool[1] cathode structure to provide the required compact anode with a large surface area for low current density and to permit a relatively free flow of new depolarizing seawater. This simple solution to the scaling and polarization problems resulted in practical cell sizes with useful long-term output voltages in the 0. 35- to 0. 7-volt bracket. Figure 2 shows how inefficient a conventional "plate" cathode cell can be. Through the use of the steel wool cathode it is possible to use current densities as low as 4mA/square foot in a practical size cell. Small steel wool cathode test cells were measured for temperature characteristics (Fig. 3), oxygen content (Fig. 4), and salinity (Fig. 5). Also of interest is the fact that test cells were operated in a pressure chamber from 0 to 10, 000 psi with no measurable change in output voltage.

Another long standing major problem solved was the relatively efficient generation of useful power supply output voltages (3 to 15 V) from these magnesium-steel cells with very low output voltages.

The use of series-connected cells to increase output voltage cannot be tolerated in an inexpensive, long-life battery, because of partial short circuiting caused by the external paths in the surrounding sea. Ideally, a single low-voltage cell must somehow be used as the primary source of electrical energy. The obvious solution to the problem is the use of an efficient dc-to-dc converter with which to convert the 0. 3- to 0. 7-volt input to any desired output voltage.

An initial survey of available dc-to-dc converter vendors showed that no converter was available with a 0. 5-volt or less input rating. In 1965, new inexpensive but efficient transistor types were obtained and built into dc-to-dc converters. The new units resulted in state-of-the-art converters that provided over 60% efficiency over the specified input voltage bracket, as shown in Figures 6 and 7, for a typical 1-watt output unit.

DESCRIPTION OF LARGE TEST CELLS

In addition to building many small laboratory test cells, a number of large cells were built based on the design data obtained from the laboratory cells.

---

[1] U. S. Patent No. 3, 401, 063.

One large 2-watt cell and several 180-milliwatt units were tested for long periods. The 2-watt unit was operated within a large laboratory salt-water tank. The output of this cell fed a solid state inverter continuously for 11 months, during which time, it developed a nominal high voltage power level of 1.5 watts.

One of the 180-milliwatt cells (Fig. 8) was fitted with a resistive 2-ohm load and a long underwater cable so that the output voltage could be monitored at a shore site.

The 180-milliwatt unit was installed in a 12-foot deep inlet on the New Jersey coast on March 26, 1965, and, with the exception of a retrieval for a 1-hour examination after 194 days, it has been in continuous operation (last data taken during June 1969).

DESCRIPTION OF TYPICAL CELL CONSTRUCTION (180-MILLIWATT TYPE)

Figure 9 shows a cutaway, perspective view of a submerged cell feeding a low input voltage dc-to-dc converter. The cell is constructed of an expanded metal (steel) annular basket (10) with inner and outer angle rings (12 and 13) at both the top and bottom of the basket to confine the webbing and control the general shape of the basket.

The basket is fully but loosely packed with a commercial grade of steel wool (14). The metal wool serves as the cathode for the galvanic cell and provides a very large surface area relative to the surface area of the magnesium anode (15).

The cylindrically shaped solid magnesium anode is centrally mounted within the opening (16) in the annular basket. A supporting insulator pin (17) projecting from one end of the anode is secured to a bracket (19) spanning the bottom of basket (10). The top end of the anode serves as a connector (20), which projects upwardly through supporting bracket (21) spanning the top of basket. A plastic spacer (22) is used on the terminal rod (20) between the anode and the support bracket (21). Bare steel spanner bolts (23 and 24) passing through brackets (19 and 21) serve to hold the basket assembly together and to maintain the anode in a centered position within the central opening through the basket.

The anode (15) and pin (20) are electrically insulated from a direct conductive path through the basket to the metal wool cathode (14) by employing a plastic pin (17) and washer (22) and by electrically isolating pin (20) from the bracket (21) with a plastic sleeve (25) surrounding the pin that passes through an oversized opening (26) in bracket (21). Insulating sleeve (25) is retained in position by clamp (27).

Terminal post pin (20) is firmly secured deep inside the magensium anode (15) by threaded engagement (28) so as to assure long-term support and minimize electrical resistance between the pin and the anode as the anode is consumed to a small fraction of the original size.

The cathode terminal (29) is affixed directly to the electrically conductive portion of the basket (10) that is in electrical contact with the steel wool cathode.

The two cell leads are coupled across the input of a dc-to-dc converter (33), the output of which is applied across a load (34). The load may take the form of any electrically operated device such as a sonar beacon or oceanographic sensing equipment.

In practice where a converter is required, it could be conveniently mounted directly on the cell, e.g., on bracket (27) so as to eliminate the need for exposed terminals (20 and 29).

180-Milliwatt Cell Characteristics (Figure 8)

Size	15 in. od, 16 in. high
Weight (dry)	54 lb
Magnesium anode	6 in. dia by 12 in. high
Weight of steel wool cathode	5 lb
Calculated current density for cathode material	4 mA/ft^2
Nominal output voltage (with 2-ohm required load)	0.6 V
Nominal continuous output power (the design load)	180 mW
Nominal design life	3-1/2 yr (min.)
Cell energy capacity (over 3-1/2 yr)	5000 plus W-hr
Watt-hours per pound of cell	100 W-hr/in.^3

A plot of the cell output voltage vs time is shown in Figure 10.

The relatively wide voltage variations shown over the 4-year period are primarily due to the wide daily and seasonal temperature variations.

This cell, which was originally designed as a 3-year life unit, was last monitored in June 1969, at which time, the output voltage was 0.6 volt.

SUMMARY

The following summary is based on the laboratory and single field test cell results:

- The feasibility of the long-lived magnesium-steel wool cell has been demonstrated (as tested in a seawater inlet).

- With loaded cell output voltages of 0.5 to 0.65 volt and a calculated cathode current density of 4 mA/square feet, it may be expected to obtain as much as 100-watt-hours of energy per pound of cell (dry weight) for a 3- to 4-year period.

- Examination of a test cell after 194 days of field operation (over the summer season in a heavy fouling area) indicated no obvious detrimental cathode corrosion or scaling of the cathode and very minor marine growth within the cell.

- Except for a high starting voltage following the initial installation, cell output voltage level may be fairly constant over the life of the cell (if the seawater temperature is constant).

- The long-term cell output voltage is primarily affected by water temperature, so that in shallow water ±10% or greater, fluctuations may be experienced.

- The magnesium-steel wool cell must be operated continuously with a quasi-fixed value for which it is designed although the cell may be "opened" or shorted for very brief periods with no apparent harm.

- A low cost, low input voltage inverter with an overall conversion efficiency greater than 60% can be used with a magnesium-steel wool cell.

- The magnesium-steel cell can be stored

on-the-shelf without deterioration if the humidity is low enough to prevent rusting of the steel wool (which is commonly protected with a water soluble grease).

- Design data are available for cells in the 10- to 250-milliwatt bracket; however, for larger cells, extrapolation of small test cell data may not hold.

- Cell life may be ultimately determined by initial current density and diameter of magnesium anode. (Possibly up to 5 years of life.)

Figure 1. Laboratory Magnesium-Steel Cells

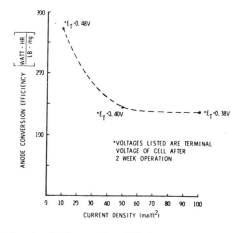

Figure 2. Anode Conversion Efficiency vs Current Density for Equal Anode and Cathode

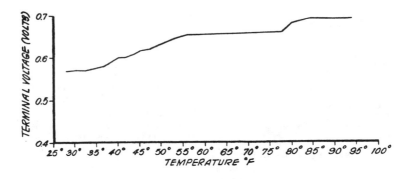

Figure 3.   Loaded Cell Voltage vs Temperature

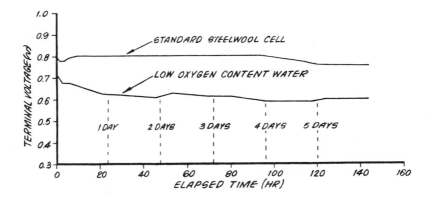

Figure 4.   Low Oxygen Content Electrolyte Voltage Test

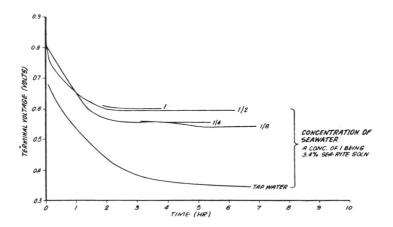

Figure 5.   Cell Terminal Voltage for Various Concentrations
of Seawater Electrolyte

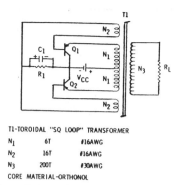

T1-TOROIDAL "SQ LOOP" TRANSFORMER

N_1	6T	#16AWG
N_2	16T	#16AWG
N_3	200T	#30AWG

CORE MATERIAL-ORTHONOL

Figure 6. Low-Voltage Inverter, Schematic Diagram

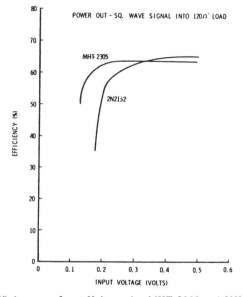

Figure 7. Efficiency vs Input Voltage for MHT 2305 and 2N2152 Inverters

Figure 8. 100-Milliwatt Field Test Cell; Left-Magnesium Anode
Center-Cathode Cage; Right-Assembled Cell

704   C. L. Opitz

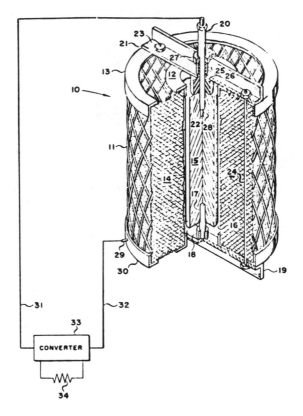

Figure 9.   180-Milliwatt Cell, Cutaway View

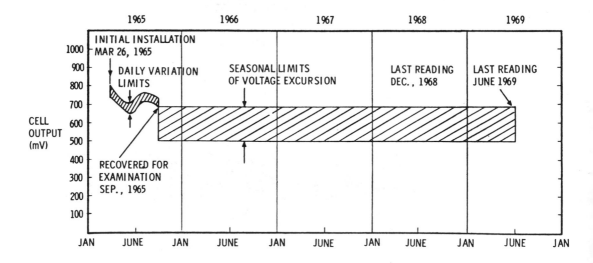

Figure 10.   Voltage-Time History of 180-Milliwatt Cell

226

17

Copyright © 1974 by the American Society of Mechanical Engineers

Reprinted from pages 665–670 of *Intersoc. Energy Convers. Eng. Conf., 9th, Proc.,* American Society of Mechanical Engineers, 1974, 1343pp.

A REVIEW OF THERMAL BATTERY TECHNOLOGY

B. H. Van Domelen and R. D. Wehrle
Sandia Laboratories
Albuquerque, New Mexico

ABSTRACT

This report reviews the evolution of thermal battery technology from World War II to the present. It discusses the first applied work with thermal cells, the transfer of this laboratory technology to the United States, the development of the initial cup technology by the U. S., and the evolution of this technology to the later pellet technology. The paper by F. Tepper will then discuss the performance of thermal batteries and compare this performance with that of other ordnance batteries.

INTRODUCTION

The electrochemical power sources used in nuclear weapons to date have been lead-acid, nickel-cadmium, silver-zinc, and thermal batteries. The lead-acid battery was used in the first atomic bomb and in later weapon systems until 1953, when it was replaced by the nickel-cadmium battery, followed shortly by the silver-zinc battery. These aqueous electrolyte batteries all had a critical deficiency: their wet stand time was so short that either the battery had to be frequently recharged or the electrolyte had to be carried separately and injected into the battery just before its use. The thermally activated battery, introduced in 1955, solved the wet stand problem. Since then, it has been the sole power source used by Sandia Laboratories for nuclear weapon systems.

The thermal battery is so called because it contains a pyrotechnic heat source that, when ignited, thermally activates the battery by melting the electrolyte, which is a salt mixture. Salt mixtures are solid and chemically inert at room temperature, but when fused become fluid and highly conductive. These properties of the electrolyte provide the long shelf life and good electrical performance of the thermal battery (1,2).

The origin of the thermal battery dates back to World War II when Dr. Ing. Georg Otto Erb, working in Germany, developed the first practical cells, using a salt mixture as an electrolyte. Dr. Erb investigated the performance of several couples and electrolytes, the effects of salts with water of hydration, and the effects of water in unhydrated electrolyte salts. He also conceived and developed batteries for several ordnance applications, including the V-I and V-II rockets and artillery fuzing systems; however, none of these batteries was produced for field use. Dr. Erb was interrogated by British Intelli-

gence at the end of the war and his studies were reported to them in a paper entitled "The Theory and Practice of Thermal Cells" (3,4). This information was conveyed by the British, and possibly through other channels, to the United States Ordnance Development Division of the National Bureau of Standards, which later became Harry Diamond Laboratories (HDL). In this paper the authors will discuss only the channel for which documentation is presently available.

Prior to August 1946 (5) Mr. Grenville Ellis, U. S. Signal Corps (now the U. S. Army Electronics Command) brought copies of Dr. Erb's report to a meeting of the Joint Battery Advisory Committee. This committee of about ten people representing the Navy, Army, Air Force, and Industry met informally to exchange ordnance battery information. Dr. A. G. Hellfritzsch, of the Naval Ordnance Laboratory and a member of this committee, recognized the relevance of Dr. Erb's work to the battery needs of the VT (Variable Time) Fuze Panel under the National Research and Development Board. Dr. Hellfritzsch conveyed this report to Roger W. Curtis, who was chairman of the VT Fuze Panel and was also in the Ordnance Development Division of the National Bureau of Standards (6). Within a few weeks Curtis and his co-workers had demonstrated the capabilities of thermal cells for ordnance applications. Combining this with the work under contract to Catalyst Research Corporation to develop gasless pyrotechnics, they developed a self-contained thermal battery (6).

In the late forties the Wurlitzer Corporation began production of HDL's first thermal battery. This battery was used in a mortar round and the total production reached over a million batteries (7).

In 1952 Curtis reported on thermal battery technology at a symposium attended by various ordnance groups (8). Because of its desirable ordnance characteristics, the thermal battery system was adopted by the military for ordnance applications and by the U. S. AEC for nuclear weapon systems.

The early thermal battery designs are frequently referred to as "cup," "pad," "cup and cover," "closed-cup," and "conventional" batteries. These early batteries had three distinguishing characteristics: heat-paper pads, electrolyte and depolarizer pads, and intercell electrical connectors, which are discussed in more detail later. These three characteristics distinguish the earlier batteries from the later pellet-type batteries which use heat powder pellets, electrolyte-depolarizer pellets, and no intercell connectors. The evolution from the earlier batteries to the present pellet-type batteries was gradual, and there were several hybrid batteries which included both technologies, such as the Naval Ordnance Laboratory/Eureka-Williams battery which was the first to use a pellet.

This review describes the early cup technology (9,10,11), one of the later hybrid systems (2,10,12), and the present pellet technology (11,13,14). Performance characteristics of the thermal battery are discussed with reference to the Ca/LiCl-KCl/CaCrO$_4$ electrochemical system and the pellet technology which is the principal system used by this Laboratory. Discussion of other electrochemical systems and references to these systems are given by Jennings (15).

THE CUP TECHNOLOGY

Figure 1 shows the typical structure of a single cell in the cup design. In the cell are the anode, the electrolyte pads, the depolarizer pads, the cup in which these are enclosed, and a heat pad to thermally activate the cell.

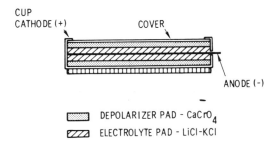

Fig. 1. - Single thermal cell - cup technology

The anode consists of calcium affixed to both sides of a metallic substrate[1] that serves both as the anode current collector and as the negative terminal by virtue of a tab extending through a slit in the side of the cell case. Although not shown in the figure, the anode is electrically insulated from the cathode cup.

The electrolyte pads are adjacent to the calcium surfaces and contain the electrolyte, the eutectic mixture of the lithium-chloride/potassium-chloride system (59 mole % LiCl-mole % KCl), which has a melting point of 352 C. The pads are formed by passing a fiberglass tape through a vat of molten electrolyte and punching pads from the tape after it has cooled and the electrolyte has solidified.

_____

1. Inconel, iron, stainless steel, and nickel are the metals most often used.

The depolarizer[2] pads, adjacent to the electrolyte pads, complete the active cell stack. These pads are punched from a paper-like sheet formed by pressing and drying a slurry containing the calcium chromate depolarizer along with glass and ceramic fibers.

The active cell stack is then placed in a metallic cup, covered with a metallic disc, and the sides of the cup are crimped over the disc as a seal. The cup and disc thus hold the cell components in place and serve as the cathode current collector and positive terminal of the cell.

The last component of the single cell is the heat pad, which activates the cell by raising its temperature to the operating range of 500 to 600 C. The pad is punched from heat paper sheets formed by a process similar to that used for the depolarizer pads, but involving a slurry containing zirconium fuel and barium-chromate oxidizer instead of the depolarizer. (The slurry also contained asbestos fibers. Later, Sandia eliminated the asbestos because it releases water of hydration when heated.)

For simplicity, only the basic components of the cup cell have been described. When one includes intercell connecting tabs, wire-mesh current collectors, electrical insulators, etc., there can be as many as 27 individual parts in a cup cell as compared to the 3 parts of the later pellet cells.

It should also be noted that with the cup technology, as well as with the pellet technology, most of the battery materials either absorb or react with water, particularly the electrolyte which is very hygroscopic. Since battery performance is severely degraded if the components are exposed to water vapor concentrations greater than 1500 ppm[3], all processes are carried out in dry rooms. At this Laboratory for example, the dry rooms are maintained at less than 300 ppm; and in addition, all battery components are vacuum dried at 100 C prior to assembly.

Figure 2 illustrates thermal battery construction using the single cell described above. The active battery stack is formed by first welding the appropriate number of cells together, the negative tab of one to the positive case of the next, and then inserting a heat pad between each pair of cells to heat them and to insulate them from each other electrically. Electrical leads are attached and a narrow fuse strip of heat paper is laid up the side of the stack and across the top cell. The stack is then capped on each end

_____

2. Depolarizer is an historic battery term used to identify the active cathode material which is reduced electrochemically during battery discharge.

3. At 25 C, 1500 ppm is 5% relative humidity.

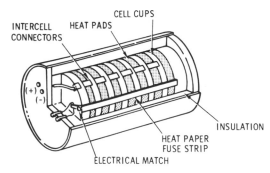

Fig. 2. - Thermal battery construction - cup technology

with an insulation disc and the entire assembly is slipped into an annular cylinder of insulation. This unit is placed inside a steel can to which a header is hermetically welded to prevent the battery materials from absorbing or reacting with water. The header has feedthroughs for the electrical leads and for an electrical match used to ignite the fuse strip[4]

The construction of a specific design is shown in Figure 3 which is a cross-section of

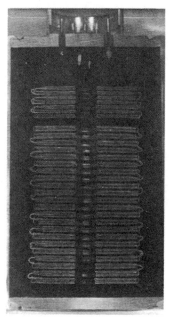

Fig. 3 - Thermal battery - cup technology

___

4. A mechanical device striking an explosive primer can also be used to ignite the fuse strip.

a battery used in a number of nuclear weapons. This battery has a center hole through the cells in lieu of a fuse strip; sparks from two electrical matches shower down the hole and ignite the heat pads. There are two battery sections, an upper 12-volt pulse section and a lower 28-volt power section (two 11-cell sections connected in parallel).

The cup thermal battery has excellent characteristics for ordnance applications. It has high current-carrying capability, wide ambient operating temperature range, high reliability, and long shelf life. Nevertheless, it has several disadvantages: an intricate cell design, short activated life, environmental limitations, heat-paper pads, high internal pressure, and complex processing and fabricating. These disadvantages are discussed below.

The numerous parts in the basic cup-type cell design, coupled with the physical properties of the depolarizer and electrolyte pads, result in a system which is not inherently reproducible. Significant performance variations occur between cells that are presumably identical (16). These variations frequently mask the effects of parameters under study and, in general, have negated a fundamental understanding of the cup-cell mechanisms.

The activated life of reliable cup-type batteries appears to be limited to about 5 minutes. Single-cup cells have been discharged for longer times, but the performance was not reproducible (16). The authors are aware of only one cup-type battery having an activated life greater than 5 minutes; but the performance was erratic, varying in an unpredictable manner from 4 to 21 minutes (17). This limitation of reliable activated life has required the use of silver-zinc systems for applications needing power for more than five minutes and has forced acceptance of the lesser performance and higher costs of the silver-zinc system.

The inability of the glass tape to immobilize completely the molten electrolyte limits the spin- and linear-acceleration capabilities of the cup battery, because of the electrolyte being displaced from the anode and cathode. The cup battery is also limited in shock along the cell-stack axis because the heat pads can deform when hot, allowing adjacent cells to contact and short.

Heat paper, although it is an excellent source of heat, has several disadvantages. It ignites so readily that it is an extremely hazardous material to handle. Its burning temperature is about 2000 C resulting in a thermal shock which sometimes damages cell components. When heat paper burns, its dimensions decrease, causing the pressure on the cells to change, which in turn affects the battery's performance. In addition, it is difficult to change the heat input to a battery because of the way heat paper is processed; i.e., a new production run must be made in order to change the heat output of the paper.

Internal pressures as high as 300 psi were observed in the earliest cup-type batteries. This required battery cases strong enough to withstand these pressures, resulting in weight and volume penalties. The sources of this pressure are: 1) impurities in the heat paper (zirconium hydride), 2) air in the void volume of the battery, and 3) moisture from the asbestos, which is often used for battery insulation.

The complex processes used in making the electrolyte, depolarizer, and heat pads, and the numerous steps required for fabricating cells are difficult to control and can cause variations in battery performance. Controlling these variations requires continuous inspection of the processing and fabrication and continuous testing of the materials produced.

Attempting to overcome these disadvantages became the motivation for developing the pellet technology.

THE PELLET TECHNOLOGY

In the mid-50's the Naval Ordnance Laboratory (NOL) and the Eureka-Williams Company developed a pellet process in which the active cell components were combined into a single three-layer pellet (2,12). The pellet was formed by compacting an anode layer of magnesium powder in a die cavity, adding and compacting a layer of lithium-chloride/potassium-chloride electrolyte and kaolin binder, then adding a layer of vanadium pentoxide and electrolyte, and finally pressing the three layers into a single pellet. The binder, a natural clay, prevents the electrolyte from flowing when molten.

Although the three-layer pellet simplified the cell design and the associated fabrication and processing, these thermal batteries still retained the other disadvantages of the cup technology; and in addition, the electrical performance of the $Ca/CaCrO_4$ system proved superior to that of the Mg/V_2O_5 system (1,2,11,13). Nevertheless, the pellet concept was a major advance in thermal battery technology.

Through NOL, Sandia Laboratories learned of the pellet concept and in 1961 began developing a pellet technology for the $Ca/CaCrO_4$ system. By 1966 the first completely pelletized thermal battery was in production, and since then pellet batteries have been the principal power source used by Sandia for nuclear weapon systems.

The pellet-type thermal cell consists of a heat pellet, a DEB pellet, and an anode (Figure 4).

The heat pellet is composed of powdered iron fuel and potassium perchlorate oxidizer pressed into a homogeneous pellet (13,14,18, 19). Unlike the heat pad, the heat pellet is

HEAT PELLET - Fe, $KClO_4$

DEB PELLET
- DEPOLARIZER, $CaCrO_4$
- ELECTROLYTE, LiCl-KCl EUTECTIC
- BINDER, SiO_2

ANODE - Ca on Fe SUBSTRATE

Fig. 4 - Cell components - pellet technology

not extremely hazardous; it ignites easily enough for reliable performance but with enough difficulty that it is safe to handle. Practically no gases are generated upon burning, eliminating the need for a heavy battery case. Another important property is that after burning excess iron in the heat pellet causes the residue to be electronically conductive. This eliminates intercell connectors, since the conductive heat pellet itself serves that function. In addition, the heat pellet is mechanically strong during and after burning, and thus does not have the environmental limitation imposed by heat-paper pads.

The DEB pellet (13,14) is composed of a mixture of depolarizer, electrolyte, and binder. The depolarizer is calcium chromate, the electrolyte is the lithium-chloride/potassium-chloride eutectic, and the binder is a finely divided silica powder whose high surface area prevents the molten electrolyte from flowing (13,14,20). These materials are fused, ground, and blended to obtain a homogeneous powder which permits the pellet to be formed in one pressing. This simplifies the processing and fabrication, and eliminates the use of additional or undesirable materials, such as glass tape, ceramic fibers, and kaolin binder.

The anode consists of either sheet calcium mechanically attached to a non-reactive metal substrate or calcium metal vapor deposited upon a metal substrate. The substrate serves both as the current collector and as the cell divider that prevents the calcium from reacting with the adjacent heat pellet.

These three pellets form a single cell, and the cells combined with end heat pellets and buffer pellets (13,14) form a battery stack (Figure 5).

The end heat pellets heat the materials at the end of the stack so that heat losses from the stack ends are reduced. The buffer pellets mitigate the thermal shock effects of the end heat pellets and act as heat reservoirs. These pellets are pressed from a homogeneous mixture of silica binder and the

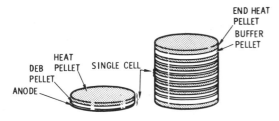

Fig. 5. - Single cell and battery stack - pellet technology

lithium-sulfate/sodium-chloride eutectic, which melts at 493 C and has a heat-of-fusion of about 84 calories per gram. Part of the heat from the end heat pellets is used to melt the eutectic salt in the buffer pellet. When the end of the battery stack cools to 493 C, the eutectic solidifies and releases its heat-of-fusion to maintain the battery temperature for a longer time.

As shown in Figure 6, the assembly of a pellet thermal battery is completed with the addition of electrical leads, match, fuse train, insulation, and battery case. The thermal insulation used is Fiberfrax for regular applications and Johns-Manville's Min-K 2002 for long-life applications. The Min-K 2002 was originally developed under AEC contract for thermoelectric generators and selected for thermal batteries after an extensive evaluation program (13,14). It is composed of silica, quartz fibers, and titanium dioxide.

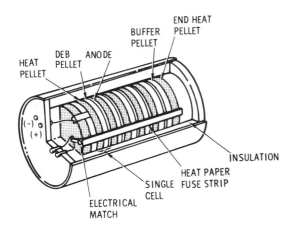

Fig. 6 - Thermal battery construction - pellet technology

Figure 7 shows the steps of battery construction and the use of center bolt construction to ensure concentricity of the pellets. In this construction, a quartz-covered steel bolt is passed through center holes in the battery stack and two steel end

plates. The result is a concentric, integral, rugged subassembly that can be easily handled and inspected.

Fig. 7 - Thermal battery - pellet technology

The pellet battery offers several improvements over the cup battery: longer activated life, more reproducible performance, greater ruggedness, no intercell connectors, lighter battery cases, better control of materials and processing, simplified fabrication, reduction of inspection and testing, elimination of hazardous heat pads, and lower cost. In addition, the pellet battery design is simpler and thus more amenable to investigation. This in turn has led to improved performance and further advancements, particularly with respect to longer life. Production has been completed on a thermal battery having an activated life of 15 minutes (21) and exploratory thermal batteries under development at Sandia have demonstrated activated lives of over 60 minutes (22).

ACKNOWLEDGEMENTS

The authors acknowledge the assistance of the individuals who provided much of the information concerning the early history of thermal batteries: A. Attewell of the Royal Air Force Establishment in the British Ministry of Defense, E. V. Forsman and C. W. Jennings of Sandia Laboratories, N. Kaplan of Harry Diamond Laboratories, R. T. Mead of Wilson Greatbatch, Ltd., R. D. Walker of the University of Florida, I. D. Yalom of the Naval Ordnance Laboratory, and particularly A. G. Hellfritzsch of the Naval Ordnance Laboratory.

REFERENCES

1. McKee, E., "Thermal Cells," <u>Proceedings of the Tenth Annual Research and Development Conference</u>, U. S. Army Electronics Command, 1956.

2. Advanced Science Division, Eureka-Williams Co., "A Discussion on Thermal Batteries," October 1971, Bloomington, Illinois.

3. Erb, G. O., "Theory and Practice of Thermal Cells," HEC 182, June 1945, Halstead Exploiting Centre, England.

4. Attewell, A., Private Communication, June 1973.

5. "Bibliography of Reports," Joint Battery Advisory Committee, August 20, 1946, p. F4.

6. Hellfritzsch, A. G., Private Communication, Feb. - May, 1974.

7. Mead, R. T., Private Communication, May 1974.

8. Curtis, R. W., "Thermal Batteries," Symposium on Batteries, Committees on Equipment and Supplies and on Guided Missiles, Research and Development Board, April 1952.

9. Thow, D. H., "Voltage Characteristics of Thermal Batteries," Proceedings of the 7th International Power Sources Symposium, Joint Services Electric Power Sources Committee, Brighton, England, September 1970.

10. Mead, R. T., "A Low-Cost Thermal Battery," Proceedings of the Twenty-Third Annual Power Sources Conference, U. S. Army Electronics Command, 1969.

11. Goldsmith, H., and Smith, J. T., "Thermal Cells in Present Use," Electrochemical Technology, Vol. 6, No. 1-2, Jan. - Feb. 1968, pp. 16-19.

12. Nielsen, N. C., "Three-Layer Pelletized Cells for Thermal Batteries," Proceedings of the Twenty-Third Annual Power Sources Conference, U. S. Army Electronics Command, 1969.

13. Bush, D. M., "Advancements in Pellet-Type Thermal Battery Technology," SC-RR-72-0218, October 1972, Sandia Laboratories, Albuquerque, N. M.

14. Bush, D. M., "Advancements in Pellet-Type Thermal Batteries," Proceedings of the Twenty-Fifth Power Sources Symposium, U. S. Army Electronics Command, 1972.

15. Jennings, C. W., "Thermal Batteries," Primary Batteries, Vol. 2, Edited by Heise, G., and Cohoon, N. C., John Wiley and Sons, New York, to be published 1974.

16. Jennings, C. W., and Bush, D. M., "Thermal Cell Studies, No. IV," SC-TM-407-60(13), January 1961, Sandia Laboratories, Albuquerque, N. M.

17. Neeld, B. G., "Characterics and Development Report for the 1264 Thermal Battery," SC-DR-65-60, July 1961, Sandia Laboratories, Albuquerque, New Mexico.

18. Bush, D. M., and Rittenhouse, C. T., "Final Report on Heat Pellet Study," SC-CR-67-2543X, Jan. 1967, Sandia Laboratories, Albuquerque, N. M.

19. Bush, D. M., and Rittenhouse, C. T., "Final Report on the Development of Pelletized Heat Sources for Thermal Batteries," SC-CR-70-6137, August 1970, Sandia Laboratories, Albuquerque, N. M.

20. Bush, D. M., "An Investigation of Materials to Immobilize Electrolyte in Pellet-Type Thermal Batteries," SC-RR-66-202A, April 1973, Sandia Laboratories, Albuquerque, N. M.

21. Baldwin, A. R., "A Long-Life, Low-Voltage, Power-Type Thermal Battery," Proceedings of the 26th Power Sources Symposium, U. S. Army Electronics Command, 1974.

22. Bush, D. M., and Baldwin, A. R., "Sixty-Minute Thermal Battery; A Feasibility Study," Proceedings of the 9th International Power Sources Symposium, Joint Services Electric Power Sources Committee, Brighton, England, Sept. 1974.

Part II

SUPERCONDUCTING MAGNETIC ENERGY STORAGE

Editor's Comments
on Papers 18 Through 22

The phenomenon of superconductivity was first observed by Kamerlingh Onnes at Leiden, Holland, in 1911, a scant three years after he had first liquefied helium. Various physical characteristics of materials near the boiling point of liquid helium, 4 K, were being tested in the laboratory at Leiden about this time. The resistivity of mercury, which was being measured near 4 K, suddenly dropped to zero on Onnes's instruments. The name *superconductivity* was given to this phenomenon. As our ability to measure resistivity has improved, the phenomenon of superconductivity or zero resistivity has persisted, and, for a direct current, the resistivity in superconductors, though still not measured, has been shown to be less than 10^{-15} $\Omega \cdot$ cm. By comparison the resistivity of copper at room temperature is 2×10^{-6} $\Omega \cdot$ cm.

Onnes very quickly recognized the possible applications of this discovery and predicted superconductivity would be used in electric power systems. Unfortunately, the first superconductors studied were mostly pure or almost pure materials that could carry only limited currents in low magnet fields, less than 0.1 T. For many years after Onnes's discovery, superconductivity was considered a laboratory phenomenon, just as batteries had been for most of the nineteenth century.

However, in the mid-1930s a new type of superconductor having magnetic properties different from those of the pure metals, such as mercury and lead, was discovered. This new type, which could carry modest currents in fields greater than 1 T, was called a type II superconductor, and the materials found by Onnes were renamed type I superconductors. Finally, in the 1950s theoretical descriptions of these two types of superconductors were proposed. These theories still appear to explain the observed phenomenon adequately.

Two superconducting materials, an alloy of niobium and titanium (NbTi), and an intermetallic compound of niobium and tin (Nb_3Sn) are the most used superconductors. They have been used in magnets, generators, and motors and have been proposed for use in superconducting power transmission lines. These two materials have transition temperatures, T_c (the temperature at which they become superconducting as the temperature is lowered), of 9.5 K and 18 K, respectively, and have critical fields, H_c (the maximum field in which the material will carry a current with zero resistance at 4 K), of 12 T and 24 T, respectively.

The largest magnet in the world is the Big European Bubble Chamber magnet at CERN in Geneva, Switzerland. This superconducting magnet was constructed between 1970 and 1974 and stores about 8×10^8 J. The conductor for this and most other superconducting magnets consists of filaments of NbTi imbedded in a copper matrix. The purpose of the copper is to carry part or all of the current for short times if the conductor is heated to a temperature above T_c.

Superconducting magnetic energy-storage (SMES) systems have been proposed for use on electric utilities for diurnal energy storage. Several economic studies have shown that if certain engineering developments are possible, SMES units will be competitive with other load-leveling and peak-shaving technologies. A brief description of the components of a SMES system is given below and then a recent cost estimate is presented. This may be compared with some cost estimates in Papers 18 through 22.

The major components of a superconducting energy-storage system are: superconducting coil, dewar, rock support, refrigerator, AC to DC to AC converter with transformer, computer, and reactive power compensation. To minimize costs, the super-conducting coil will be a simple solenoid consisting of many turns of a superconductor wrapped in a helical pattern. The coil may consist of one or more layers, depending on the desired current and field. The superconductor will be cooled either by being placed in a bath of liquid helium or, if it is fabricated as a hollow conductor, by the forced flow of liquid or other low-temperature helium.

The dewar is a structure consisting of an outer vessel, which has air and rock outside it and a vacuum within, and an inner vessel, which is surrounded by the vacuum and which contains the superconducting coil.

The outward forces of small superconducting coils are usually taken up by stainless steel bands. Special stainless steels that remain ductile at cryogenic temperatures are used instead of carbon steels, which become brittle. These forces in a large magnet for diurnal storage, which might be 300 m across and 100 m high, could be contained by stainless steel bands, but the cost would make SMES uneconomical. Instead it is proposed that the SMES unit be constructed underground and the forces be transmitted to the surrounding rock. The rock thus will contain the magnetic forces just as it usually contains the forces of the same magnitude in a dam. The structure for transmitting the forces from the magnet at 4 K to rock at some ambient temperature near 300 K must have a high compression strength and a low thermal conductivity. A fiber glass-epoxy composite is the most likely candidate for the strut material at present.

Heat that flows to the coil, either along the struts or by radiation from the rock walls, will be removed by a cryogenic refrigerator. Refrigerators operating over the large temperature difference, 300 K to 4K, are limited in their cooling capacity by the laws of thermodynamics and some unavoidable inefficiencies. These refrigerators require 200 to 500 times as much energy at 300 K as they can remove at 4 K. Thus, it is imperative to limit the heat flow into the coil at 4 K.

The ac to dc to ac converter will convert the ac power into dc for charging and discharging the magnet. The converter will consist of one or more three-phase Graetz bridges, which are each composed of six silicon-controlled rectifiers (SCRs).

The current in the superconducting coil always flows in one direction so that only the voltage on the bridge need be changed to change the direction of energy flow. The output voltage on the Graetz bridge is changed by simply adjusting the instant during the 60 Hz (50 Hz for Europe) cycle that a firing or turn-on pulse is supplied to each SCR. The power reversal associated with the change from charge to discharge and vice-versa has been accomplished in less than 10 ms in laboratory tests designed to simulate large storage systems. Thus a SMES unit should be able to follow rapid power changes in addition to meeting the diurnal-load variations. A step-down transformer will be required to convert the ac power that may be 500 kV and 5000 A to perhaps 50 kV and 50 kA for the superconducting coil.

The computer is for data acquisition and control. It will monitor all the components of the SMES system and also the external power system.

The first three papers in this section describe large superconducting magnet systems proposed for electric utility applications. Paper 18 examines the costs of superconducting magnet systems and gives a cost comparison with other load-leveling systems.

Papers 19 and 20 also describe an underground superconducting coil, discuss possible construction and operating costs, and cite the advantages of SMES over other systems. Paper 21 looks at our present understanding of superconducting wires and also addresses the technology of designing superconductors to meet various energy-storage-coil requirements. The last paper in this section, Paper 22, focuses on the use of magnetic energy to drive a pulsed resistive load. The system described will be used for fusion research.

BIBLIOGRAPHY

Brechna, H. 1973. *Superconducting Magnet Systems.* Springer-Verlag, New York.

Cohen, M. H., ed. 1968. *Superconductivity in Science and Technology.* University of Chicago Press, Chicago.

Foner, S., and B. B. Schwartz, eds. 1974. *Superconducting Machines and Devices: Large System Applications.* Plenum Press, New York.

Gregory, W. D., W. N. Mathews, Jr., and E. A. Edelsack, eds. 1973. *The Science and Technology of Superconductivity,* vols. I and II. Plenum Press, New York.

Herbstschüle über Anwendung der Supraleitung in Electrotechnik und Hochenergiephysik, 1972, Proc. Ges. Kernforsch., Karlsruhe.

Lynton, E. A. 1964. *Superconductivity.* Methuen & Co., London.

Meyerhoff, R. W., ed. 1977. *Int. Conf. Manuf. Supercond. Mater.,* Port Chester, New York, Nov. 1976, *Proc.* American Society of Metals.

Saint-James, D., G. Sarma, and E. J. Thomas. 1969. *Type II Superconductivity.* Pergamon Press, Oxford.

Shoenberg, D. 1965. *Superconductivity*. Cambridge University Press, Cambridge.

Summer Study on Superconducting Devices and Accelerators, 1968, Proc. Brookhaven National Laboratory, New York.

Tinkham, M. 1975. *Introduction to Superconductivity*. McGraw-Hill, New York.

[*Editor's Note:* The proceedings of the following meetings frequently have up-to-date information on superconductivity, superconducting magnets, and cryogenic systems: International Conference on Magnet Technology (biannual meetings), International Cryogenic Engineering Conference (biannual meetings), Applied Superconductivity Conference (ASC, biannual meetings) published by IEEE in their *Transactions on Magnetics*, Cryogenic Engineering Conference (meetings held on alternate years with ASC) published as *Advances in Cryogenic Engineering* by Plenum Press, and Symposia on Engineering Problems of Fusion Research (biannual meetings).]

18

Copyright © 1975 by the Institute of Electrical and Electronics Engineers, Inc.
Reprinted by permission from IEEE Trans. Magn. **MAG-11**:482–488 (1975)

WILL SUPERCONDUCTING MAGNETIC ENERGY STORAGE BE USED ON ELECTRIC UTILITY SYSTEMS?*

William V. Hassenzahl[†]

ABSTRACT

As the cost of fossil fuel has increased and the load factors on electric utilities have decreased, the need for efficient, reliable energy storage systems has increased. Although pumped hydro storage is now used extensively on those utility systems having the appropriate resources nearby, it is only 65% efficient. Superconducting magnetic energy storage which promises to be more than 90% efficient and easily sited may become a competitive energy storage technology. A comparison of the various energy storage systems is presented in terms of performance on electric power systems, and cost. Emphasis is given to the various technologies involved in the development of large superconducting magnets. A brief review of the Los Alamos Scientific Laboratory program on superconducting magnetic energy storage is included.

I. INTRODUCTION

The cost of fossil fuels[1,2] has increased at an alarming rate in the past year. Simultaneously the capital cost of nuclear power plants has increased to $400-$500/kW. To make the electric power situation even more difficult the load factor, the ratio of average power to peak power or the fraction of available power capacity being used, has dropped to about 50%.[3] Energy storage has been used on a limited scale in the past to increase the load factor and thereby utilize more of the relatively inexpensive offpeak power.[4] The installations in the past have been in the form of pumped hydro facilities such as the unit at Ludington, Michigan.[5]

The need for energy storage systems which have few geographical constraints, have high efficiencies (pumped hydro has an efficiency of only 65-70%) are needed at the present time. The demand for this type of system will continue to increase in the future as electric power systems grow.

Superconducting magnetic energy storage is in many ways an ideal energy storage system for the electric utilities. It is relatively small in size, it can be sited in many geographical areas, in particular near the load centers such as major metropolitan areas, and it is expected to have an efficiency of nearly 95%. An additional benefit lies in the characteristics of the SCR converter-inverters which are the interface with the electric power system. These will provide a complete range of power control and can respond to system demands in a few milliseconds. This feature will be effective for system stability and automatic generation control. The pumped hydro storage units now in use typically have discharge times greater than 8 hours. Any storage unit which would be used for system or inter system stability could have excess converter capacity which would reduce the discharge time to tens of minutes.

The costs of various superconducting magnetic energy storage systems appropriate for electric utility systems are described in section II. Systems with a variety of energy storage capacities have been considered. These range in size from an 11,000 MWh (3.9 x 10[7] MJ) unit which has a storage capacity

Manuscript received September 30, 1974

*Work performed under the auspices of the U.S.A.E.C.

approximately equal to the pumped hydro installation at Ludington, to an 8 MWh (3 x 10[4] MJ) unit which might be considered a prototype and which could be effectively used to study the effect of these devices on system stability. In section III these are compared to the costs of other energy storage and load leveling systems.

The Los Alamos Scientific Laboratory is exploring the development of superconducting magnetic energy storage systems for electric utility applications. A summary of the LASL program[6,7] is given in section IV.

II. COSTS OF SUPERCONDUCTING MAGNETIC ENERGY STORAGE SYSTEMS

The costs of the various components of superconducting magnets have been described by Lubell, et.al.[8] and more recently by Hassenzahl, et.al.[9] The component costs described in these references are still approximately correct, with the major cost items being

    a) Superconductor
    b) Reinforcing and support structure
    c) Dewar
    d) Refrigerator
    e) Converter-Inverter
    f) Fabrication

The cost of each of these items for toroidal and solenoidal energy storage systems is briefly described in this section. The cost of fabrication is not independently considered. However, it has been included in the dewar and structure costs.

Because of the present rate of inflation, it is quite difficult to predict the costs of fabrication and materials for a construction project which will be carried out some 10 or more years in the future. Thus, to make a relevant comparison with other technologies, the cost of materials and processes which have been determined in the last year are used in the cost comparisons.

II-1 Superconductor

The superconducting materials cost which may be expected in the future when large scale magnet systems are being routinely constructed has been described by Powell.[10] These costs are considerably lower than present day costs of small orders. However, they appear to be reasonable when the efficiencies of large scale, mass production are considered. (At the current rate of inflation superconductor costs may not decrease, in an absolute sense, but should decrease relative to other costs.) A summary of the cost of various superconducting materials from Powell is given in Table I. The first three entries in Table I are for 4.2 K operating temperature, and the last entry is for 1.8 K. The smaller figure of merit for NbTi at 1.8 K implies, for a given amount of stored energy, that only 58% as much NbTi will be required at 1.8 K as at 4.2 K.

The calculation of the quantity of superconductor, Q_{sc}, required for toroidal magnets with circular cross section is straightforward if the winding thickness is small compared to the minor radius. The value of Q_{sc} is given by

[†]Los Alamos Scientific Laboratory, Los Alamos, New Mexico, 87544.

$$Q_{sc}(kAm) = 5 \times 10^3 \left| \frac{E_m^2}{B_m} \right|^{1/3} \times \left\{ \frac{1}{\left[(\frac{r}{a} - 1) \left[\frac{r}{a} - (\frac{r^2}{a^2} - 1)^{1/2} \right] \right]^2} \right\}^{1/3}$$

where

E_m = Maximum stored energy (MJ)
B_m = Maximum field (T)
r = Major radius
a = Minor radius

Calculations of the quantity of superconductor required for toroids with noncircular cross sections have been carried out by Kaiho.[11]

One general result which applies to all geometries of magnetic energy storage systems is that Q_{sc} is proportional to $E^{2/3}$. Thus the unit cost of superconductor, in \$/MJ, will be less for large magnets than for small magnets.

Although the most reliable cost estimates made to date have been for toroidal magnetic energy storage systems, this geometry is inherently more expensive than the solenoids due to the inefficient utiliza-tion of superconductor. The main disadvantage of solenoids is the fringe field which may extend for very long distances. However, it is possible to shield solenoids or to locate them in regions where the stray field is not a problem. One geometry which has been considered for the solenoid consists of a cylinder with its diameter and height approximately equal and with a relatively thin winding of superconductor. For ex-ample, a 31-m mean diameter, 30-m long solenoid is 80% as efficient in terms of energy stored per unit volume of superconductor as the most efficient rectangular geometry, a Brooks coil. A torus with r/a = 3 is only ≈ 36% as efficient.

The quantity of superconductor required for sole-noidal energy storage magnets may be estimated by a variety of techniques. A general source of informa-tion on inductances is the book by Grover.[12]

The cost of superconductor for several energy storage magnets are given below in Table II

TABLE I

The cost of superconducting materials, from Powell[3]

Superconductor	Field (T)	Cost (\$/kAm)*	Cost/operating field (\$/kAmT)
NbTi	5.0	0.36	0.072
Nb₃Sn	10.0	0.87	0.087
V₃Ga	16.0	3.9	0.24
NbTi @ 1.8 K[†]	7.0	0.29	0.041

*Costs of superconductor are based on 30% inflation from Powell's values using 1971 dollars

[†]Calculation of costs for 1.8 K superconductor are based on measurements of existing materials. With optimized heat treatment etc. the costs for 1.8 K material could be even less.

TABLE II

Quantity of Superconductor Required for Energy Storage Magnets

Configuration	Quantity of S.C. (kAm)	Cost[a] \$ x 10⁶	Cost[b] \$ x 10⁶
3.0 x 10⁴ MJ Torus r/a = 3	6.2 x 10⁶	4.0 x 10⁰	2.5 x 10⁰
10⁶ MJ Torus r/a = 3	6.4 x 10⁷	4.2 x 10¹	2.5 x 10¹
10⁶ MJ Solenoid d = h	2.9 x 10⁷	2.0 x 10¹	1.2 x 10¹
3.9 x 10⁷ MJ Torus r/a = 3	7.4 x 10⁸	4.8 x 10²	2.9 x 10²
3.9 x 10⁷ MJ Solenoid d = h	3.3 x 10⁸	2.2 x 10²	1.3 x 10²

a) Superconductor used in low field region identical to superconductor in high field region.
b) Superconductor adjusted for local maximum field intensity, 40% cost savings assumed.

II-2 Structure

The structure for a superconducting magnet consists of the reinforcing material which resists the $J \times B$ forces, and the material which supports the weight of the coil and the reinforcing structure. The support material is a small part of the total structure and is included in the dewar costs.

The reinforcing material in small magnet systems is the copper and superconductor in the coil windings. However, for larger coils additional material under tension, typically in hoop stress, is required. For very large coils where the strength of the conductor is insignificant, the minimum mass of reinforcing material[13] is given by

$$M = \kappa \frac{\rho E}{\sigma}$$

where

ρ = density of the material (kg/m^3)
E = maximum stored energy (J)
σ = tensile strength of the material (Pa)
$\kappa \geqslant 1.0$

The factor κ is a geometrical constant which would be 1.0 if all the reinforcing material were in biaxial tension but which is greater than 1.0 for all practical cases.

For stainless steel used at cryogenic temperatures, the fabricated cost based on $\kappa = 1.1$ will be on the order of $26/MJ. For a system with an 8 hr discharge this corresponds to a cost of $900/kW, which is greater than the installed cost of a generation plant. Thus it will not be economical with the present state of the art to use magnetic energy storage systems having cold reinforcement for load leveling. For systems which are to be charged and discharged over shorter time periods the cost per kW will decrease. For example, if the converter-inverter capacity is such that the system can be discharged in 30 min, the reinforcement will only contribute $56/kW to the cost of the facility.

A technique which appears quite promising for the systems which could be used for load leveling on a daily cycle is to construct the magnet underground and use rock as the reinforcing material.[6,14,15] One possible application of magnetic energy storage using this technique is shown in Fig. 1.

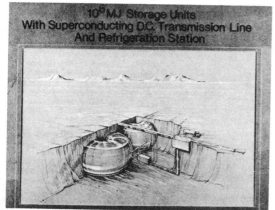

Fig. 1.   Two 10^6 MJ superconducting magnetic energy storage systems installed underground using rock reinforcement. One possible application as shown here is to use SMES systems in conjunction with dc superconducting transmission lines.

The cost of the materials required to transmit forces to a rock support structure has been estimated to be about 25% of the cost of cold stainless steel support. This value has been used in the estimates of warm reinforcement costs which are listed along with other costs in section II-6 below.

II-3 Dewar Costs

At the present time a large liquid helium dewar will cost between $2500 and $3000 per square meter of surface area. This depends to a certain extent on size, shape, boiloff rate, etc. An estimate of dewar costs for large toroidal magnet systems was made by Cryogenic Consultants Inc.[16] The dewar costs given in Table III are based on the CCI calculations. These include the cost of support structure, intermediate cooling stations, etc.

TABLE III

Dewar costs for large toroidal superconducting magnets. Warm reinforcement costs include structure to transmit forces from the magnet at 4 K to the rock support. Total dewar cost and cost per square meter of surface are given.

Dewar Type	Dewar Cost			
	3×10^4 MJ		3.9×10^7 MJ	
	$/m^2	$\times 10^6	$/m^2	$\times 10^6
Conventional conductor immersed in a helium bath				
i) Cold reinforcement	3030	4.6	2800	400
ii) Warm reinforcement	3220	4.8	6000	720
Forced helium flow in hollow conductor				
i) Cold reinforcement	1890	2.7	1770	220
ii) Warm reinforcement	2090	3.0	3400	500

II-4 Refrigerator Costs

The refrigerator requirements depend on the operating characteristics of a system as well as the structural characteristics. The three contributions to the refrigeration load are radiation heating, thermal conduction along the support structure and eddy current and hysteresis losses in the windings. For small units using cold reinforcement the refrigerator capacity is roughly proportional to the surface area of the dewar, with the radiation heat leak being about 0.05 W/m^2. Thermal conduction is only significant for the warm reinforcement systems and eddy current and hysteresis losses will become important for the units with cycle times of less than one hour.

The costs of helium refrigerators were estimated by Strobridge.[17] These costs have remained almost constant for the last 5 years, with increased production costs being offset by savings associated with a greater production volume. The refrigerator capacities and costs associated with the various energy storage magnets are presented in Table IV.

II-5 Converter-Inverter Costs

The cost of the converter-inverter system for high capacity dc transmission lines is approximately $30/kW.[18] A transformer to convert the high voltage, low current power associated with the electric power system to relatively low voltage, high current power for the energy storage device is included in this cost. The value $30/kW is a small fraction of the cost of any magnetic energy storage system. Thus

TABLE IV

Refrigerator capacities and refrigerator costs for
large superconducting magnets operating at 4.2 K. Refrigerator capacities are based on 8 hr discharge times.

Energy storage capacity (MJ)	Reinforcement type	Refrigerator capacity*(kW)	Refrigerator cost ($x10⁶)
3×10^4	cold	0.1	0.17
10^6	cold	0.8	0.5
10^6	warm	10.0	4.5
3.9×10^7	cold	8.2	1.3
3.9×10^7	warm	100.	12.0

*
Based on calculations in Ref. 16.

increasing converter capacity and decreasing cycle time
will decrease the cost-per-kW. In fact for small systems the relative cost of the converters is so small
that a tenfold increase in converter capacity results
in a factor of 9 <u>decrease</u> in per kW costs.

II-6 Total System Costs

The total cost of solenoidal energy storage systems
based on the component costs given earlier in this section, are included in Table V. A miscellaneous cost of
10% has been added to cover items such as a control
computer, personnel facilities, etc.

TABLE V

Total costs of magnetic energy storage systems having
solenoidal geometry.

Maximum Stored Energy (MJ)	$3x10^4$	10^6	$3.9x10^7$
Reinforcement	Cold	Warm	Warm
Superconductor Cost ($x10⁶)	1.2	12	130
Dewar Cost ($x10⁶)	2.7	30	400.
Structure Cost ($x10⁶)	0.8	(a)	(a)
Refrigerator Cost ($x10⁶)	0.17	4.5	12.
Converter Inverter Cost ($x10⁶)	0.1	0.9	41.
Miscellaneous Cost ($x10⁶)	0.5	4.7	58.
Total Cost ($x10⁶)	5.5	52.1	641

(a) Included in dewar cost.

The costs of magnetic energy storage systems having toroidal geometries are given in Ref. 9.

III. COMPARISON OF ENERGY STORAGE SYSTEMS

There are a variety of ways to compare the many
energy storage systems which can be used on electric
power systems. These include such simple methods as
comparing the initial capital costs in $/kW. However,
for a realistic comparison the operation of the device
on a utility system must be considered. For example
pumped-hydro installations have been used in the past,
but available sites are diminishing and those which do
exist are at large distances from power generation
facilities and load centers. These distances necessitate additional transmission lines and imply reduced
flexibility.

A few of the factors which must be considered for
proper comparison of energy storage technologies are
a) Capital cost
b) Efficiency of operation
c) Ease of siting and environmental considerations
d) Effectiveness of the storage device for "sys-
     tem stability"
e) Operation and maintenance costs
f) Expected life
One method of comparing the various technologies
which includes all of these factors is to calculate
the cost of delivered energy from the various devices
used for load leveling. This type of comparison was
first carried out by Kyle et.al.[18] and was extended
to include magnetic energy storage in Ref. 9. In this
technique a monetary credit is applied to those systems
which may be sited close to load centers, or at distribution stations because additional transmission
facilities are not required. Additional credits may
be assigned for those systems which have large power
delivery capacity or which can change power flow in
tens of milliseconds.

The results of calculations of the cost of delivered energy are included in Table VI below. The details of the calculations leading to the entries may
be found in Ref. 9 and 18.

TABLE VI

A comparison of delivered energy, in mills/kWh, for
various energy storage systems based on off peak energy
costs of 10 mills/kWh.

Operating time hours/yr	2000	3000
Pumped-hydro, 0.67 eff Capital cost = $220/kW	33	29
Gas turbine Capital Cost = $150/kW Fuel Cost = 300¢/10⁶ Btu	52	49
Batteries[a] 10h/day discharge, 0.7 eff Capital Cost = $15/kW		
5 year life	30	26
3 year life	45	38
SMES Solenoid warm reinforcement[a,b] 3.9 x 10⁷ MJ	30	26
SMES Solenoid cold reinforcement[a,b,c] 3 x 10⁴ MJ	60	51

a) Units given $100/kW credit since fewer geographi-
    cal restrictions allow siting near load centers.
b) SMES unit has additional converter capacity for the
    discharge. Capital cost increased appropriately,
    and system stability credit of $15/kW/yr given.
c) Small unit for system stability, converter capabili-
    ty allows 30 min discharge, with continuous charge-
    discharge at half of power capacity, 75% efficiency
    assumed.

The cost of delivered energy given in Table VI
indicates that for large systems SMES will be competitive with other alternatives for load leveling. However, for smaller systems the only economical application will be for system stability and peak shaving.

There has been a major change in the costs of
delivered energy since Refs. 9 and 18 were completed,
in that the cost of oil and other petroleum products
on the international market has increased by about a
factor of three in the last year. This mainly affects
the economy of gas turbine peak shaving systems.
Two years ago, when the cost of #2 fuel oil was 75-
100¢/10⁶ BTU, no one believed it could reach 150¢/10⁶
BTU by 1980, and as recently as 15 months ago the

extrapolation to 175¢/10⁶ BTU was thought to be outrageous. Today the cost of #2 fuel oil is about 300¢/10⁶ BTU and will inevitably increase by 1980 if for no other reason than inflation. The cost of delivered energy increases by 12 mills/kWh for each increase in fuel costs of 100¢/10⁶ BTU.

An additional advantage of an energy storage system which has excess power capacity is that it may be used to defer the installation of additional generation. This is due to the utility requirements for spinning reserve. In fact the response time and control capability of the SCR converters make this device more effective as spinning reserve than the inertial characteristics of generators.

IV. THE LOS ALAMOS SCIENTIFIC LABORATORY SUPERCONDUCTING MAGNETIC ENERGY STORAGE PROGRAM.

In the process of developing the technology of superconducting magnetic energy storage it is useful to develop model systems which test the principle of energy storage and its application to electric power systems. These models also provide information about the relevant component technologies and where improvements or advances are required. The LASL program has been a combined effort a) to analyse future power system requirements and how magnetic energy storage will compete with other storage technologies and b) to develop an understanding of the performance of magnetic energy storage systems by model construction.

The initial steps in the former stage led to the concept of underground construction with the use of natural rock for reinforcement, which has been followed by analyses of costs of various systems and the cost comparisons described in section II and III.

The studies of model systems began some time ago with the operation of an electric train from a 10-kJ energy storage magnet through a transistorized converter unit. The second stage, which is not yet completed, is a storage system which will eventually be composed of 16 pancake solenoids which may be assembled in toroidal or solenoidal configurations. The coils and the dewar in which the tests are being performed are shown in Figs. 2 and 3. At the present time 9 coils have been fabricated. The maximum stored energy achieved to date was 91-kJ when five of the coils shown in Fig. 4 were tested.

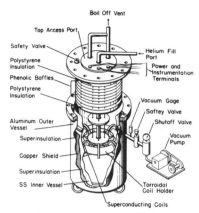

Fig. 3.   Artists conception of toroidal configuration of pancake solenoids in liquid helium dewar.

Fig. 4.   Stack of six pancake solenoids. The lower five coils were tested at 4.2 K and reached 65 A at 48 kG equivalent to 91 kJ stored energy.

The next step in the LASL program will be to construct a 100-MJ energy storage system as shown in Figs. 5, 6, and 7. This will consist of a 3 meter diameter solenoid in a liquid helium cryostat, a liquid helium refrigerator, an SCR inverter converter system, and the necessary control and monitoring instrumentation. The specifications of the system are given below in Table VII.

There are several technical problems which must be considered before large superconducting magnetic energy storage devices will be used on electric utility systems. Some of these are

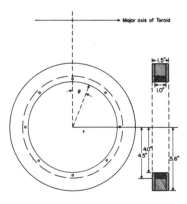

Fig. 2.   Pancake solenoids used for model energy storage magnet. Nine units have been constructed with approximately 3000 turns of 0.53 mm diameter wire.

TABLE VII

Preliminary design parameters of the Los Alamos Scientific Laboratory 100 MJ Energy Storage Magnet

Coil		
Inner Diameter	3.0	m
Outer Diameter	3.3	m
Height	2.25	m
Weight	10,000	kg
Maximum Stored Energy at 4 K	100	MJ
Dewar		
Inner Diameter	2.2	m
Outer Diameter	4.2	m
Height	4.65	m
Converter capacity		
Power	1.5	MW
Voltage	500	V
Current	7000	A

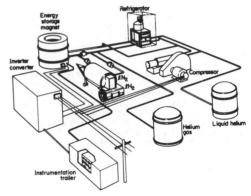

Fig. 7. Installation of 100 MJ energy storage coil including inverter, converter, instrumentation trailer, refrigerator and other equipment.

a) Advanced conductor design
b) Electrical insulation at low temperatures
c) Load bearing insulators
d) High current leads
e) Cyclic stress characteristics of materials at low temperatures
f) Operation of a small SMES unit on utility systems
g) Rock containment structures.

The LASL 100 MJ project will address most of these problems except the last, the use of rock structure, which is not feasible with such a small unit.

Advanced conductor design will be investigated in two separate areas. The first will be to use a NbTi, Cu, CuNi superconductor for the 3.0 m diameter coil. This material will have low hysteresis and eddy current loss characteristsics. Two conductor geometries have been considered for this purpose. The first is a solid conductor having a cross section of 1.25 x 0.8-cm and the second is a conductor composed of three separate wires each having a cross section of 0.5 x 0.5-cm. These are shown in Fig. 8. Space is provided in the high field region in the cold volume for testing a second advanced conductor design, a hollow superconductor using forced helium flow. Tests will be made with forced flow of both superfluid and supercritical helium.

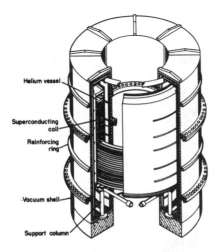

Fig. 5. Artists concept of 100 MJ energy storage coil in liquid helium cryostat.

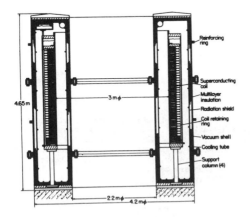

Fig. 6. Cross section of 100 MJ energy storage coil.

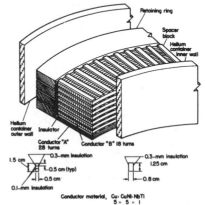

Fig. 8. Conductor and windings for 100 MJ superconducting magnetic energy storage system. Helium container and retaining ring support structure are shown.

The load bearing insulator between the coil and the reinforcing stainless steel, as shown in Fig. 8, will provide an excellent test of b), c) and e) above. The material will be subjected to a cyclic stress at one cycle per 5 minutes which will be simultaneous with the charging voltages.

High current leads with a steady state current carrying capacity of greater than 7000 A will be required. Only a few leads with a current capacity of this magnitude have been constructed in the past.

The 100 MJ unit will be installed at Los Alamos and tied to the electric power system which receives a large fraction of its power from New Mexico Public Service Co. The unit has been designed to be small enough to be transported, making tests on other power systems possible after the testing phase at Los Alamos, or to allow the magnet to be used for other purposes.

SUMMARY

Superconducting magnetic energy storage appears to be an alternative to other energy storage technologies. The capital cost will be greater than for some of the competitive technologies; however, the expected high efficiency makes it very attractive with the decreasing supplies and rising costs of fossil fuels.

The first application of this technology will be small units on utility systems for the purpose of peak shaving and system stability. To be economical, or nearly so, units with a capacity of at least 3×10^4 MJ will be required. Units of this size could be constructed and installed on utility systems in the 1980's.

REFERENCES

1. R. A. Fernandez, "Hydrogen Cycle Peak-Shaving for Electric Utilities," Proc. Intersociety Energy Conversion Engineering Conference, 9th, San Francisco, 1974, Paper 749032, p. 413-422.

2. Philip N. Ross, "Implications of the Nuclear-Electric Economy," Westinghouse Electric Corp. publication, presented at Conference on Research for the Electric Power Industry, Washington, D.C., Dec. 1972.

3. "Power Generation at the Top 100 Electric Utilities," Electric Light and Power, p. 30-31, July 1973.

4. S. R. Knapp, "Pumped Storage — the Handmaiden of Nuclear Power," IEEE Spectrum, p. 46-52, April 1969.

5. "Huge Pumped-Hydro Nears Completion," Electric World, p. 60-63, Sept. 1973.

6. E. F. Hammel, W. V. Hassenzahl, W. E. Keller, T.E. McDonald, and J. D. Rogers, "Superconducting Magnetic Energy Storage for Peakshaving in the Power Industry," Los Alamos Scientific Laboratory report LA-5298-MS, June 1973.

7. W. V. Hassenzahl, R. D. Turner, "Progress Report USAEC Division of Applied Technology Superconducting Magnetic Energy Storage Project at LASL," Los Alamos Scientific Laboratory report, LA-5588-PR, April 1974.

8. M. S. Lubell, H. M. Long, J. N. Luton Jr., and W. C. T. Stoddart, "The Economics of Large Superconducting Toroidal Magnets for Fusion Reactors," Oak Ridge National Laboratory Report, ORNL-TM-3927, August 1972.

9. W. V. Hassenzahl, B. L. Baker, W. E. Keller, "The Economics of Superconducting Magnetic Energy Storage Systems for Load Leveling: A Comparison with Other Systems," Los Alamos Scientific Laboratory Report, LA-5377-MS, August 1973.

10. James R. Powell, "Costs of Magnets for Large Fusion Power Reactors: Phase I, Cost of Superconductors for DC Magnets," Brookhaven National Laboratory Report, BNL 16580, Feb. 1972.

11. Katsuyuki Kaiho, "Consideration on the Superconducting Coil for Energy Storage," Bulletin of the Electrotechnical Lab., 37, 1091-1100, Tokyo, Japan, 1973.

12. F. W. Grover, Inductance Calculations, D. Van Nostrand Co., Inc., New York, 1946.

13. R. H. Levy, ARS Jour.,p. 787 (1962).

14. J. R. Powell and P. Bezler, "Warm Reinforcement and Cold Reinforcement Magnet Systems for Tokamak Fusion Power Reactors: A Comparison," Brookhaven National Lab. Report BNL 17434 (Nov. 1972).

15. R. W. Boom, G. E. McIntosh, H. A. Peterson and W. C. Young, "Superconducting Energy Storage," presented at the Cryogenic Engineering Conference, Atlanta (August 1973).

16. Henry F. Daley Jr., "Cost Estimates for Large Liquid Helium Dewars for Containing Superconducting Magnetic Energy Storage Devices." Study prepared under contract to the Los Alamos Scientific Laboratory (June 1973).

17. T. R. Strobridge, Trans. IEEE, Nucl. Sci. NS-16, p. 1104, (1969).

18. M. L. Kyle, E. J. Cairns, and D. S. Webster, "Lithium/Sulfur Batteries for Off-Peak Energy Storage: A Preliminary Comparison of Energy Storage and Peak Power Generation Systems." Argonne National Laboratory report ANL-7958.

19

Paper presented at *Energy Storage: User Needs and Technology Applications*,
Eng. Found. Conf., Technical Information Center, ERDA, 1976, 16pp.

MAGNET DESIGN FOR SUPERCONDUCTIVE ENERGY STORAGE
FOR ELECTRIC UTILITY SYSTEMS*

R. W. Boom, B.C. Haimson, M.A. Hilal, R. W. Moses,
G.E. McIntosh, H.A. Peterson, R.L. Willig, and W.C. Young
College of Engineering
University of Wisconsin-Madison
Madison, Wisconsin 53706

ABSTRACT

The design of large superconductive single layer segmented solenoids for
energy storage in large electric utility systems is discussed. Such a unit
would be charged at night when demand is low and discharged during the peak
demand daytime hours. Typical uses are 2-10 h discharges at rates of
500-2000 MW. The coil is mounted in bedrock with forces carried by the rock.
The structural and cryogenic aspects of designing a low loss cryogenically
stable solenoid limited by cost optimization are presented. With conservative
cost estimates it is shown that such units might add only approximately
0.018$/kWh to the low nighttime power cost for competitive daytime resale.

INTRODUCTION

Superconductive energy storage has been considered since the discovery of
high field superconductors in 1961 [1,2,3,4,5,6,7,8]. The Wisconsin studies
began in 1970 [9] with the realization that a three-phase ac/dc Graetz bridge
converter provides an ideal interface between a dc energy storage coil and a
large three-phase ac utility system; together forming an inductor-converter
(I-C) unit. The next important suggestion was the use of bedrock instead of
expensive cold structure [10]. The Wisconsin group has conducted conceptual
feasibility studies for the past three years [11,12,13,14,15,16] which have
resulted in conservative cost estimates. A typical representative cost was
$500 x 10^6 for a 10,000 MWh unit. Recent engineering design improvements,
primarily to reduce the amount of material needed, give promise of reducing
this cost to about $300 x 10^6.

During the early period the emphasis was on engineering solutions to
such problems as force containment, force transmitted from low temperature
conductors to ambient temperature rock faces, composite conductor stability,
conductor manufacture and winding, magnet assembly and system analysis. A
10,000 MWh solenoid at 5 T was about 100 m radius by 100 m long. The preference was a one layer solenoid 25 cm thick with radial ripples in the conductor
and dewar to allow for thermal contractions and magnetic force cycling without
mechanical motion at the nodes or support points. The major energy loss is
the power required to run the refrigeration system to cool off the solid epoxy
fiberglass support struts between the conductor and rock.

*Presented in part at the International Conference on Magnet Technology -
Frascati (Rome), Italy, April 21-25, 1975.

246

TiNb-aluminum composite conductors operated at 1.8 K are best; TiNb because of its ease in fabrication and aluminum because it is about 1/7 the cost of copper for equivalent usage and is easier to form. Superfluid helium was selected because of its superior heat transfer characteristics. Epoxy supports impose unusual conditions on superconductor optimization; for example, a composite conductor at 1.8 K will require only about twice the total refrigeration compared to a composite conductor operated at 20 K, since the major loss is through the epoxy fiberglass struts. The result is a gross insensitivity to improvements in T_c of the superconductor or to reduction of temperatures from 4.2 K to 1.8 K. The improvements in J_c and heat transfer predominate in optimization.

The above points were discussed previously [12] and, to some extent, are amplified in subsequent sections of this paper. The main points in this paper deal with magnet design and introduce newer concepts, such as a three-tunnel solenoid, a conceptual "hour-glass" solenoid with forces at right angles to the rock walls, optimized strut cooling schemes, and an engineering design optimization which leads to the promise of 40% cost reductions.

ELECTRICAL

The dc solenoid buried in bedrock is charged and discharged with a three-phase Graetz bridge converter such as those used throughout the world for dc power transmission. Such converters using solid state controllable thyristors complete with transformers, connectors, filters and control circuitry cost about 40$/kW at rated power. The simplest description of the converter operation is that the large dc irreversible current in the storage solenoid is commutated by the converter into three 120° phase segments into or out of the three-phase power lines. The converter applies a voltage to the inductor which can be + for charging, – for discharging, near zero for idling, or any ± V as demanded. The control between $\pm V_{max}$ and 0 volts is accomplished by adjusting the control pulse on the thyristor grids; power flow direction can be changed very quickly, even within a fraction of a cycle, to store or return energy alternately from the utility system upon command without switching.

Discussions in the previous papers cover supplementary uses for I-C units in power systems, such as AGC (automatic generation control), transient stability regarding major disturbances and voltage regulation [12,14,15,16]. The major use, of course, is the diurnal storage and release of energy.

MAGNET DESIGN

The virial theorem regarding the required structure to confine a magnetic energy W is such that cold stainless steel structure is prohibitively expensive [11,12,17]. The same virial theorem applies for bedrock support, except that an essentially infinite mass of structure is available for the cost of excavation. Magnets must still be held together by structure in tension; for rock, tension is obtained by relieving the residual compression in the rock. Such residual compressions occur in the bedrock due to overburden and rock formation stresses. It is not necessary that the bedrock be a single intact mass but only that joints remain closed under compression. If the tension in the rock in an annular region would exceed residual compression, then, of course, no tension could exist because of joints and the outward pressure would be transmitted to larger radii (reduced by $1/r^2$) and contained where tension can exist.

Optimization for magnets with warm support is completely different than for magnets with cold support, primarily because the storage unit cost has little relationship to the mass of the structural rock needed. The three largest cost items, the conductor, the epoxy struts, and the dewar, are related respectively to: (1) the ampere-meters IS required, (2) the total force F carried from cold to warm structure, and (3) the surface area A of the magnet. Each of the three quantities should be as small as possible for a given stored energy. Moses [12,18] has shown that for solenoids:

$$W = \frac{\pi}{2} \mu_o n^2 R^2 \ell K(\beta) I^2 = \frac{\pi R^2 \ell K B_M^2}{2\mu_o K'^2} \tag{1}$$

$$R = (\frac{\mu_o K'^2}{\pi \beta K})^{1/3} \; W^{1/3}/B_M^{2/3} \equiv G(\beta) W^{1/3}/B_M^{2/3} \tag{2}$$

$$F_r = 2(1 - \frac{\beta}{2K} \frac{\partial K}{\partial \beta})(\frac{\pi \beta K}{\mu_o K'^2})^{1/3} \; W^{2/3} B_M^{2/3}$$

$$\equiv Q_{fr}(\beta) \; W^{2/3} B_M^{2/3} \tag{3}$$

$$IS = 4(\frac{\pi \beta K'}{\mu_o K^2})^{1/3} \; W^{2/3}/B_M^{1/3}$$

$$\equiv Q_{is}(\beta) \; W^{2/3}/B_M^{1/3} \tag{4}$$

$$A = 8(\pi \mu_o^2 \beta K'^4/K^2)^{1/3} \; W^{2/3}/B_M^{4/3}$$

$$\equiv Q_a(\beta) \; W^{2/3}/B_M^{4/3} \tag{5}$$

where, in MKS units, W is total stored energy, F_r total radial force, I conductor current, B_M maximum field at midplane, R solenoid radius, ℓ solenoid length, $\beta = \ell/2R$, K a shape factor which approaches one for long solenoids [19] and K' a shape factor from the equation $B_M = \mu_o nIK'(\beta)$, where n is turns per meter. Several values of K and K' are given in Table I.

Table I. K Factors

β = 0.1	0.2	0.3	0.4	0.5	1	2
K = 0.203	0.320	0.405	0.472	0.526	0.688	0.818
K'= 0.570	0.617	0.656	0.690	0.719	0.821	0.917

Q_f for the axial force, the radial force, the linear force sum, and the vector force sum are plotted in Fig. 1. The ordinates are β and Q_f. Similar curves not shown for Q_{is} and Q_a [12,18] lead to the following conclusions: (1) Q_f vector is 20% less than Q_f scalar with either quite satisfactorily close to minimum below β= 0.5, (2) Q_{is} has a broad minimum around β= 0.3 and (3) Q_a has a broad minimum below β = 0.3. Therefore solenoids should be short solenoids with $\beta \cong 0.3$ and not above 0.5, see Table II for approximate values.

248

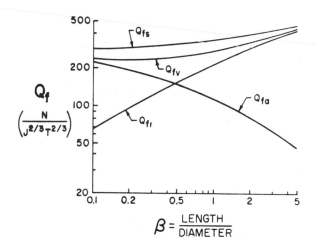

Fig. 1. Q force factors for solenoids.
$F = Q_f W^{2/3} B_M^{2/3}$. Q axial, Q radial, Q scalar
sum and Q vector sum shown.

Table II. Quality Factor Fluctuations

$0.1 < \beta < 0.5$	$\Delta Q/Q$ (%)
$Q_{fv} \leq 251$	5
$Q_{fs} \leq 327$	10
$Q_{is} \leq 604$	5
$Q_a \leq 1.07 \times 10^{-3}$	25

As mentioned earlier, we have become more interested in three tunnel
solenoids in which the central tunnel is twice as high and at a smaller radius
than the end tunnels, see Fig. 2. This unit is an "hour-glass" shaped
solenoid which reduces end field problems and provides six basic rock floors
and ceilings for mounting and axial force bearing surfaces. Without frequent
opportunities to carry axial forces to the rock surfaces it would be necessary
to employ a "virial theorem" amount of internal cold structure under compression.

In case the segmented solenoid is used the same Q factors apply except
that the maximum field B_M in eqs. (1-5) should be replaced by

$$[\frac{2K'\gamma}{1 + \gamma(2K' - 1)}] B_M \tag{6}$$

where γ is the fraction of the length occupied by conductors and β is an average β for the total solenoid. Replacing B_M by expression (6) is more accurate as the number of tunnels increases. The new term, eq. (5), is now multiplied by γ.

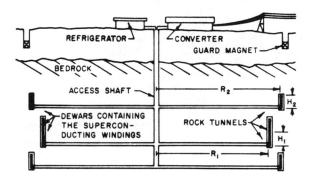

Fig. 2. Segmented solenoid. One layer coil in each dewar. Guard coil partially cancels external dipole moment; guard coil radius should be ~ 5 x R_2.

In reference [12] we have derived Q values for toroids and dipoles. In all respects a toroid is inefficient requiring twice as many ampere-turns. A dipole is quite similar to a low β solenoid and may be competitive following future detailed design.

STRUCTURE

In order to use static bedrock support it is necessary to maintain positive contact between the bedrock, dewar wall and conductor with epoxy fiberglass struts which are fixed at both ends. If the conductor and close fitting dewar are rippled or corrugated in shallow circular arcs, see Fig. 3, then the nodes can be fixed. Cool down contractions and magnetic field expansions are then taken up by shortening and lengthening the "free standing" part of the conductor between nodes. Magnetic tension on a conductor is T = BIR, with a ripple radius $R \sim 1.2$ m. Even with the reduced stress due to rippling it is necessary to reinforce the conductor with strong aluminum alloy occupying 1/3 of the cross-section. The conductor is 2 x 15 cm and carries a tensile load of 4.7×10^5 N at 5 T, almost all of which is in the reinforcing material. The conductor, epoxy-fiberglass struts and assembly generalizations are shown in Fig. 3.

ROCK MECHANICS

The need for inexpensive rock structure was presented above to result from the virial theorem. Excavation and wall preparation at 50$/m^3 for a three tunnel 10,000 MWh storage unit, Fig. 2, cost only 37$/kW for 10 h discharges. Recall that stainless steel cold structure for the same unit costs 6400$/kW. The selection of location can be determined by rock quality, freedom from ground water, generator, and load centers locations. A survey has shown that a great majority of the states within the U.S. (possibly excluding the Gulf Coast and Florida) contain rocks of suitable strength and extent at reasonable depths, which could be used for housing the magnet.

For fields of 5 T the average pressure on the rock is 9.95×10^6 N/m^2 (1443 psi), although the maximum pressure under foot-pads might be 10 times as

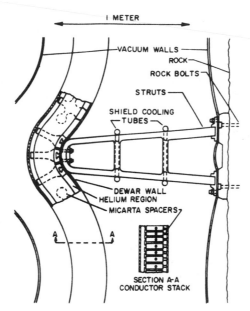

Fig. 3. Composite reinforced 2 cm x 15 cm conductor mounted with 1.2 m radius of curvature. Epoxy-
fiberglass strut is cooled at 11 k and 70 K. Super-
insulation not shown.

much at 9.95×10^7 N/m^2. The compressive strength of most igneous and meta-
morphic rock is about 3.0×10^8 N/m^2 while sedimentary rock is typically half
as strong at about 1.5×10^8 N/m^2. Some rock properties measured in Wisconsin
at potential rock structure sites are given in Table III.

Table III

Wisconsin Rock Properties in N/m^2

Rock Type	Compressive Strength	Tensile Strength	Young's Modulus	Poisson's Ratio
Granite	3.1×10^8	1.4×10^7	7.5×10^{10}	0.2
Rhyolite	4.1×10^8	1.4×10^7	7.5×10^{10}	0.2
Quartzite	3.3×10^8	1.7×10^7	8.3×10^{10}	0.1
Dolomite	1.4×10^8	0.8×10^7	6.9×10^{10}	0.3

The force generated in the conductor flows through the dewar wall, along
epoxy-fiberglass struts to foot-pads on the rock and then compresses rock out
to a radius where a circumferential tensile ring of rock confines the total
forces. The tensile load is taken by the relief of residual compression,
which could vary along the circumference from as high as 2×10^7 N/m^2 to as
low as 0. N/m^2. Forces will be balanced and contained with rock structure
within a few radii of the solenoid. The relatively low tensile strength of
rock is not always a factor since joints, which are often prevalent, would
always open up first.

A major geological engineering activity is to map out joint configurations using core drilling, seismic methods and surface mapping. The integrity, frequency and direction of joints must be known in order to orient the structure and the resulting forces to advantage.

The surface of a rock excavation can be left smooth and without overbreak by controlled blasting. Rock bolts will be used to prevent surface slippage and also to help relieve stress buildup around corners and edges of the structure. Finite element stress analysis around 3 tunnel structures has shown that surface tensions develop in a few locations, up to about 2×10^6 N/m^2, which can easily be reduced with surface shaping and rock bolting.

Access for repairs may require a scalloped surface made up by placing concrete ridges on the rock surface. Mounting foot pads could be mounted on the ridges with crawl holes between.

One final advantage of rock support is that, after consolidation on the first cycle, the rock could remain substantially rigid due to its tightly packed infinite mass. In contrast metallic structure sufficiently thick to remain elastic at all times would still deflect about 10 cm radially each cycle, which is an unnecessary daily flexure for all the components.

CRYOGENICS

Forced flow cooling with supercritical helium in hollow conductors was discarded [12] because of the excessively high fluid friction power loss. Pool boiling at 4.2 K is well understood. As applied to the closely mounted stack of conductors in Fig. 3 some disadvantages are: (1) narrow horizontal channels between turns rate only very low heat transfer values, possibly less than 0.1 W/cm^2, (2) narrow passages might induce vapor locking, and (3) the helium quality (entrained vapor bubble percentage) would be low in tall narrow annular regions possibly 10-20 meters high. Each of these problems could be overcome but only at the expense of more dispersed turns, larger coils, more helium volume and other size related cost increases without proportionate increases in energy stored.

Superfluid cooling seems to overcome all the above problems provided a critical heat flux of 0.5 W/cm^2 is not exceeded [20]. There is no bubble problem and cooling in narrow horizontal channels is possible. Completely stable magnet design is obtained with a minimum helium inventory. Pool cooling vs. hollow tube cooling for 1.8 K superfluid helium was considered but again hollow tube cooling was rejected because only 0.4 m of conductor could be reasonably stabilized at a heat flux of 0.5 W/cm^2 in the cooling tubes. A major concern has been possible "super-leaks." However, for such large units a "super-leak," which is just a small hole $\sim 1\mu$ in diameter, is also unacceptable at 4.2 K. Almost absolute integrity of all weld joints is necessary at any temperature since over 10,000 meters of dewar weld joints might be needed. The plan is to x-ray and helium leak test every weld before proceeding with construction.

The major energy loss of the total operating system is the heat leak along the solid epoxy struts. The heat would be intercepted by a baffle system of fixed temperature shields. In the idealized case [21] for an infinite number of shields each cooled with a perfect carnot refrigerator the minimum total power is:

$$P_{min} = T_H \frac{A}{\ell} \left[\int_{T_C}^{T_H} \frac{\sqrt{k}}{T} dT \right]^2 \qquad (7)$$

where P_{min} is the room temperature refrigeration power, T_C is the cold end temperature (1.8 K here) T_H is the hot end temperature (300 K), A/ℓ is the area/length of the epoxy struts and k is the thermal conductivity of the epoxy-fiberglass.

An optimization procedure has been devised by Hilal [12,21] to best select the number, location and intermediate temperatures of the shields. The results for T_C = 1.8 K and T_H = 300 K are compared with the ideal minimum power requirement. The values for 1, 2 and 3 shields cooled at carnot efficiency are $P\ell/A$ = 31.7, 21.8 and 19.1 W/cm respectively, while $P\ell/A$ for an infinite number of shields from eq. (7) is 16.3 W/cm as calculated for Narmco No. 570 epoxy-fiberglass [12]. On the basis of this comparison we note that the 2 shields are much better than 1 shield but not much worse than 3 shields. Therefore 2 shields are chosen as less complicated than 3 or more. The final temperatures and locations of the two shields are shown in Table IV. The length x_1 is the distance between the 1.8 K dewar wall and the 11.1 K shield, x_2 is the next interval to the 70.3 K shield and x_3 the final epoxy length between 70.3 K and 300 K. The above method of optimizing the station temperatures and locations simultaneously yields the best selections possible without uncertainties. The two shield design is seen to be only about 25% worse than the absolute perfection of an infinite number of shields.

The refrigerator for the 1.8 K system will be a five-engine design for maximum efficiency, which is about 800 watts/watt $\sim$ 20% carnot. Helium will be liquified and reduced in temperature to 1.8 K by pumping. The magnet will be immersed in helium with heat removed via the vapor while the 1.8 K helium makeup will be supplied from a reservoir.

Table IV

Thermal Shields for Mechanical Supports

T_C = 1.8 K, T_H = 300 K, Narmco 570 epoxy-fiberglass	Carnot Cycle	Actual Cycle
Conductor temperature	1.8 K	1.8 K
1st Shield temperature	13.4 K	11.1 K
2nd Shield temperature	69.7 K	70.3 K
Ambient (end) temperature	300 K	300 K
1st Shield location x_1/ℓ	.319	.305
2nd Shield location x_2/ℓ	.294	.324
x_3/ℓ	.387	.370
Total 300 K Refrigeration Power, $P\ell/A$, W/cm	21.8	94.5
Ideal Carnot, infinite no. of shields, $P\ell/A$, W/cm	16.3	--

Heat removed by pumping vapor through a pipe can be expressed [22] by:

$$Q' = 1.4 \times 10^5 \left(\frac{P_I^2 - P_O^2}{\ell}\right)^{.56} D^{.67} = 7.4 \text{ W/cm}^2, \tag{8}$$

where Q' is the flux in W/cm^2, $P_I = 1.65 \times 10^4$ dynes/cm^2 the input pressure (12.5 torr), $P_O = 1.31 \times 10^4$ dynes/cm^2 (10 torr) the output pressure, the pipe length $\ell = 10^4$ cm and pipe diameter $D = 20$ cm. This method of heat removal is preferred because the tube is full of low pressure helium weighing only $\sim 2 \times 10^{-4}$ g/cm^3 and still transmits a high heat flux (7.4 W/cm^2). Recall that a pipe full of superfluid helium at a density of ~ 0.125 g/cm^3 could transmit only 0.5 W/cm^2. Losses are listed in Table V [12].

Table V

Losses for a 10,000 MWh I-C Unit

10 h discharge - 14 h charge, $I_O = 157,000$ A, $\Delta I = .68\ I_O$ per cycle	
Source	ΔE (MWh)
Magnetic hysteresis	7.4
Epoxy-fiberglass struts	
1.8 K	235.0
11 K	233.4
70 K	200.5
Radiation	3.4
Current leads	14.8
Mechanical hysteresis	.4
Eddy currents	1.0
Total at 300 K	696

A refrigeration system rated at 29 MW at the compressors will handle the above losses. An additional refrigerator capacity is required for redundancy and cool down, which is taken as 20%.

The calculation of losses has been detailed before [12]. A change here is that the number of TiNb filaments has been increased to 6420 in order to reduce the hysteresis loss and the current leads have been optimized to the loss shown. The major change is to take the forces to the rock face directly from each turn with epoxy fiberglass struts which largely eliminates the internal cold structure. The losses quoted above are for 1.1 m struts, Fig. 3, which were selected for a three-tunnel I-C unit by a cost optimization given below.

COST ESTIMATES

The basic I-C unit described here is a three-tunnel 10,000 MWh design (Fig. 2), with 2 cm x 15 cm composite conductor, with forces carried directly to the rock from each turn (Fig. 3) to avoid axial structure and with the scaler force quantity of epoxy-fiberglass strut material, Q_{fs} in Fig. 1. Table VI is a list of parameters for 10,000 MWh and 1,000 MWh I-C three-tunnel segmented solenoids.

254

Table VI. Segmented Solenoid Three-Tunnel I-C Units

W_o	10,000 MWh	1,000 MWh	Comments
B_o	5 T	5 T	midplane field
b_o	6 T	6 T	local max. at conductor
T	1.8 K	1.8 K	pool cooling
R_1	146.5 m	68 m	middle segment (Fig. 2)
R_2	160.5 m	74.5 m	end segments (Fig. 2)
H_1	16.2 m	7.5 m	half middle (Fig. 2)
H_2	16.2 m	7.5 m	end (Fig. 2)
Δ	14.0 m	6.5 m	separation (Fig. 2)
Al	2 cm x 15 cm	2 cm x 15 cm	conductor (Fig. 3)
TiNb	6240 filaments	6240 filaments	0.01 cm dia. each
ℓ_o	1.1 m	1.1 m	epoxy fiberglass strut
N	2675	1240	turns
I	157,000 A	157,000 A	current
L	2920 H	292 H	inductance
Forces	10^{11} N	10^{11} N	Comments
F_{r1}	1.33	.29	radial force, half middle segment
F_{r2}	.38	.08	radial force, end segment
F_r	3.41	.73	total radial force
F_{a1}	1.51	.33	axial force, half middle segment
F_{a2}	1.20	.26	axial force, end segment
F_a	5.40	1.16	total axial force
F_s	8.81	1.90	total scaler sum carried to rock
F_v	6.81	1.46	total vector sum

The costs in Table VII can be used to scale to any other B, ℓ and W. The power required for the refrigerator is almost independent of charge-discharge cycles since, according to Table V, the steady strut loss is 96% of the total loss. The cost of stored energy can be found from the yearly owning and operating cost as follows:

$$c'E_o \ 365 = r[C_1+C_2+C_3+C_4+C_5+C_6] + C_7 \tag{9}$$

where:

C_1--C_5 and P_r are listed in Table VII

C_6 = converter bridge cost

= 40 P_m = peak power rating x 40\$/kW

C_7 = purchased energy per year

= c" [E_o + 24P_r + .02 $\bar{P}$ 24] 365

c" = \$/kWh purchase rate

c' = \$/kWh production cost

r = (0.16)(1.2) = 0.19
(16% of capital costs for interest, taxes, dividends, maintenance, etc., and 20% for interest during construction)

E_o = fixed energy delivered daily

= 9000 MWh, this example

$\bar{P}$ = average converter bridge power per 24 day at average energy loss of 2%

P_d = average discharge power

t_d = discharge time

= 10 h

The total stored energy W is determined from:

$$0.9W = E_o + (P_r + 0.02P_d)t_d, \tag{10}$$

where the factor 0.9 is taken in order that 0.1 W is always kept in storage; then the structure will never relax mechanically. With eq. (10) as a restriction, eq. (9) is optimized for c' by varying ℓ and B to get the costs shown in Table VIII for a daily energy delivery of 9000 MWh in a 10 h period. Cost C_5 and power P_r are slightly more complicated than shown in Table VII due to losses in leads, hysteresis, etc. which account for ∿4% of the refrigerator power. The actual computer optimization properly accounts for all losses.

Several observations can be made from Table VIII:

(1) The strut length, is always about 1.1 m, independent of B. Flared struts, Fig. 3, are stressed at 3.45 x 10^8 N/m² at the cold end and half as much at the warm end.

(2) As B increases, W increases since more energy must be stored in order to offset the larger refrigerator load.

(3) The aluminum conductor, made up of 2/3 high purity aluminum and 1/3 high strength aluminum alloy, is the dominant cost item. One half the cost quoted is allowed for fabrication, which might be too conservative.

Table VII. Cost Factors

Reference Unit, Table VII, W_o = 10,000 MWh, B_o = 5 T,

ℓ_o = 1.1 m, α = bJ_c, α_o = 1.82 x 10^{10} AT/m^2 at 2.2 K and b_o = 6,

T = conductor = 1.8 K, shields at 11 K and 70 K

---

cost component	cost in 10^6

IS related

1. Aluminum conductor $\qquad c_1 = 95.7 (\frac{B_o}{B})^{1/3} (\frac{W}{W_o})^{2/3}$

2. TiNb + Cu $\qquad c_2 = 33 (\frac{\alpha_o}{\alpha}) (\frac{B}{B_o})^{2/3} (\frac{W}{W_o})^{2/3}$

A related

3. Rock excavation

4. Helium $\qquad\qquad c_3 = 49.3 (\frac{B_o}{B})^{4/3} (\frac{W}{W_o})^{2/3}$

5. Dewar + shields
   + food pads + walls

F related

6. Epoxy struts $\qquad c_4 = 31.8 (\frac{\ell}{\ell_o}) (\frac{B}{B_o})^{2/3} (\frac{W}{W_o})^{2/3}$

7. Refrigerator $\qquad c_5 = 27.5 [(\frac{\ell_o}{\ell}) (\frac{B}{B_o})^{2/3} (\frac{W}{W_o})^{2/3}]^{.8}$

Refrigerator Power

P_o = 2.85 x 10^4 kW (reference) $\qquad P_r = P_o (\frac{\ell_o}{\ell}) (\frac{B}{B_o})^{2/3} (\frac{W}{W_o})^{2/3}$

257

Table VIII. Yearly Optimized Costs, E_o = 9000 MWh

	1	2	3	4	5	6	7	8
B, tesla	1.3	1.2	1.1	1.1	1.1	1.1	1.1	1.1
ℓ, (m), strut	1.1	1.1	1.1	1.1	1.1	1.1	1.1	1.1
W, (MWh), stored	10349	10388	10453	10500	10545	10587	10628	10714
E_o, (MWh), delivered in 10 h	9000	9000	9000	9000	9000	9000	9000	9000
rC_1, (10^6), Al conductor	31.8	25.3	22.2	20.2	18.8	17.8	16.9	16.3
rC_2, (10^6), TiNb	3.9	4.0	4.5	5.3	6.5	8.1	10.0	12.7
rC_3, (10^6), area related	81.9	32.6	19.1	13.0	9.7	7.6	6.2	5.2
rC_4, (10^6), struts	2.5	3.6	4.4	5.3	6.2	7.0	7.8	7.8
rC_5, (10^6), refrigerator	2.2	3.1	4.0	4.6	5.2	5.7	6.2	7.2
rC_6, (10^6), converter	10.4	10.4	10.4	10.4	10.4	10.4	10.4	10.4
C_7, (10^6), purchased energy	17.5	17.8	18.0	18.2	18.4	18.6	18.7	19.1
Yearly capital cost $rC_1 .. + rC_6$	132.7	79.0	64.6	58.8	56.8	56.6	57.5	59.6
Owning and operating yearly cost (10^6)	150.2	96.8	82.6	77.0	75.2	75.2	76.2	78.7
c' selling rate mills/kWh	45.7	29.5	25.1	23.4	22.9	22.9	23.2	24.0
c" purchase rate, mills/kWh	5	5	5	5	5	5	5	5
Charge for storage, mills/kWh	40.7	24.5	20.1	18.4	17.9	17.9	18.2	19.0

(4) The second most expensive component is area related, 60% of which is for rock excavation at 50$/m^3. Lower costs are often found for excavation and wall preparation.

(5) Struts, refrigerator and TiNb seem to have been optimized to relatively small costs; improvements would not gain very much in these areas.

(6) The converter equipment at 40$/kW is unavoidably an expensive item.

(7) The power cost, C_7, is sensitive to the purchase rate. Table VIII shows costs of $18 x 10^6 at 5 mills/kWh for the energy to offset the refrigerators and for resale, which would be $7 x 10^6 if 2 mills/kWh power could be purchased from nuclear stations. The number 5 mills was taken since that has recently been the maximum nighttime cost in Wisconsin.

(8) The overall yearly costs show that fields above 4 T are satisfactory.

(9) The total capital cost estimates are about $57 x 10^6/.19 $\sim$ $300 x 10^6.

(10) A storage cost of 18 mills/kWh is acceptable in some cases since peaking power on occasion costs as much as 40-60 mills/kWh.

CONCLUSIONS

The cost reductions obtained by eliminating most of the axial cold structure and by using smaller conductors reduced the estimated cost of stored energy by 40%. The amount of epoxy-fiberglass strut material is selected for scalar force summing, that is, the radial forces and the axial forces are separately taken to the rock face. An improvement is shown in Fig. 4 in which the rock face is circular and two extra end coils adjust the forces to be perpendicular to the rock. Such a configuration promises to save at least 20% in strut material (and losses) and would tend towards the conceptual optimum design.

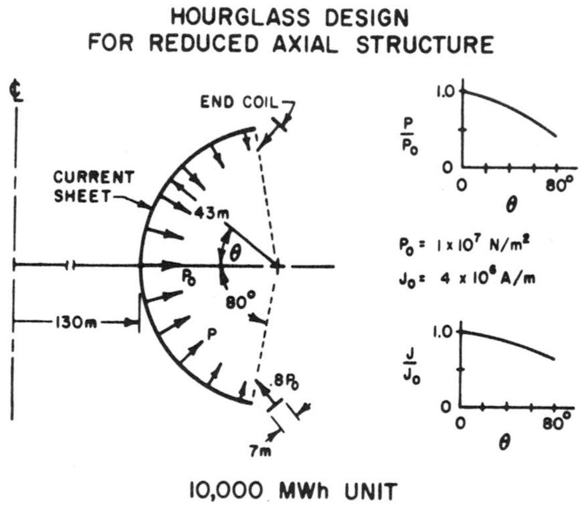

HOURGLASS DESIGN
FOR REDUCED AXIAL STRUCTURE

10,000 MWh UNIT

Fig. 4. Semi-circular solenoid with end coils. The force vectors are largely perpendicular to the supporting wall.

259

Magnet Design for Superconductive Energy Storage

The cost of storing energy in the above 9000 MWh-10 h discharge example is approximately 18 mills/kWh. Future improvements, such as the circular I-C unit in Fig. 4, and detailed total system design offer promise for cost improvements. The estimations used are conservative, about twice the costs expected for modern automated production, and must remain so until actual cost experience is obtained.

ACKNOWLEDGEMENTS

The authors thank E. Ibrahim, J. Saarivirta and A. Khalil for substantial assistance with the cryogenic design. The project is supported by the National Science Foundation, the University of Wisconsin and the Wisconsin Utilities Research Foundation.

REFERENCES

1. R. Carruthers, in: High Magnetic Fields (H. Kolm, B. Lax, F. Bitter, and R. Mills, eds.), MIT Press, Cambridge, Massachusetts (1962), p. 307.

2. Z. J. J. Stekly, "Magnetic Energy Storage Using Superconducting Coils," IDA/HQ 63-1412, Proceedings Pulse-Power Conference (February 1963), p. 53.

3. J. Sole, "Stockage D'Energie Possibilities Supraconducteurs en Vue des Decharges de Grandes Puissances," Rapport CEA--R3243, Comm. Energie Atomique, Limeil, France (June 1967).

4. M. Ferrier, in: Low Temperatures and Electric Power, Intern. Inst. Refrigeration, London (March 1969), p. 425.

5. F. Irie and K. Yamafuji, "Some Fundamental Problems with Superconducting Energy Storage," in: Low Temperatures and Electric Power, Intern. Inst. Refrigeration, London (March 1969), p. 411.

6. H. Brechna, F. Arendt, and W. Heinz, in: Proceedings of 4th Intern. Conf. on Magnet Technology, Brookhaven, New York (1972), p. 29.

7. W. V. Hassenzahl, "Will Superconducting Magnetic Energy Storage be Used on Electric Utility Systems," IEEE Trans. Magnetics, MAG-11(2):482 (1975).

8. F. E. Mills, "The Fermilab Cryogenic Energy Storage System," IEEE Trans. Magnetics, MAG-11(2):489 (1975).

9. R. W. Boom and H. A. Peterson, "Superconductive Energy Storage for Power Systems," IEEE Trans. Magnetics, MAG-8(3):701 (1972).

10. J. R. Powell and P. Bezler, "Warm Reinforcement and Cold Reinforcement Magnet Systems for Tokamak Fusion Power Reactors: A Comparison," BNL-17434, Brookhaven National Laboratory, New York (November 1972).

11. R. W. Boom, G. E. McIntosh, H. A. Peterson, and W. C. Young, "Superconducting Energy Storage," Advances in Cryogenic Engineering, Vol. 19, Plenum Press, 1974.

12. R. W. Boom, H. A. Peterson, W. C. Young, B. C. Haimson, G. E. McIntosh, et al., "Wisconsin Superconductive Energy Storage Project," Feasibility Study Report, Volume 1, July 1, 1974, written by interdisciplinary team of College of Engineering, University of Wisconsin.

13. R. W. Boom, B. C. Haimson, G. E. McIntosh, H. A. Peterson, and W. C. Young, "Superconductive Energy Storage for Large Systems," IEEE Trans. Magnetics, MAG-11(2):475 (1975).

14. H. A. Peterson, N. Mohan, W. C. Young and R. W. Boom, "Superconductive Inductor-Converter Units for Pulsed Power Loads," Int. Conf. on Energy Storage Compression and Switching, Torino, Italy, Nov. 5-7,1974.

15. H. A. Peterson, N. Mohan and R. W. Boom, "Superconductive Energy Storage Inductor-Converter Units for Power Systems," IEEE Winter Power Meeting, New York, 1975.

16. H. A. Peterson, R. W. Boom, W. C. Young and N. C. Storck, "A Look at Superconductive Storage," Electrical World, March 1, 1975.

17. R. W. Moses, "The Virial Theorem and Flywheel Energy Storage," submitted for publication in Physics Today.

18. R. W. Moses, "Cost Related Parameters of Superconductive Energy Storage Magnets," submitted for publication in Advances in Cryogenic Engineering, Vol. 20, Plenum Press, 1975.

19. F. W. Grover, Inductance Calculations, Van Nostrand Pub. Co., 1946 (Dover, 1962).

20. C. Linnet, V. Purdy, Y. W. Chang, and T. H. K. Frederking, "Unsaturated Helium Cooling Limits," Rept. DAAK02-68-C0064, UCLA (October 1970).

21. M. Hilal and R. W. Boom, "Optimization of Mechanical Supports for Large Superconducting Magnets," submitted for publication in Advances in Cryogenic Engineering, Vol. 20, Plenum Press, 1975.

22. M. Hilal and G. E. McIntosh, "Cryogenic Design Elements for Large Superconducting Energy Storage Magnets," Submitted for publication in Advances in Cryogenic Engineering, Vol. 20, Plenum Press, 1975.

20

Reprinted from pages 367–382 of *Energy Storage: User Needs and Technology Applications*, 1976 Eng. Found. Conf., Proc., Technical Information Center, ERDA, 1977, 424pp.

SUPERCONDUCTIVE ENERGY STORAGE INDUCTOR-CONVERTER UNITS
FOR POWER SYSTEMS

H. A. Peterson
University of Wisconsin-Madison

Abstract

The objective of our Superconductive Energy Storage Research and Development Program at the University of Wisconsin is to establish a sound technical and economic basis for arriving at a decision to build or not to build an I-C (Inductor-Converter) unit experimental model with warm support in bedrock probably in the range from 10 to 30 MWh of energy storage capability. Bedrock appears to provide the only possible means for realizing the required low structural costs. Furthermore, the inherent advantage of economy of size so unique for magnetic energy storage (but so characteristic of most other major components in large bulk power transmission systems) provides strong motivation for establishing the technology essential to ultimate realization of the envisioned longer range benefits in very large sizes, say 10,000 MWh. I-C units are shown to have cost advantages in larger sizes representative of bulk power transmission system concentrated storage needs. Projected steady-state as well as dynamic power system benefits and credits are shown to be substantial. Foreseeably, I-C units would be logical companions for base load generation regardless of the prime mover energy source (coal, nuclear fission, or fusion).

Introduction

The objective of our Superconductive Energy Storage Research and Development Program at the University of Wisconsin is to establish a sound technical and economic basis for arriving at a decision to build or not to build an experimental I-C unit model with warm support in bedrock probably in the range from 10 to 30 MWh of energy storage capability.

It must be emphasized that I-C units show greatest long range promise of economic advantage only if they can be built in very large sizes, say 10,000 MWh. Bedrock appears to provide the only possible means for realizing the required low costs (refs. 4,5,6). Furthermore the inherent advantage of economy of size so unique for magnetic energy storage (but so characteristic of most other major components in large bulk power transmission systems) provides strong motivation for establishing the technology essential to ultimate realization of the envisioned longer range benefits.

Cost estimates for energy storage in useful amounts (in a power system engineering sense) by means resting upon new and untried technological advancements are at best hazardous. Nevertheless, the purpose of this paper is two-fold:

1. To present some results of recently completed optimized cost studies,

2. To comment upon power system benefits unique to I-C unit applications.

Basis for Cost Calculations

Long range power system engineering planning including effects of energy storage on electric power production costs is a highly complex problem. For the present purpose of cost analysis of I-C units, only two criteria are examined:

I. Annual Operation Cost in $/kW-yr,
II. Energy Storage Cost in mills/kWh.

It is believed that on the basis of these criteria the basic long range application possibilities of I-C units will be clarified. Only the daily cycle of energy storage and recovery is considered.

Nomenclature

For cost calculation purposes, the following nomenclature is useful:

C_C = Capital cost of converter in $/kW including all costs

α_C = Fixed cost annual rate factor for converter

C_I = Capital cost of inductors in $/kWh

α_I = Fixed cost annual rate factor for inductors

F_I = Fractional portion of inductor energy cycled

$\overline{P}_C$ = Converter average power rating in kW

T_D = Discharge time at average rated power $\overline{P}_C$ per cycle in hours

N = Number of days used per year

$T_Y = NT_D$ = hours used per year

C_T = Total capital cost in $

C_{TA} = Total annual operation cost in $/yr

$\dfrac{C_{TA}}{\overline{P}_C}$ = Total annual operation cost in $/kW-yr

W_Y = Energy recovered in kWh/yr

$\dfrac{C_{TA}}{W_Y}$ = Energy storage cost in $/kWh

$C_E = 1000 \dfrac{C_{TA}}{W_Y}$ = Energy Storage cost in mills/kWh

Cost equations for I-C units may be written using the foregoing nomenclature as follows:

$$C_T = C_C \overline{P}_C + \frac{C_I \overline{P}_C T_D}{F_I} \tag{1}$$

$$C_{TA} = \alpha_C C_C \overline{P}_C + \frac{\alpha_I C_I \overline{P}_C T_D}{F_I} \tag{2}$$

In these equations C_I is not a constant. It depends upon the energy storage capability, i.e., the size of the inductor. C_I decreases as

$\dfrac{\overline{P}_C T_D}{F_I}$ (which is a measure of size) increases. This has been emphasized

before [3,4,5,6]. This inherent nonlinearity has special significance in the optimization process and in evaluating comparisons among energy storage alternatives.

In inductor storage units, it is anticipated that all of the energy stored will not be cycled ($F_I < 1$). Retaining F_I, equations (1) and (2) become:

$$\frac{C_{TA}}{\overline{P}_C} = \alpha_c C_C + \frac{\alpha_I C_I T_D}{F_I} \tag{3}$$

$$\frac{C_{TA}}{N\overline{P}_C T_D} = \frac{\alpha_c C_C}{NT_D} + \frac{\alpha_I C_I}{NF_I} \tag{4}$$

from which the annual costs per kilowatt year and the energy storage costs in mills per kilowatt hour can be calculated.

Converter Rating and Cost

An I-C unit operating over the current range from 0.316 to 1.0 per unit corresponds to cycling 90% of the total energy W_o stored. Such an operating range imposes a limitation in converter power at $I_d = 0.316$ per unit which is only slightly above 0.316 of the converter power available for $I_d = 1.0$ per unit. For a representative value of commutating reactance $X_c = 0.20$, curves of maximum dc voltage E_d and power P_d are shown in Fig. 1. The slightly rising voltage E_d with decrease in current I_d is because of the commutating reactance X_c. Since dc power is $E_d I_d$, the decrease in power with I_d is obvious. For an average charging power $\overline{P}_C$ of 1.0 per unit, the range of power is from 1.5 at $I_d = 1.0$ to 0.5 at $I_d = 0.316$. In order to effect in service an average power of 1.0, over the specified current range it is necessary to purchase a converter with a peak rating of 1.5 in per unit. This amounts to saying that instead of the \$40/kW assumed in our earlier studies, a cost of 1.50(40) or \$60/kW should be used, reflecting the basic fact that for average base power during the discharge cycle, excess converter rating must be purchased in the amount of 50% for the specified current range 0.316 to 1.0 per unit.

To show the more general converter rating requirement for constant average power during discharge as a function of the fraction F_I cycled of total energy stored, Fig. 2 has been prepared. It is to be emphasized that the converter maximum rating actually required is consistently taken at \$40/kW.

Figure 3 shows an arrangement of converters operating in parallel which should be helpful for visualization. L_o is the large superconductive energy storage inductor. It is anticipated that it will be a high current device, say 100,000 amperes. Thus it is similar in current rating to that of a dc superconductive underground cable transmission link. For an average power of 1000 MW, the converter voltage rating would be 1.5(10) kV or 15 kV consistent with the foregoing criterion relating average power to maximum rating. The total inductor current I_d would be subdivided inside the Dewar and brought through the temperature barrier in conductor pairs as shown, $I_1 + I_2 + \ldots + I_n = I_d$. Currents I_1, I_2, ...I_n are supplied by converters C_1, C_2, ... C_n^* respectively,

---

*Each of these converters might be made up of two or more series converted sub-units so that reduced rates of charging or discharging could be achieved without excessive VAR requirements.

264

all operating in parallel for I_d = 1.0. At reduced currents, one or more converters could be taken out of service, the criterion being that I_d should be low enough so that the converters remaining in parallel operation should not be overloaded. Those converters thus freed from parallel operation could be used for voltage regulation purposes if required. Alternatively they could simply be switched out of use completely in order to minimize losses. Each converter is analogous to a reversible motor-generator in a pumped storage hydro plant. An additional converter could be provided for increased reliability to serve as a replacement in case any one of the units is unavailable.

It should be evident from Fig. 1 that over the specified operating range of I_d from 0.316 to 1.0, average power on charge or discharge is not always available. This is true for I_d < 0.67. On the other hand, for I_d > 0.67, more than average power is available up to a maximum of 1.5 at I_d = 1.0. However, damping capability for small oscillations is not critically dependent on the magnitude of I_d over this range when the unit is large enough to provide significant storage capability in relation to system size.

Optimized Design Calculations

Results of optimized design studies by computer techniques are summarized in Fig. 4. This is a very important figure because it summarizes results of carefully optimized design of magnets in bedrock following methods outlined in references 4 and 8. Converter costs are included since the optimization included all costs of the I-C unit. Capital cost curves are shown (solid lines) for 10, 5, and 2 hour discharge times T_D. Also shown are curves (dashed lines) for constant values of energy cycled daily: 500, 1000, 2000, 5000, and 10,000 MWh all of which are net values returned to the system. All losses are included and are a factor in the optimization process which involves tradeoffs between capital and operating costs. The curves of Fig. 4 are based upon the assumption that off peak energy for charging and supplying losses is available at zero cost. The storage inductor size is taken as equal to 1/.9 times the sum of the energy delivered plus the losses during discharge. The optimization is extended to include the effect of the cost of off peak energy (other than zero) later in this paper.

While the equations (1) to (4) inclusive are helpful in gaining insight into the mechanism of cost structure, and relative magnitudes of fixed and variable components, it is clear that over ranges of size such as are being discussed here the parameters in general are not constants. For that reason, we abandon the equations (1) to (4) at this point and use the information in Fig. 4 to calculate the annual operation cost in $/kW-yr shown in Fig. 5. As a comparison, the corresponding energy storage costs are shown in Fig. 6. For comparison purposes, the "Future Batteries" curve from reference 2 is shown in Fig. 5 and from it, the energy storage cost curve was calculated and is shown in Fig. 6. A careful study of these two figures reveals the nature and magnitude of the challenge of superconductive energy storage.

The influence of the cost of off peak energy is important, as is shown in Figs. 7, 8, and 9 for 100, 1000, and 10,000 MWh respectively. The optimization process includes the subtle influence of several interrelated design parameters. However, as a first approximation, the total cost of energy storage is simply increased by the amount paid for off peak energy. Thus, if we wish to know the approximate effect of off peak energy cost on the results shown in Fig. 6, we have only to add that off peak cost to the results shown in Fig. 6.

The effect of off peak energy cost on annual operation cost is a bit more elusive. For an off peak cost of 10 mills per kWh, the curves of Fig. 10 have been calculated. The effect is to increase the annual operation cost quite markedly. For lower off peak costs, annual operation costs would lie between those shown in Fig. 5 and Fig. 10.

Because all of the optimization was developed with minimized mills/kWh costs as the objective, efficiency became an issue of lesser importance. However, so that we may know how it is influenced under the optimization procedure, the results shown in Table I are helpful.

Table I

Net Energy Delivered to System MWh	% Efficiency Max	Min
10,000	91.0	87.0
1,000	90.2	80.5
100	70.5	66.5

As a general comment, it is clear that high efficiency is not always the most worthy objective. Furthermore, from an engineering design point of view, the resultant structure at the lower efficiency would appear to be more readily realizable and acceptable.

Power System Benefits

It has been shown [3] that suitably controlled I-C units may be uniquely beneficial to the dynamic performance of a power system. These benefits are in addition to the primary benefits accruing from the daily cycling of energy into and out of storage. Benefits are significant from the standpoint of small load disturbances and relate therefore to the AGC (Automatic Generation Control) problem. Because the response of an I-C unit may be very fast, such small disturbances can be rendered inconsequential because the electrical load change is in effect cancelled by an electrical load change of opposite sign so quickly that machine rotor torques are essentially unchanged through such disturbances. During major disturbances, which lead to the problem of transient stability, benefits are also quite significant. It is essentially impossible to assign "credits" to these benefits in any general way. Each case in turn must be studied and decisions made based upon the findings attending that specific case.

Transmission credit, pollution control credit, and spinning reserve credit are also difficult to assess numerically. In our studies, transmission credits have been found to range from about $10/kW to $60/kW of converter rating. Pollution control credits range from $5 to $125 per kilowatt depending intimately upon the alternative considered for satisfying the peaking requirement. The spinning reserve credit range similarly was quite large, from $20 to $100 per kilowatt approximately, depending upon alternatives and the specific problem at hand.

It should be emphasized that all of these benefits and credits, and comments pertaining thereto in this paper relate to bulk power transmission system engineering concepts rather than to distribution system engineering. Energy storage in distribution systems is a different system problem. In general, superconductive energy storage would not be economical in distribution systems.

Conclusion

While there are several modifications in engineering design under consideration, there is no foreseeable prospect of a major change in the basic economic trends presented in this paper. Actual costs or improved cost estimates may justify cost reductions of up to 50% once the actual assembly equipment has been designed in detail. All costs used in this paper have been upper bound, highest estimates based on conservative principles. In the past we have suggested that cost reductions of 33% might be warranted. The principal need

266

at this point is experimental results from a well conceived, coordinated, and executed program of basic and project oriented meaningful research. In time, results of these efforts will provide a firmer foundation for more rigorous economic analysis.

We have no illusions as to the problems and difficulties ahead. The path to ultimate realization of the substantial benefits envisioned, with the high degree of reliability required for electric power industry acceptance, is demanding and tortuous. With appropriately muted feelings of optimism, we shall continue our journey.

ACKNOWLEDGEMENTS

For this research, support received from the University of Wisconsin, the National Science Foundation, the Wisconsin Utilities Research Foundation and the General Electric Education Foundation is gratefully acknowledged.

References

1. Peaking Power Batteries for Electric Utilities, O.G. Berkowitz, J.T. Brown. *American Power Conference*, April 21-23, 1975.

2. Energy Storage, Fritz R. Kalhammer, Vance R. Cooper. *Research Progress Report* FF-2, EPRI, January 1975, pp. 11-18.

3. Superconductive Energy Storage Inductor-Converter Units for Power Systems, H.A. Peterson, N. Mohan, R.W. Boom. *IEEE PES Transactions Paper T 75 152-4*, Winter Power Meeting, January 26-31, 1975.

4. Wisconsin Superconductive Energy Storage Project, H.A. Peterson, R.W. Boom, W.C. Young. *American Power Conference*, April 21-23, 1975.

5. A Look at Superconductive Storage, H.A. Peterson, R.W. Boom, W.C. Young. *Electrical World*, March 1, 1975.

6. Wisconsin Superconductive Energy Storage Project. Feasibility Study Report, Volume I, July 1, 1974, written by interdisciplinary team of College of Engineering (Professors R.W. Boom, H.A. Peterson, W.C. Young, B.C. Haimson, G.E. McIntosh, et al.).

7. AC/DC Power Conditioning and Control for Advanced Conversion and Storage Technology. EPRI 390-1-1. Report prepared by Westinghouse Electric Corporation for the Electric Power Research Institute, August 1975.

8. Magnet Design for Superconductive Energy Storage for Power Systems. R.W. Boom, M.A. Hilal, R.W. Moses, G.E. McIntosh, H.A. Peterson, R.L. Willig, and W.C. Young. *International Conference on Magnet Technology*, Rome, Italy, April 21-25, 1975.

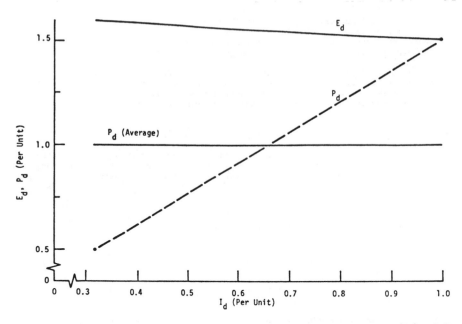

Fig. 1. Upper Limits of Voltage and Power as Functions of I_d for $P_d(Av) = 1.0$
Per Unit. Ignition Delay Angle $\alpha = 0$, $\beta = 0.90$.

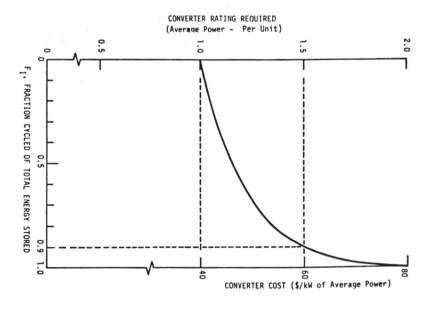

Fig. 2. Converter Rating Requirements and Cost.

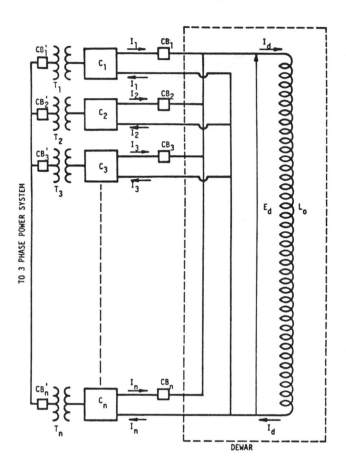

Fig. 3. Converters Arranged for Parallel Operation to Supply Total Inductor Current I_d.

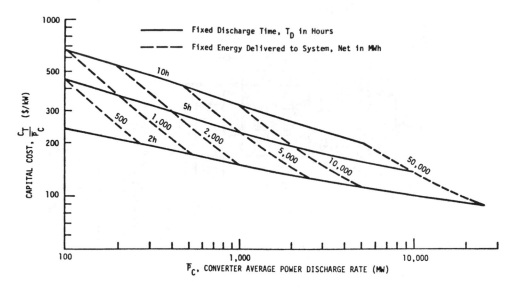

Fig. 4. Capital Cost of I-C Storage Units. Optimized Design for 90% Discharge of Energy; Converter Cost Included at $40/kW of Max. (or $60/kW of Av) Rating.

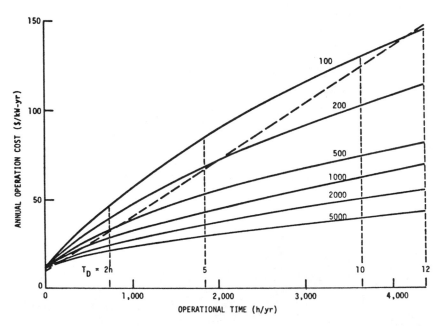

Fig. 5. Annual Operation Cost of I-C Units. Optimized Design with Off Peak Energy Cost 0 Mills/kWh; Converter Cost $40/kW of Max. (or $60/kW Av) Rating; 90% Energy Cycled Daily; 19% Annual Rate of Return. Numbers on Curves are I-C Unit Average Ratings in Megawatts (Dashed Curve for Future Batteries with Converters at $40/kW).

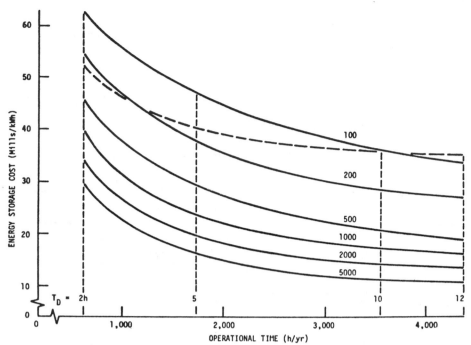

Fig. 6. Cost of Energy Storage in I-C Units. Optimized Design with Off Peak
Energy Cost 0 Mills/kWh; Converter Cost $40/kW of Max. (or $60/kW Av)
Rating; 90% Energy Cycled Daily, 19% Annual Rate of Return. Numbers
on Curves are I-C Unit Average Ratings in Megawatts (Dashed Curve for
Future Batteries with Converters at $40/kW).

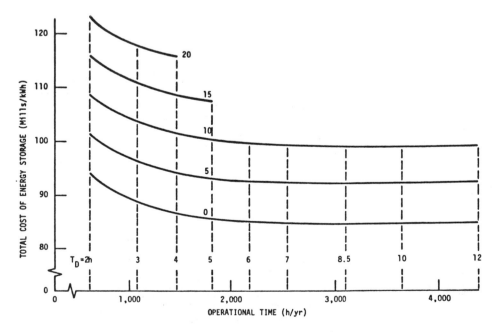

Fig. 7. Cost of Energy Storage in I-C units with 100 MWh Net Delivered.
Optimized Design; Converter Cost $40/kW of Max (or $60/kW Av) Rating;
90% Energy Cycled Daily, 19% Annual Rate of Return. Numbers on
Curves are Off Peak Energy Costs in Mills/kWh.

271

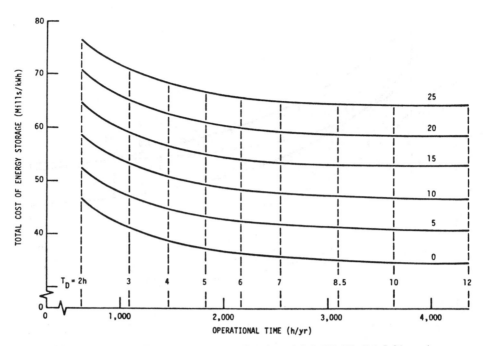

Fig. 8. Cost of Energy Storage in I-C Units with 1,000 MWh Net Delivered. Optimized Design; Converter Cost $40/kW of Max. (or $60/kW Av) Rating; 90% Energy Cycled Daily; 19% Annual Rate of Return. Numbers on Curves are Off Peak Energy Costs in Mills/kWh.

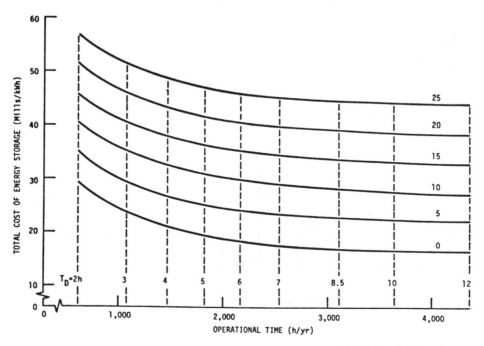

Fig. 9. Cost of Energy Storage in I-C Units with 10,000 MWh Net Delivered. Optimized Design; Converter Cost $40/kW of Max. (or $60/kW Av) Rating; 90% Energy Cycled Daily; 19% Annual Rate of Return. Numbers on Curves are Off Peak Energy Costs in Mills/kWh.

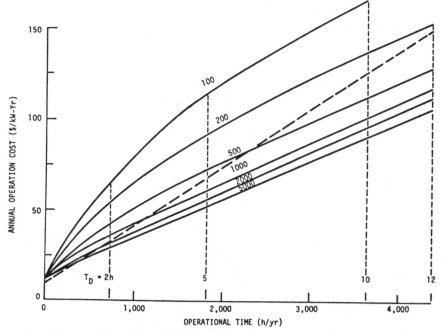

Fig. 10. Annual Operation Cost of I-C Units (Including Charging and Losses).
Optimized Design for 10 mill/kWh Off Peak Energy Cost; 90% Energy
Cycled Daily; 19% Annual Rate of Return. Numbers on Curves are I-C
Unit Average Power Ratings in MW (Dashed Curves for Future Batteries
with Converters at $40/kW).

273

21

Copyright ©1975 by The Science Research Council

Reprinted from pages 615–627 of *Int. Conf. Magnet Technology, 5th, Proc.*, The Science Research Council, 1975, 757pp.

SUPERCONDUCTING MATERIALS: SOME RECENT DEVELOPMENTS

M. N. Wilson

SUMMARY

   This review concentrates on four subject areas which have seen significant changes in recent years; AC losses, stability, multifilamentary Nb_3Sn and multifilamentary V_3Ga. Several applications are now being considered in which a small amplitude AC field is superposed on a larger DC field; fine filaments may not always give the lowest hysteresis loss under these conditions. The theory of self field instability has recently been improved and extended to the point where it can provide a good picture of the instability process under realistic thermal conditions. Some points of disagreement with experimental observation remain however, a possible explanation is offered for this. Multifilamentary Nb_3Sn and V_3Ga are now reaching the stage of meriting serious consideration as fully practical magnet conductors. The remaining problems of production and utilization are being solved and composites have been tested with at least twice the overall current density of filamentary NbTi at all field levels.

1.   INTRODUCTION

   The whole field of superconducting magnet materials has become so wide and diverse that a comprehensive review would now require very much more space than is available here. Such an all-embracing review has not therefore been attempted, but instead four subject areas have been chosen which have seen significant changes since the last Magnet Technology Conference. Each of these subjects has then been dealt with in some depth. For a more complete coverage of the field, the reader is referred to the recent review by Fietz and Rosner(1).

   Two of the subject areas are theoretical: AC losses and stability. Although the theory of AC losses has not basically changed, several of the more recent applications have imposed different AC conditions and the conventional wisdom that 'finest filaments are best' may not always be true in this case. Stability theory has progressed in recent years and we now have a much better picture of the self field instability although there are still some areas of uncertainty and disagreement with experiment; a possible explanation for this is offered.

   The two remaining subject areas are concerned with materials technology - more specifically with the production and utilization of multifilamentary composites of Nb_3Sn and V_3Ga. Good progress has been made in recent years and both materials are rapidly advancing to the stage where they must merit serious consideration by the magnet technologist as fully practical magnet conductors.

2.   AC LOSSES

   The multifilamentary composites which are now in general use were originally developed to eliminate flux jumping instabilities in the superconductor and thus promote stable operation in a magnet. The filaments were subsequently made finer and the coupling between them was reduced by the use of cupronickel barriers etc. so that the composites could be pulsed with a minimum hysteresis loss. The objective of this work was, and still is to a large extent, the development of a superconducting synchrotron. In a synchrotron, the magnets are pulsed from a low field to their maximum field and then to the low field again. Under these conditions, it is easy to show that the hysteresis loss in a superconducting magnet may be minimized by making the filaments as small as possible. In fact the loss per unit volume is simply proportional to the filament diameter, provided the filaments are sufficiently de-coupled from each other. Any coupling between the filaments will tend to increase the loss and so the coupling must be kept to a minimum by tightly twisting the composite and interposing resistive barriers between the filaments. A range of sophisticated composites has thus been developed to have minimum hysteresis loss in a synchrotron magnet or any other application (e.g. energy storage) where the field must be pulsed from zero to maximum or vice versa.

   Recently however, another class of applications has emerged in which the superconductor is subject to alternating fields of a different kind. In this case, the major component of field is a steady DC level, but there is superposed on this an AC component of much smaller amplitude. Such conditions can occur for example in the DC rotor windings of a superconducting alternator or in the lift magnets of a magnetically levitated vehicle as it encounters irregularities in the guide way. They may also occur in some kinds of proposed fusion reactor where a single pulse of field may be applied, in addition to the

DC containment field, to ignite the plasma. It has often been assumed that the low hysteresis composites developed for the synchrotron application will also be the best choice here, but it has recently been pointed out(2) that this is not necessarily so.

Adopting the simple one dimensional slab model of a superconducting filament, illustrated in Figure 1, we may distinguish two cases. Firstly, as shown in 1(a), the case where the disturbance caused by the fluctuating component of the field does not penetrate to the centre of the filament and there is in fact no loss in the centre. Secondly, as shown in 1(b), the disturbance penetrates the whole filament.

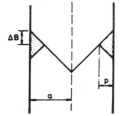

(a) below full penetration

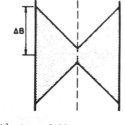

(b) above full penetration

FIG. 1. Slab model of AC field patterns (shaded area depicts flux change)

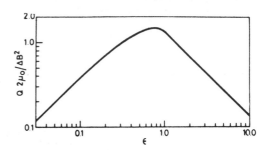

FIG. 2. Dimensionless loss per cycle $Q.2\mu_o/\Delta B^2$ versus dimensionless half width $\epsilon = a/p$.

In the first case, the loss per unit volume for a complete cycle may be shown to be

$$Q = \frac{2\Delta B^3}{3\mu_o^2 J_c a}$$

where ΔB is the amplitude of the AC component of field, J_c is the critical current density and a is the half width of the slab. If the penetration depth of the AC field is p i.e. $\mu_o J_c p = \Delta B$, we may define a dimensionless half width $\epsilon = a/p$ and obtain the loss per cycle as Q

$$Q = \frac{\Delta B^2}{2\mu_o} \left\{ \frac{4}{3\epsilon} \right\}$$

Similarly, in the fully penetrated case we find

$$Q = 2\Delta B a J_c - \frac{4\mu_o J_c^2 a^2}{3}$$

which may be written

$$Q = \frac{\Delta B^2}{2\mu_o} \left\{ 4\epsilon - \frac{8\epsilon^2}{3} \right\}$$

We may thus plot a 'dimensionless loss' $Q.2\mu_o/\Delta B^2$ as the simple function of ϵ shown in Figure 2.

It may be seen that, for a given field amplitude, the loss is a maximum when the superconductor half width is slightly less than the penetration depth of the superconductor. In the synchrotron case, where the penetration depth is large, ϵ is small and one is working at the left hand side of the peak. The most effective way of reducing losses will thus be to make ϵ as small as possible i.e. fine filaments.

In the second class of applications however the penetration depth may be small, one is now at the right of the peak and the best way to reduce losses may be to increase ϵ i.e. large filaments. For example, in a magnetically levitated vehicle, the ΔB caused by guideway irregularities is expected to be of the order of 0.02T. If we assume a DC field of 5T i.e. $J_c = 1.5 \times 10^9$ A m^{-2} and a filament diameter of 50μ i.e. a = 25 ×10^{-6}m, we find $\epsilon = 2.36$, a dimensionless loss of 0.57 and an actual loss Q of 90 J m^{-3} (0.09 mJ/cm^3) per cycle. At a frequency of 50 Hz this would dissipate 4500W m^{-3} (4.5 mW/cm^3) which is very similar to the loss in a synchrotron magnet. This loss may be halved by simply making the filament twice as big i.e. 100μ dia. In order to halve the loss by reducing diameter however it would be necessary to make $\epsilon = 0.075$ i.e. a filament diameter of only 1.6μ or roughly 4000 times as many filaments as the 100μ case! Clearly the most economical way to reduce losses in this case will be to make the filaments as large as possible. Even at 50μ diameter however there is some danger of instability and at 100μ diameter there will probably be flux jumping. Stability thus provides the upper limit to filament size and the exact value of this will depend on other parameters such as operating field, proportion of copper in the conductor etc.

Superconducting Materials: Some Recent Developments

It is dangerous to generalize about filament size however. In the fusion reactor application for example, one might have a DC background field of 7.5T and a pulsed field of 0.5T(3). In this case we have $J_c \approx 5 \times 10^8$ A m^{-2} and hence p = 800μ; even with a 100μ diameter we find ε = 0.0625 and are thus far to the left of the peak in Figure 1. Any reduction in filament size will therefore reduce the losses; any feasible increase in filament size will increase losses.

Similar considerations also apply to the coupling of filaments through the matrix of the conductor. Roughly speaking, when the filaments are well coupled, they behave co-operatively as one single large filament. Thus, in certain cases the loss may be reduced by increasing the coupling between filaments. For example Satow, Tanaka and Ogama(4) have found that under some conditions the losses in untwisted composites can be less than in twisted composites. Stability could be a problem in this case of course. Again it is impossible to generalize and each condition must be treated individually.

3. STABILITY

Magnetic instabilities or flux jumps can drastically degrade the performance of a magnet although of course they are not the only cause of degradation; mechanical movement is quite certainly another. In recent years the mechanisms of cryostatic stabilization and of intrinsic adiabatic stability have become rather well understood. The coupling of filaments in a composite in response to an external field and its effect on stability are also understood. The self field coupling and consequent self field instability is still however less well understood. Self field effects are important because they limit the overall size of a composite. Significant advances in the theory of self field effects have recently been made however by Turck and Duchateau and their work will be referred to frequently in this section.

Very briefly, the self field effect is the tendency for the transport current in a filamentary composite to flow preferentially in the outer regions of the composite and to only penetrate inwards when critical current density is reached at the outside. The current and field distributions in the composite are therefore as shown in Figure 3. This 'piling up' of critical current in the outermost filaments tends to be unstable. It is not possible to encourage a more uniform sharing of current by simply twisting the composite. To achieve this, a full transposition of the filaments is needed whereby the inner filaments pass to the outside and vice versa as one moves along the composite.

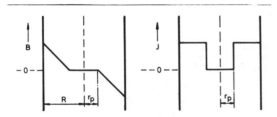

(a) field distribution. (b) current distribution.

FIG. 3. Self field patterns in twisted composites

On applying a simple adiabatic stability theory to the situation illustrated in Figure 3, one is forced to conclude that a typical filamentary niobium titanium composite will be unstable at composite diameters greater than ~ 0.5 mm. Practical magnet experience shows that this is clearly not the case. Two main reasons may be advanced for the failure of the simple theory; firstly the self field flux jump is not adiabatic and secondly, the current distribution is not as simple as the one shown in Figure 3. The newer theories take account of both these effects. In order to illustrate their scope and limitations, they will now be applied to two different conductors which have been tested in several medium sized solenoids at the Rutherford Laboratory.

3.1 Experimental Tests. The first conductor tested was a 1000 filament (IMI B1000), 1 : 1 matrix: superconductor ratio, 2 mm diameter composite, twisted at a pitch of 25mm. Three solenoids were fabricated from this conductor, each of 100 mm bore, resin impregnated and producing a peak field of about 6T. All the magnets showed some training - between one and four quenches - but quickly reached their short sample current. The subsequent quench currents were surprisingly constant to better than 1 in 10³: a good indication that no instabilities were present. At each quench, the whole magnet was heated above critical temperature so that the current distribution in the superconductor started each time from the virgin state. This means that the training which was seen could not have been primarily electromagnetic but was probably mechanical.

The theory of Turck and Duchateau treats a cylindrical composite carrying a current distribution like that shown in Figure 3. Full account is taken of the magnetic damping in the matrix which slows down the progress of the flux jump and of the thermal conduction processes which remove the heat generated by flux motion in the superconductor. The thermal conditions at the boundary may either correspond to the composite being cooled by liquid or gaseous helium or immersed in a resin impregnated winding. The major factor determining stability is the parameter β defined in (5) or (6) as

$$\beta = \frac{\mu_0 (\lambda J_c)^2 R^2}{c\,\theta_c}$$

276

where λ is the space factor of filaments in the composite, R is the outer radius of the composite, c is the specific heat and θ_c is the critical temperature rise defined by

$$\theta = -J_c \cdot \frac{d\theta}{dJ_c}$$

(note that some of Turck's symbols have been changed to avoid conflict with the rest of this paper) For the composite in question, at a field of 6T we find β = 48. Referring now to Figure 12 in ref. (6), we may infer that the composite should become unstable in a resin impregnated magnet when the current flowing in the outer region reaches 70% of the critical current of the composite, i.e. when the inner boundary of this region r_p has moved in to a radius of 0.55 mm.

This is not the whole story however because a more detailed consideration of the current distribution within the composite will show that one might also expect additional currents to flow in the region $r < r_p$. This is brought about by the way in which the self field and external field can interact with the magnetization and transport currents within the individual filaments to cause a shift in the 'electric centre line' of the filament, (7)(8). In this way, a current distribution will be induced to flow in the central region. Because these currents are subcritical, they will be quite stable and the composite will thus be able to carry this additional transport current with unimpaired stability.

In order to determine the magnitude of the internal currents we refer to Figure 9 of ref. (7). This shows how the internal subcritical currents in the composite increase with radius until at a certain radius they reach critical. The boundary conditions between the two regions will then be perfectly matched if this radius, a, is set equal to the penetration radius r_p of the outer region. Thus we may substitute $a = r_p$ into equation (9) of ref. (7).

$$\beta_w = \left\{ \frac{2\mu_o \lambda J_c}{B_e d} \right\}^{\frac{1}{2}} a$$

where the original β has been written as β_w to distinguish it from Turck's β, B_e is the external field, d the filament diameter. For the composite in question with λ = 0.5, $J_c = 1.03 \times 10^9$ A m^{-2}, B_e = 6T, d = 45μ and a = r_p = 0.55 mm we find β_w = 1.20. This value of β_w may now be used to find the total current induced in the region $r < a$. Figure 10 of ref. (7) shows f as a function β_w where f is the total induced current expressed as fraction of the critical current of the inner region. For β_w = 1.20 we find f = 0.84.

The final result of all this is that we would expect the 2 mm composite to stably carry up to 70% of its critical current in an outer sheath flowing at critical density. The inner region of the composite would carry stable sub-critical currents up to 84% of the critical current of the inner region or 0.84 x 30% = 25% of the entire composite. The maximum stable current for the composite would thus be 70% + 25% = 95% of critical. The fact that the magnets actually reach critical current is therefore perhaps not too surprising.

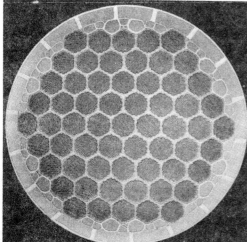

FIG. 4. The 3.5mm dia. 14701 filament composite used in self field rest solenoid.

The second magnet result is surprising however. This was a solenoid of similar size which was made to test the large three component composite shown in Figure 4. The composite was 3.5 mm diameter containing 14701 filaments with 39% of NbTi, 39% of copper and 22% of cupro-nickel. After some training the magnet reached 93% of its 'short sample' field, the current carried was 87% of critical.

Applying the theory of Turck and Duchateau to this magnet we find β = 133, predicting an instability in the outer sheath when it carries a current of ∿ 50% of the critical of the whole composite, i.e. $r_p = R/\sqrt{2}$ = 1.24 mm. Because the filaments are much finer in this composite, the 'electric centre' effect is less than before. We find β_w = 7.2 and hence f = 0.26. This inner region will thus only carry 26% of its critical current in a stable sub-critical fashion before it reaches critical density at its outer boundary. We therefore expect the composite to become unstable at 50 + 0.26 x 50 = 63% of its critical current. In this case it is clear that, in spite of its many corrections and greater thoroughness, the theory is still too pessimistic.

3.2 Decay of the Self Field Current Distribution.

A possible explantation for this which does
not seem to have been mentioned so far is that the
ideal current distributions assumed in (5) and (6)
might decay rather rapidly in a magnet. This
could be brought about by the 'resistive transition'
effect illustrated in Figure 5. Although it is not
well understood, this effect is thought to be shown
by all multifilamentary superconductors. As
critical current is approached the composite does
not switch sharply to the resistive state but
rather undergoes a gradual transition in which a
measureable resistance in progressively developed
from $\sim$ 80% of critical current density upwards. It
seems reasonable to suppose therefore that any array
of filaments, carrying critical current density will
develop a resistance. The outer sheath of critical
current would thus be resistive.

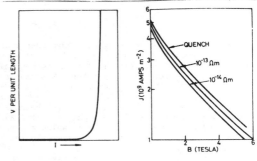

(a) voltage - current   (b) lines of constant
     curve   resistivity

FIG. 5. Resistive transitions in filamentary
          NbTi composites

The usual mechanism postulated for the decay of the self field current distribution is that the current,
as it is fed into the composite at the magnet terminals, must cross the matrix. The transverse resistance
of the matrix will thus cause the purely inductive current distribution which is initially set up to decay
with a time constant

$$\tau \simeq \frac{\mu_o}{\rho_m} \cdot \frac{\ell^2}{4}$$

where ρ_m is the mean transverse resistivity of the matrix and ℓ is the length of the composite. The
composite shown in Figure 4 had an unusually high transverse resistivity of $\sim 10^{-8}$ Ωm and the coil was wound
from a very short length of conductor - only 36 M. Even in this case however the decay constant is
$\sim 4 \times 10^4$ secs. The magnet was actually taken to its quench current in 26 secs. It is clear that the
matrix resistivity, even though it was exceptionally high, could not have caused any significant decay of
the self field current distribution. Most magnets, with lower matrix resistivities and longer lengths of
conductor will have much longer time constants.

Much shorter time constants are found however if it is assumed that the filaments in the outer sheath
are developing an effective longitudinal resistivity. For the 3.5 mm diameter composite carrying 50% of I_c
in a critical outer sheath we may roughly calculate that the time constant for this to decay into the inner
region is

$$\tau \simeq \frac{\mu_o R^2}{4\rho_e} \cdot \ell n\left[r_1/r_2\right]$$

where r_1 and r_2 are the mean radii of the outer and inner regions and ρ_e is the effective longitudinal
resistivity of the array of filaments. Note that this time constant does not depend on the length of
conductor. As shown in Figure 5(b), a typical composite might develop a resistivity of $10^{-13}\Omega$ m at 95%
of its ultimate quench current and $10^{-14}\Omega$ m at 90% of I_c. For $\rho_e = 10^{-13}\Omega$ m we thus find $\tau \simeq 5$ secs.

It is clear therefore that within a rather short space of time, the potentially unstable outer current
sheath will have decayed to a more stable sub-critical level - perhaps to 95% of the critical current
density. This process will continue with a gradually increasing time constant until the distribution of
current is uniform over the cross section. One would therefore expect it to be stable, at least against
small disturbances.

Note that this in no way contradicts the theory of Turck and Duchateau. Nor is it in disagreement with
the results of their experiments which were performed on fairly short samples of conductor. The experiments
did in fact indicate that the non-uniform current distribution decayed with time. It was assumed however
that this would not happen in magnets involving long lengths of conductor.

An interesting possibility arises from the idea of a partial decay of the self field current distribution. Let us assume that the magnet containing the 3.5 mm composite is being charged and has reached, say,
70% of its critical current. The composite would be unstable and would quench if its outer current sheath
were flowing at critical density as assumed in the theory. This outer sheath has however decayed resistive-
ly to 95% of critical density and it is stable. Suppose now that a local energy release in the magnet

suddenly raises the temperature of the composite by just the amount needed to reduce the critical current density of the superconductor by 5%. The outer sheath will now be at critical current density, it will be unstable and will quench. Our small (5%) temperature rise has thus been able to quench a conductor which is only at 70% of critical. In other words, large diameter filamentary composites, although they may not actually be self field unstable, will be more than usually vulnerable to small temperature fluctuations within the magnet, e.g. mechanical movement effects.

It is possible that such an effect has already been seen by the workers at Fermi Laboratory where magnets wound from a large twisted composite have been seen to train much more than similar magnets made from transposed cable(9).

4. NIOBIUM TIN

Niobium tin tape has been available commercially for several years and has been used in the construction of all the large high field superconducting magnets currently in use. It does however have the disadvantage that it is basically unstable in the presence of magnetic fields perpendicular to the broad face of the tape. This instability may be controlled by soldering copper strip to the face of the tape (dynamic stability) and arranging for the edge of the strip to be cooled by liquid helium. The charge times of such magnets can be long however (∿ 1 hour) and the need for edge cooling can cause difficulties in some types of coil geometry.

Filamentary niobium tin can be expected to be much more stable than the tape and is now being actively pursued in many places. An unusual 'filamentized' niobium tin tape has recently been made by IGC(10) and is shown in Figure 6. The subdivision of the tape has been achieved by a chemical milling process. It is interesting to note that, whereas conventional tapes showed a strong tendency to flux jump in normal fields, no flux jumps were observed in the filamentized tape. It is not clear however whether tapes of this kind can be produced economically in long lengths and to a consistent standard of quality.

FIG. 6.   'Filmentized' Nb$_3$Sn tape

A more popular approach to the production of filamentary Nb$_3$Sn has been the use of the 'bronze route'. In this process, filaments of pure niobium are drawn down in a tin copper bronze. At its final size, the wire is heat treated for typically ∿ 100 hours at 550°C - 750°C so that the tin in the bronze reacts with the niobium filaments to form filamentary niobium tin. Not only does this process offer a very convenient way of producing the brittle Nb$_3$Sn in filamentary form but it can also ensure that the Nb$_3$Sn so produced is of good quality. At reaction temperatures below 930°C the only compound of niobium and tin to form in the bronze process is Nb$_3$Sn (cf. the liquid tin process where Nb$_6$Sn$_5$ and NbSn$_2$ normally form below 930°C). Because the bronze process is a solid state reaction, the resulting filaments of Nb$_3$Sn can be very smooth and uniform. Because the reaction takes place at low temperature, the grain size of the Nb$_3$Sn so produced can be small and its critical current density large.

An interesting variation on the bronze method has recently been demonstrated by Tsuei(11). In this process, a filamentary structure is produced metallurgically by first chill-casting an alloy of Cu,Sn and Nb and then working it down to finished size, when a final heat treatment is applied to form Nb$_3$Sn. Unfortunately the overall current densities exhibited so far by these composites are very low. If they could be improved, the method would clearly be a very cheap way of producing multifilamentary Nb$_3$Sn.

4.1 Current Density. Figure 7 shows a typical overall current density for a bronze matrix niobium tin composite. Considerable effort is currently being put into the optimization of this process and there is probably scope for the improvement of J_c. It is generally felt that the basic problem is one of obtaining good stoichiometry throughout the layer while at the same time keeping a fine grain size. Flux pinning is thought to occur at the grain boundaries so that a small grain size should mean a high density of pinning points.

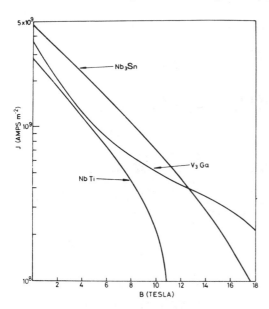

FIG. 7. Overall current densities of filamentary composites

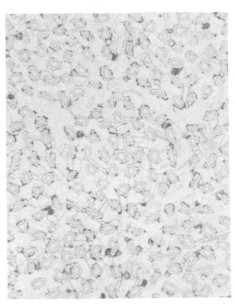

FIG. 8. Reacted filaments of Nb₃Sn in bronze

If the heat treatment is terminated before complete reaction, the resulting filament will have a pure niobium core as shown in Figure 8. This is thought to confer good mechanical properties on the composite but is thought to not produce the best superconducting properties because there will be a variation in stoichiometry across the Nb₃Sn with a niobium rich layer adjacent to the unreacted core. A hint that this is indeed the case can sometimes be seen by measuring the critical temperature of the composite by both inductive and resistive methods. It is often found(12) that, whereas the resistive measurement shows a sharp transition at ∿ 18 K, the inductive measurement indicates a gradual transition spread over several degrees up to 18 K. One might thus infer that the filament contains several layers of different T_c but that most of the current is carried in the stoichiometric high T_c layer. The longer times or higher temperatures needed to produce a more stoichiometric compound might however also produce grain growth with a consequent reduction in J_c.

On the basis of the above arguments, it would appear that the best way to obtain a high current is to use very fine filaments of niobium which can be reacted to completion at lower temperatures and in shorter times. This has yet to be conclusively demonstrated in practice. Another approach is to dope the niobium with small quantities of zirconium or zirconium oxide which is expected to retard the rate of grain growth in the Nb₃Sn. It has been found(13) that, although the Zr additions are effective in raising the current density of thick layers of Nb₃Sn, the highest current densities are still found in the thinnest layers of undoped material.

When seen from the point of view of the magnet technologist of course the important current density is not the J_c of the Nb₃Sn compound but the overall J_c of the composite. It is not practical to draw down a bronze which contains more than ∿8 at % of tin because of the brittle intermetallics which form above this level. It has been found that the Nb₃Sn reaction will continue until about 0.3 at. % of tin remains in the bronze. From these two figures we may calculate that the maximum filling factor of the reacted Nb₃Sn filaments in the composite can only be about 33%. In practice it has so far been nearer 10%-20%. It is interesting to speculate on what the highest possible overall current densities might be. If we take the highest J_c in the Nb₃Sn observed in filamentary composites(13) as ∿ 5.5 × 10⁹ A m⁻² at 10T and multiply this by the filling factor of 33% we obtain an overall J_c of 1.8 × 10⁹ A m⁻², a factor of ∿ 2½ better than the 7.5 × 10⁸ A m⁻² plotted in Figure 7. It is however by no means clear whether this performance can be achieved in a practical conductor. All the above figures neglect the diluting effect of the additional regions of pure copper which will be needed in any large magnet conductor.

The surface diffusion process offers a way of increasing the final proportion of Nb₃Sn in the composite by allowing a higher tin content in the bronze so that more niobium may be reacted to completion. After processing to final size the composite is given a surface coating of tin and is heated to an intermediate temperature to allow the tin to diffuse into the composite. It doesn't matter if the bronze becomes brittle at this stage because all the drawing processes have been finished and higher tin concentrations are permissible. Indeed it is not even necessary to draw down bronze at all; a pure copper matrix may be used and all the tin finally added by coating. This has manufacturing advantages because pure copper is much easier to work than bronze. Using this process, McInturff and Larbalestier(14) have achieved the highest overall current densities so far reported for filamentary Nb_3Sn : 3×10^9 A m^{-2} at 5T.

Another approach, recently described by Hashimoto, Yoshizaki and Tanaka(15) makes use of a very tin rich bronze to provide a 'reservoir' of tin. A single rod of Sn-20 at % Cu is placed inside a copper tube which has many thinner rods of pure niobium embedded in its wall. The whole composite is drawn down to finished size and then heated to $\sim$ 750°C. During this heat treatment the tin diffuses into the copper and then reacts with the niobium. One advantage of this process is that all the starting materials are much more ductile than Cu-8 at.% Sn bronze and the drawing down to final size can be performed without the need for frequent heat treatments. Another advantage could be that, like surface diffusion, this process could allow a greater proportion of Nb_3Sn in the finished composite.

4.2 Stabilization and Protection.

It is an unfortunate fact of life that the Nb_3Sn reaction appears to tail off when the proportion of tin in the bronze falls to $\sim$ 0.3%, giving a residual resistivity of $\sim 10^{-8}$ Ω m. If it were possible to 'clean out' the bronze and leave only pure copper with a resistivity of $\sim 10^{-9} - 10^{-10}$ Ω m, the matrix would be able to perform the two very useful functions of stabilizing the composite and protecting it from burn out at quench in a magnet. Instead, it has so far been necessary to add pure copper to the matrix as a separate item. Because the tin in the bronze is very mobile at the reaction temperature, it is essential that the pure copper is protected from the bronze by a diffusion barrier of tantalum or some other metal. One may choose to have islands of copper surrounded by a barrier in a matrix of bronze, like the Harwell composite(16) shown in Figure 9. Alternatively the bronze and superconductor may be divided into islands, surrounded by a diffusion barrier and immersed in a copper matrix, like the Airco composite(17) shown in Figure 10. The former arrangement normally gives the lowest AC losses whereas the latter allows a higher proportion of copper in the matrix. A third approach, used in the IMI composite(18) shown in Figure 11 and also being pursued by Supercon and Westinghouse(19) is to use the niobium itself as the barrier. Each filament is now a hollow tube of niobium containing bronze and immersed in a matrix of copper. Reaction takes place on the inner surface of the tube and, provided it does not go too far, the copper outside should remain uncontaminated.

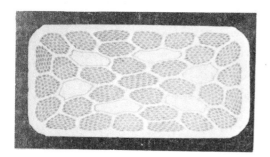

FIG. 9. Harwell multifilamentary Nb_3Sn with islands of pure copper in bronze matrix

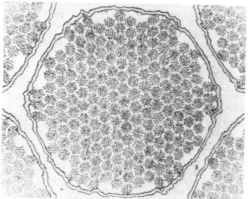

FIG. 10. Region of an Airco multifilamentary Nb_3Sn with clusters of filaments in bronze, grouped in copper matrix.

If the composite is to be made into a stranded cable, the copper can be added as separate strands in the cable, as for example in the cable made by IGC(20). To avoid mechanical damage to the Nb_3Sn it is preferable to make the cable before reaction, so that the copper strands must again be protected by a barrier. Finally, it is possible to simply plate copper onto the reacted composite. Both the plating and the cabling techniques will certainly help to protect the composite from burn out at quench but the copper may be too far away from the filaments to provide much dynamic stability. The plating technique also precludes any possibility of reacting the composite after winding the magnet.

FIG. 11. Region of an IMI multifilamentary composite, niobium tubes containing bronze in a pure copper matrix (unreacted)

4.3 Dynamic Stability. Copper in the matrix will help stability by providing magnetic damping to slow down a flux jump and thus allow more time for the heat generated to diffuse away. For this to happen, the copper must be sufficiently close to the superconductor to slow down the motion of flux within it. We may make a very rough estimate of how close this should be by applying the dynamic stability criterion(21)

$$d < \left\{ \frac{8\theta_c k(1-\lambda)}{\rho \lambda \, J_c^2} \right\}^{\frac{1}{2}}$$

where d is the stable filament diameter, k is the thermal conductivity of the superconductor, ρ is the copper resistivity. For composites like the one shown in Figure 10 we are concerned with clusters of filaments rather than individual filaments. In this case, provided the filaments are relatively fine, it is probably a fairly good approximation to consider the whole cluster as a single 'macro filament'. The values of J and k should then be taken as a suitable weighted mean between bronze and Nb_3Sn. The filling factor λ is now the fraction of the whole composite occupied by the Nb_3Sn and bronze together. Using figures for a typical composite at 5T, we find $d \sim 400\mu$. This is the maximum cluster size at which the copper will be effective in damping down instabilities, e.g. self field flux jumps. For the other geometries we may perhaps adopt a general rule of thumb that no Nb_3Sn filament in bronze should find itself more than 200μ away from a region of pure copper.

4.4 Mechanical Properties. Niobium tin is a brittle material; this poses severe problems in coil winding and also in supporting magnetic field stresses in the finished magnet. In bulk form, Nb3Sn breaks at a tensile strain of about 0.2%. It is certainly better than this in filamentary form but opinion is still somewhat divided as to how much better it is. Several workers(12,19,20) report that strains of up to 1% can be imposed before there is any noticeable reduction in current carrying capacity. Others have seen damage at strains as low as 0.3%(22). At the Rutherford Laboratory we have found that the amount of degradation seen can depend very strongly upon the sensitivity of the measurement(23). Although a strain of 0.67% caused only a small reduction in the quench current of a short sample, it reduced the current at the 10^{-14} Ω m detection level by almost a half. In a fully impregnated magnet, the composite would be expected to quench at somewhere between the 10^{-13} Ω m and 10^{-14} Ω m short sample currents. The exact level will depend on the local thermal environment at the high field point in the magnet. It is therefore important, when assessing the amount of damage caused by straining a wire, to carry out a proper resistive transition measurement. The quench current of a well cooled short sample can be a very poor guide to the quench current of an impregnated magnet.

It is almost certain that filaments which are only partly reacted will exhibit better mechanical properties than fully reacted filaments. In this case, the unreacted niobium core probably serves to halt the propagation of cracks originating at the surface of the filament. It also seems likely that the filaments are generally under a compressive strain of $\sim$ 0.2% caused by their differential contraction with respect to the matrix on cooling down from the reaction temperature. This can only help the mechanical properties of the composite.

Opinion is also divided on the question of whether it is better to react the wire first and then wind the magnet or wind first and then heat the whole magnet. At Rutherford, we are especially interested in the construction of dipoles and quadrupoles for beam transport. In these magnets, the minimum bending radius at the coil ends can be as small as 5 mm so that, even at 1% strain, a reacted conductor could only be 0.1mm in diameter. We have therefore chosen to react the whole magnet after winding and feel that the need to use heat resisting insulation and coil formers is a small price to pay for the avoidance of risks associated with winding reacted conductor. We use E glass fibre insulation, either braided or lapped. In order to protect the glass from abrasion during winding, it is desirable to use some kind of binder to glue the fibres together. We are presently trying to find the best binder; one which can be easily applied to the conductor and which will volatilize cleanly at the heat treatment stage without leaving behind a conducting residue of carbon, or oxidizing the conductor. The most promising candidate at present seems to be perspex.

Many other groups have opted for the alternative 'react first and wind later' technique. In this case the diameter of the wires must be small to avoid filament damage on bending. If such fine wires are wound singly into a magnet of reasonable size, there will be protection problems. For example, if the 0.1 mm wire already mentioned were used to make a dipole magnet of stored energy $\sim$ 1 M Joule, the peak internal resistive voltage at quench might be $\sim$ 1 MV, even if the composite contains $\sim$ 20% of pure copper. It is clearly necessary to increase the operating current and reduce the coil inductance by using many strands in parallel as a braid or cable. The cable may be impregnated with indium or similar metal after reaction. It would appear that the resulting cable can be almost as flexible as its component strands, i.e. that the minimum beinding radius is determined by the diameter of the individual wires rather than the diameter of the cable(20)(24).

If the conductor is intended for the fabrication of a large magnet with large bending radii, such as a bubble chamber, there seems to be little doubt that it can be reacted before winding. Care will still be needed however to avoid any local kinking of the conductor.

5. VANADIUM GALLIUM

The critical current density of vanadiun gallium falls off with field much less rapidly than that of niobium tin. When viewed in terms of overall current density as shown in Figure 7, the two curves cross over at around 12.5 T, giving V_3Ga the advantage at high fields. At 17 T for example, a V_3Ga composite has twice the overall current density of Nb_3Sn. The flatter shape of the J_c curve is also generally to be preferred on grounds of stability and hysteresis loss. Against these advantages however must be set the lower T_c of V_3Ga : 14.5 K as against 18 K for niobium tin and the significantly higher cost of vanadium and gallium in comparison with niobium and tin.

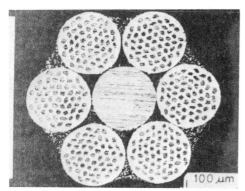

FIG. 12. Furukawa multifilamentary V_3Ga cable, six composite twisted around a tungsten wire.

V_3Ga tapes were probably the first superconductors to be made using a bronze process. It was found(25) that the rate of growth of the V_3Ga layer and also its critical current could be markedly increased by copper plating the gallium coated vanadium wire before heat treatment. The bronze technique was subsequently applied to the first commercial production of multifilamentary A15 compound superconductors: the V_3Ga composites and cables made by the Furukawa Electric Company(26). Figure 12 shows a cross section of one of these cables. The centre strand is a tungsten wire, for strength; the six outer strands are composites of 55 filaments of vanadium in gallium bronze. After reaction the cables are impregnated with indium. Overall densities are as plotted in Figure 7.

5.1 Current Densities.   Although the overall current density of V_3Ga composites at medium fields is presently rather disappointing, this situation could be quite dramatically changed by the recent work of Howe and Weinman(27). These workers have evolved a technique in which both the vanadium core and the copper matrix are alloyed with gallium, e.g. 9 at.% Ga in the vanadium and 17.5 at.% Ga in the bronze matrix. In this way, the reaction temperature may be reduced from $\sim$ 600°C to $\sim$ 550°C. As a consequence of the lower temperature of formation, the critical temperature has been found to increase by $\sim \frac{1}{2}$ K and the critical current density by a factor of $\sim$ 5. The maximum current density observed at 10 T was in fact 10^{10} A m^{-2}.

In spite of their high local current densities, the filling factors of these experimental composites is low (as shown in Figure 13) and their overall J_c would not be very interesting to the magnet constructor. At present, one can only speculate on what the maximum overall current density in a practical magnet conductor using this process might be. If, for example, one could start with 20 at.% of gallium in the bronze and then react until this level had fallen to 10 at.% and all the vanadium was consumed, the resulting filaments of V_3Ga would occupy $\sim$ 25% of the composite cross section. If the highest current densities could be obtained over the whole cross section of each filament, the current density at 10 T over the composite would be 2.5 x 10^9 A m^{-2}. A spectacular figure, but there are many 'ifs' in the argument leading up to it.

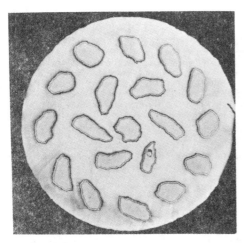

FIG. 13. Experimental high J_c composite of V_3Ga

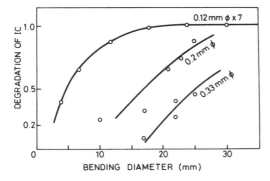

FIG. 14. Bending properties of V_3Ga composites

It has also been found by Tachikawa Itoh and Tanaka(28) that the addition of 1-5 at.% of aluminium to the vanadium filament can markedly increase the rate of formation of V_3Ga and raise its critical current density.

5.2 Stabilization and Protection. The V_3Ga reaction is thought to stop when the gallium content of the bronze falls to about 10 at.%. At this concentration, the resistivity of the bronze is reported to be $\sim 10^{-7}$ Ω m at 4.2 K. This is almost as high as the cupro-nickel used to provide resistive barriers in AC NbTi composites. It is not surprising therefore that the pulsed behaviour of multifilamentary V_3Ga is good.

If the composite is to be used in the construction of a magnet of reasonable size however, some pure copper or other good normal conductor will be needed for protection from burn-out at quench. Any composite or cable of diameter greater than, say, $\frac{1}{2}$ mm would also benefit from the stabilizing influence of a high conductivity component. The indium coating on the Furukawa cable is thought to confer a degree of stability but for large magnets and cables, something of lower resistivity will be needed. To the author's knowledge, no such cable or composite has yet been produced.

5.3 Mechanical Properties. The mechanical properties of V_3Ga appear to be very similar to those of Nb_3Sn. Figure 14 shows the degradation of critical current in two different wires and a cable as they are bent around various diameters(29). From this data, it would appear that the filaments can be strained to $\sim$ 0.7% without damage and also that the minimum permissible bending radius of a seven strand cable does not differ appreciably from that of its component strands. One should again beware however of the difference in quench current between a well cooled short sample and a fully impregnated magnet.

6. DISCUSSION AND FUTURE PROSPECTS

It seems likely that multi-filamentary composites will continue to be the most common type of super-conductor used in magnet construction. New requirements for low losses in small amplitude AC fields may however demand different designs for such applications as magnetic levitation. The theory of magnetic instabilities in filamentary composites has been improved and extended in recent years and is now fairly well understood. The major cause of degradation in magnets however is now mechanical - small movements, slippage, cracking etc. causing a local release of energy. This is not very well understood at present and more work is needed in this area. Magnetic stability theory should at least be able to tell us which conductors might be especially vulnerable to these mechanical effects.

Filamentary Nb_3Sn and V_3Ga are advancing to the stage where they must be considered as useful magnet conductors. Many of the problems of production and utilization have been sorted out and we may confidently expect to see these materials used in the construction of several medium sized magnets before the next Magnet Technology Conference. These is undoubtedly scope for a further improvement in the current carrying capacity of these materials; it will be disappointing if a further factor of two in overall J_c is not forthcoming in the fairly near future.

In the longer term, perhaps the most exciting prospect is the promise shown in Figure 15 - Nb_3Ge: critical temperature 23 K, upper critical field 38 T. Even at a temperature of 18 K, the critical current density is quite respectable and at lower temperatures it is probably very high indeed.

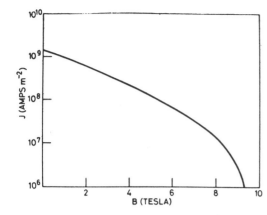

FIG 15.   Nb$_3$Ge current density at 18 K

It is very difficult to form stoichiometric Nb$_3$Ge and the low critical temperatures measured in the past can be attributed to non-stoichiometric compound. Recently however, Gavaler(30, 31) has shown that it is possible to make stoichiometric Nb$_3$Ge using a sputtering process under very carefully controlled conditions of gas pressure, voltage and temperature. It would be difficult to maintain such strict control in any kind of continuous process. Present indications are that Nb$_3$Ge cannot be made by a bronze process but this is by no means fully established, possibly a matrix other than copper may yield a satisfactory result. However, in view of the very recent nature of these discoveries, it is really not very useful to speculate on production techniques at this point in time. One can probably say with some confidence however that in the next few years, while the magnet technologists are grappling with the problems of using the present 'exotic' materials Nb$_3$Sn and V$_3$Ga, the more basic research work is going to find several new materials offering greatly improved H$_c$, T$_c$ and J$_c$ and that at least one of these will prove to be technologically useful.

ACKNOWLEDGEMENTS

    It is a pleasure to acknowledge the help of

Y Furuto, D G Howe, J A Lee, I L McDougal, R L Stoecker, K Tachikawa, B Turck, T A de Winter, J Wong and many others who supplied valuable information for this paper and also D C Larbalestier who contributed useful information and comments to the Nb$_3$Sn section and C A Scott who supplied the results of solenoid tests quoted in section 3.

REFERENCES

1. Fietz W A, Rosner C H, 'Advances in Superconductive Magnets and Materials' IEEE Trans MAG 10 2 p. 239 (1974).

2. Hunt T K, 'AC Losses in Superconducting Magnets at Low Excitation Levels' Jnl. Appl. Phys. 45 2 p. 907 (1974).

3. Cornish D N and Khalafallah K, 'Superconductors for Tokomac Toroidal Field Coils', Proc 8th Symposium on Fusion Technology.

4. Satow T, Tanaka M and Ogama T, 'AC Losses in Multifilamentary Superconducting Composites for Levitated Trains Under AC and DC Magnetic Fields', Adv. Cryo. Eng. 19 p. 154 (1973).

5. Duchateau J L and Turck B 'Theoretical and Experimental Study of Magnetic Instabilities in Multifilamentary NbTi Superconducting Composites' Proc. 1974 Applied Superconductivity Conference. Oakbrook, U.S.A.

6. Duchateau J L and Turck B "Dynamic Stability and Quenching Currents of Superconducting Multifilamentary Composites in Usual Cooling Conditions', Saclay Report SEDAP/75-03 to be published in Jnl. Appl. Phys.

7. Wilson M N, 'Filamentary Composite Superconductors for Pulsed Magnets' Proc. 1972 Applied Superconductivity Conference IEEE 72CH0662-5-TABSC.

8. Duchateau J L, Turck B, 'External Field Effect on the Current Distribution in Superconducting composites' Cryogenics Oct. 74 p. 545.

9   Strauss B P, Sutter D F, Ioriatti E, Habrylewicz W, 'Results of Magnet Prototype Evaluation for the Fermilab Energy Doubler Project. Proc 1974 Applied Superconductivity Conference

10. Brisbin P H, Coles W D, 'Multifilamentary Niobium Tin Superconductor Tape' Proc. 1974 Applied Superconductivity Conference.

11. Tsuei C C, 'Ductile Superconducting Cu-Rich Alloy Containing A15 Filaments' Proc. 1974 Applied Superconductivity Conference.

12. Suenaga M, Sampson W B, Klamut C J, 'The Fabrication and Properties of Nb$_3$Sn Conductors by the Solid Diffusion Process', Proc. 1974 Applied Superconductivity Conference.

13. Suenaga M, Luhman T S, Sampson W B, 'Effects of Heat Treatment and Doping with Zr on Critical Current Densities of Multifilamentary Nb$_3$Sn Wires' Jnl. Appl. Phys. __45__ 9 p. 4049 (1974).

14. McInturff A D, Larbalestier D C, 'Effect of Metallurgical History on J$_c$ (5T) in Surface Diffused Multifilamentary Nb$_3$Sn' Proc. 1974 Applied Superconductivity Conference.

15. Hashimoto Y, Yoshizaki K, Tanaka M, 'Processing and Properties of Superconducting Nb3Sn Filamentary Wires' Proc. ICEC5 (1974).

16. Larbalestier D C, Madsen P E, Lee J A, Wilson M N, Charlesworth J P, 'Multifilamentary Niobium Tin Magnet Conductors' Proc. 1974 Applied Superconductivity Conference.

17. Gregory E, Marancik W G, Ormand F T, 'Composite Conductors Containing Many Filaments of Nb$_3$Sn' Proc. 1974 Applied Superconductivity Conference.

18. McDougal I L, 'Superconducting Solenoids Using High Current Density Nb$_3$Sn Filamentary Composites' paper H11 at this conference.

19. Randall R, Wong J, Deis D W, Shaw B J, Daniel.M R, 'Fabrication and Properties of Multifilament Nb$_3$Sn Conductors' Proc. 1974 Applied Superconductivity Conference.

20. Scanlan R M, Fietz W A, 'Multifilamentary Nb$_3$Sn for Superconducting Generator Applications' Proc. 1974 Applied Superconductivity Conference.

21. Superconducting Applications Group, Rutherford Laboratory, 'Experimental and Theoretical Studies of Filamentary Superconducting Composites' J. Phys. D3 p. 1517 (1970).

22. Deis D W, Gavaler J R, Shaw B J, 'Multifilament Nb$_3$Sn Conductors' J. Appl. Phys. __45__ 10 p. 4594 (1974).

23. Larbalestier D C, Edwards V W, Lee J A, Scott C A, Wilson M N, 'Multifilamentary Niobium Tin Solenoids' Proc. 1974 Applied Superconductivity Conference.

24. Tanaka Y, Furuto Y, Ikeda M, Dobashi N, 'Multifilamentary Superconducting Nb$_3$Sn Wire' 13th Convention, Cryo. Soc. Japan, Nov. 1974.

25. Tachikawa K and Tanaka Y, 'Superconducting Critical Currents of V$_3$Ga Wires Made by a New Diffusion Process' Japan J. Appl. Phys. __6__ (1967) p. 782.

26. Furuto Y, Suzuki T, Tachikawa Y and Iwasa Y, 'Current Carrying Capacities of Superconducting Multifilamentary V$_3$Ga Cables' Appl. Phys. Lett __24__ 1 p. 34.

27. Howe D G and Weinman L S, 'Superconducting Properties of V$_3$Ga in the V$_3$Ga/Cu-Ga Composite System' Proc. ICEC5 (1974) p. 326.

28. Tachikawa K, Tioh K and Tanaka Y, 'Studies on Multifilamentary V$_3$Ga Wire' Proc. 1974 Applied Superconductivity Conference.

29. Ikeda M, Furuto Y, Tanaka Y, Inoue I, Ban M, 'V$_3$Ga Stranded Wire Wound Superconducting Magnet' Unpublished Report, Furukawa Electric Co. 1974.

30. Gavaler J 'Superconductivity in Nb-Ge Films Above 22K, App Phys. Lett. __23__ 8 P. 480.

31. Gavaler J, Janocko M A, Jones C K, 'Preparation and Properties of High T$_c$ Nb-Ge Films' J. Appl. Phys. __45__ 7 p. 3009.

[Editor's Note: The discussion has been omitted.]

22

Copyright © 1976 by IPC Science and Technology Press Ltd

Reprinted from *Cryogenics*, Dec. 1976, pp. 699–704

Pulsed energy conversion with a dc superconducting magnet

M. Cowan, E.C. Cnare, W.B. Leisher, W.K. Tucker, and D.L. Wesenberg

A generator system for pulsed power is described which employs a dc superconducting magnet in a magnetic flux compression scheme. Experience with a small-scale generator together with projections of numerical models indicate potential applications to fusion research and commercial power generation. When the system is large enough pulse energy can exceed that stored in the magnet and pulse rise time can range from several microseconds to tens of milliseconds.

A generator system called PULSAR is being developed to meet pulse power requirements of future fusion research.[1-4] The system also has applications in commercial power generation to improve efficiency and reduce capital cost.[5,6] The main components of a PULSAR system are a superconducting magnet, a generator coil of normal conductor, and an armature which may be a metallic conductor or a plasma. Energy conversion involves magnetic flux compression which is achieved by using chemical energy to increase forcibly the mutual inductance between the armature and generator coil. This is done either by expanding the armature radially within or by driving it axially through the coil. The superconducting magnet provides the operating magnetic flux but develops only a small current transient during pulse generation.

As pulse generators for fusion research, full-scale systems are projected to produce pulses with energy ranging from slightly less to substantially greater than that stored in the associated magnet and with rise times ranging from several microseconds to several tens of milliseconds. Thus it should be possible to produce pulses with peak power above the terawatt level. Employed as a topping stage in a coal-fired power plant, PULSAR is projected to increase plant efficiency with a lower temperature and a much smaller magnet than MHD systems require.[6]

A small-scale system has been used in an experimental programme which has confirmed the predictability of electrical performance and delineated the engineering approach to full-scale systems. This paper will summarize experimental results and give size and performance projections of full-scale systems for specific applications.

Modelling generator performance

A computer code which models operation of PULSAR systems with metallic armatures has been written for a CDC 6600 digital computer. It solves a set of time-dependent,

The authors are with the Sandia Laboratories, Albuquerque, NM 87115, USA. Received 13 September 1976.

mutually coupled, lumped parameter circuit equations together with the equation of motion of the moving armature. Fig. 1 shows an example of a greatly simplified circuit model for PULSAR with only the three basic elements shown. For solutions of practical value, additional circuits must be included. For example, a steel sleeve is used to provide structural support to the generator coil and the superconducting wire of the magnet is wound on an aluminium bobbin. Both of these parts form circuits which act to minimize the current transient in the magnet. The superconducting and generator circuits are modelled as single loops since their current distributions are controlled by the winding configuration. However, current in the armature can drift axially and diffuse radially. This is accounted for by conceptual dividing of the armature axially and radially into many pieces and representing each piece by a circuit containing resistance and both self and mutual inductance. Proper time variation of the mutual inductances accounts for the motion of the armature, and produces a current

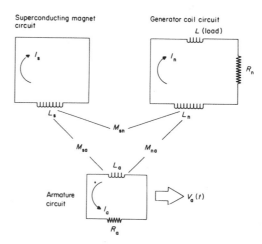

Fig. 1   Circuits used to model PULSAR operation

Fig. 2   First layer-turn of generator coil winding

pulse in the load. Provisions are also made for dividing the superconductor winding into layers so that layer-to-layer voltages can be calculated These are smaller than 1 kV in the present system.

Additional code work is underway to simulate PULSAR operation with a plasma armature. The armature must have sufficient velocity for flux compression to occur; that is, time for significant magnetic diffusion into the armature must be long compared to the time required for either its axial passage through or its radial expansion within the generator coil. Large systems meet this requirement with lower armature velocities than small ones. Large size also makes it easier to achieve a higher maximum coupling coefficient between the armature and generator coil. This is important for efficient operation. It is difficult to obtain close coupling in small scale because of construction tolerance, the necessity of multilayers in the windings and finite skin depth. These differences due to scale result in performance projections for full-scale systems greatly exceeding that achieved in small-scale experiments. However, experimental results are accurately predicted by the same code used for projections.

Experimental systems

Experimental systems have used a superconducting magnet with a clear bore of 460 mm and a 267 mm length of winding. It has 6656 turns in 26 layers wound on an aluminium bobbin 8.3 mm thick, and stores 200 kJ at peak current of 150 A (10^6 A-turns). The winding is potted in a filled epoxy and restrained axially by nylon spacers. The wire is rectangular, 0.51 mm x 0.15 mm, and has 132 x 36 μm NbTi filaments in copper with 2.6 copper to NbTi ratio. The twist pitch is 25.4 mm. Short sample critical current is 250 A at 3 T. The magnet requires ~60 litres LHe to cool from 77 K to 4.2 K and fill, after which it may be operated for about 2 hours before refilling is required.

The copper generator coil has a bore of 290 mm, a length of 350 mm, and incorporates six layers at one turn per layer for an inductance of 7 μH. It is wound on an epoxy-impregnated fibre-glass core. Fig. 2 shows the lay-up of the first layer-turn which is composed of 26 individually insulated wires of rectangular cross-section. A precast MOCA-cured Adiprene sleeve was slip-fit over the winding and Adiprene resin was injected under pressure to fill the voids. An Inconel 718 stainless steel cylinder was then slipped over the Adiprene to resist radial forces produced during operation. This coil

structure is nested co-axially within the bore of the super-conducting magnet with a radial clearance of 4.5 mm. Fig. 3 shows this arrangement with the coil and magnet axes vertical. Notice that the magnet fill tube is mounted at 45° so that operation is also possible with the axes horizontal. The co-axial cables from the generator coil connect to the load, which is not shown. This magnet and coil arrangement has been used for experiments in both the radial and axial modes of operation.

In full-scale systems the radial mode will be required for pulse rise times less than 1 ms. For this, the armature is either an expendable thin-walled aluminium tube or a plasma which expands radially within the generator coil. When relatively slow pulses are adequate the axial mode may be used. The armature in this case is a heavy-walled free piston which is driven axially through the nested coil and magnet and which is designed to survive repeated use. Fig. 4 shows radial and axial mode armatures which have been used in the experimental programme along with the two principal parts of the generator coil assembly.

The radial expansion of thin-walled aluminium tubes was first studied using magnetic pressure created by capacitor bank discharge. This study found that 1100-0 aluminium gave best expansion at the low strain rates required. Finally, armatures large enough for generator experiments were developed for expansion by high explosive. A typical radial armature is 1.3 mm thick, 220 mm in diameter and 330 mm long. Multipoint detonation of a 0.42 kg charge distributed along the armature axis produces nearly cylindrical expansion at 575 ms^{-1}. For a radial mode experiment the magnet and coil are mounted on a firing stand as shown in Fig. 3.

The axial-mode armature was designed to withstand the high pressures associated with both the acceleration by chemical propellant and the magnetic fields produced in flux compression. It was made of a 6061-T6 aluminium tube with 12.7 mm wall and a 25 mm thick steel back plate. Its diameter and length were 290 mm and 450 mm, respectively, and it was propelled through the generator

Fig. 3   Set-up for an experiment in radial mode with coil and magnet axes vertical

Fig. 4 Coil parts and aluminium armatures. Left to right are the steel sleeve, the winding structure, an axial armature (with interior plywood stiffener) and a radial armature

coil with velocities ranging from 300 to 750 ms^{-1} by a recoilless cannon. Fig. 5 shows the magnet and generator coil mounted on the cannon for an axial mode experiment. The cannon is 20 m long with a 290 mm smooth bore. Propellant is fired at the centre of the barrel to accelerate in opposite directions the armature and a reaction projectile of equal mass.

Plasma armatures, which are still in an early development stage, allow generation of pulses with the shortest rise times. They have been produced experimentally by shock heating deuterium with the expanding products of the high explosive fuel. This has been done with and without pre-ionizing the deuterium, however, without pre-ionization, geometric conversion must be used in the high explosive system to increase shock velocity to more than 15 km s^{-1}. The plasma conductivity is typically $\sim 10^4$ mhos m^{-1}. Thus, compared to the plasma flow in a conventional MHD generator the plasma armature of PULSAR has much higher velocity and conductivity. In fact, magnetic Reynolds numbers greater than unity have been achieved.

The first experimental generator to incorporate a plasma armature is shown in Fig. 6. For convenience it was designed to be self-destructive. The generator coil was a single turn of aluminium, 600 mm in diameter and 65 mm long. Instead of the superconducting magnet a slow electromagnet (Helmholtz pair) was used which was powered by a capacitor bank discharge. Plates of lucite were used to form a chamber within the single turn, and this was evacuated and then backfilled with deuterium gas to 10 torr (1.3 kN m^{-2}). Two cylinders of high explosive were purged of air and filled with deuterium at atmospheric pressure. These were located on axis at opposite sides of the chamber and detonated so that the two high-speed flows collided at the centre of the chamber and expanded radially into the single turn. The flow entered the chamber through thin Mylar diaphragms. Detonation of the explosive charges produced a radial ionizing shock of about 17 km s^{-1} in the low density deuterium within the chamber.

Experimental results

The present generator coil design has survived several radial and axial tests and remains in good condition. Comparisons between measured and calculated results are made in Fig. 7 and 8 for typical radial and axial mode experiments with

aluminium armatures. The measured quantity is time derivative of current I in a 0.6 μH load. The experimental $\dot{I}$ curve is integrated to obtain a current-time curve and calculated points of current at selected times are shown for comparison. For the radial mode test of Fig. 7 the magnet was charged to its maximum current of 150 A, while for the axial test of Fig. 8 it was charged to 100 A. Calculated results agree well with measurements except near peak current in the radial test where they appear to be too high. This results from a 13 mm thick rubber cushion used to line the generator coil. No allowance was made in the code for the slowing of the armature by the rubber. Also the armature rebounds after colliding with the coil and often tears which accounts for the appearance of the $\dot{I}$ record after peak current. No calculations are attempted to simulate this behaviour. However, in the axial case comparison is made for the complete passage

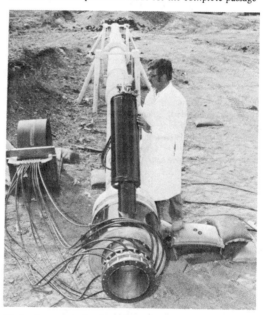

Fig. 5 Arrangement for an experiment in axial mode with coil and magnet axes horizontal. Cables run from coil to 0.6 μH load

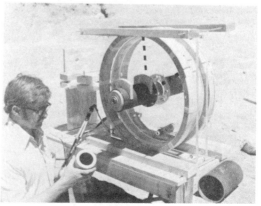

Fig. 6 Self-destructive generator used to test plasma armature. Worker holds one of the two explosive elements used in this test

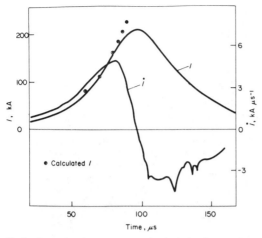

Fig. 7   Comparison between measured and calculated current for radial test with magnet at peak current

different conductivities assumed for the plasma until the calculated and measured currents compared as shown. This established an estimated plasma temperature of about 4 eV. Since this is higher than shock heating alone could produce, it is concluded that ohmic heating was responsible for the good conductivity. The last 90% of the current was produced in 11 μs and the final e-folding time was about 5 μs. Peak field was 1.5 T, about three times higher than the original field established by the Helmholtz coil.

Magnet response

The present experimental magnet has been involved in five radial and four axial experiments. Radial tests have been conducted at 30, 100, 125, and 150 A while all axial ones were at 100 A. Twice during radial tests the vacuum system of the magnet opened to the atmosphere resulting in rapid quenching. Damage was minor and easily repaired. The cause was finally found to be misalignment between the armature and the magnet which developed during charging and which was caused by an asymmetric array of ferrous bolts used in

of the armature through the coil and agreement has always been as close as that indicated in Fig. 8. The axial mode armature is somewhat longer than the coil and this accounts for the flat portion of the $\dot{I}$ record at about 2 ms. This tends to make a flat top on the current record and demonstrates that the geometry of the armature can be used to shape the pulse. Table 1 summarizes generator performance for these two experiments.

Notice the relatively large kinetic energy for the axial armature. This illustrates the problem of achieving efficiency in small scale for this mode. The mass becomes large for structural strength yet the speed must be high because of the scale. Even so, axial systems can be quite efficient, if they are large enough.

To illustrate the potential for plasma armatures Fig. 9 shows the electrical record from operation of the experimental generator shown in Fig. 6. For this experiment the calculated points shown were determined in a relatively artificial way. The thickness and expansion speed (17 km s^{-1}) for the conducting plasma were determined from framing camera pictures of the event. Since the mass-velocity distribution of the radial flow was not known the conducting plasma was arbitrarily assigned a mass so that calculated velocity would match the measured one. Then, code runs were made with

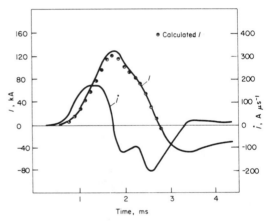

Fig. 8   Comparison between measured and calculated current for axial test with magnet at 2/3 peak current. Note that $\dot{I}$ scale is A μs^{-1}

the set-up. Otherwise the magnet has not quenched during radial mode operation. Quench has occured after every axial experiment. Unlike that resulting from loss of vacuum, this has been relatively mild, taking several seconds after pulse generation to develop. When a non-conducting armature was used in an axial test in which all other experimental conditions were unchanged the magnet did not quench, eliminating purely mechanical causes. Additional axial experiments will be carried out in an effort to establish the reason for quench in axial mode operation.

Structural system

The present experimental generator coil (with its steel sleeve) is designed to withstand field intensities of 20 T. This is probably near a practical design limit and projections of large scale generator performance have assumed no more than 25 T. In fact, the present experimental generator cannot generate fields above 12 T: but, in the radial mode, the

Table 1.   Generator performance for radial and axial mode tests

	Radial (150 A)	Axial (100 A)
Pulse rise time	60 μs	700 μs
Armature speed	575 ms^{-1}	300 ms^{-1}
Armature kinetic energy	0.15 MJ	1.46 MJ
Peak power in load $(L\dot{I})$	460 MW	9.5 MW
Pulse energy in load $(1/2\,LI^2)$	12.4 kJ	4.6 kJ
Pulse energy/magnet energy	0.061	0.051

armature impacts the generator coil wall, producing compression shocks in the range of 300–500 MPa. In large radial systems it should be possible to avoid this physical impact and the resulting shocks by achieving magnetic rebound at 20 to 25 T.

Since the radial PULSAR system will be used in an enclosed building, shock waves, detonation products and sound vibrations must be properly mitigated. This can be done with a muffler system which will be designed to fit the generator coil. Within this system several layers of shock-mitigating material sandwiched between perforated baffle plates will be used. This same type of system has been used in blast containment chambers and shock wave diffusers. In this manner radial PULSAR can be safely operated in an enclosed building.

Applications

High energy electron beam accelerators

A radial PULSAR system with a plasma armature has the potential to replace the traditional Marx generator for high energy electron beam accelerators, achieving substantial reduction in cost. It may be possible to operate directly into either a water capacitor or an advanced diode without additional power amplification. It will not be possible to design with confidence large generators which use plasma armatures until a proper MHD code has been developed. However, crude extrapolations made from experimental results are encouraging.

Flash lamps for upgraded SHIVA

The SHIVA laser at the University of California, Lawrence Livermore Laboratory, will produce a beam with a power of more than 20 TW for study of fusion by inertial confinement.[8] The possibility of an upgraded version for the 1980s is under consideration which would increase laser power by an order of magnitude or more. A PULSAR system is one of the alternatives to a 90 MJ capacitor bank being considered for the prime power source in the upgraded version. The required pulse rises to peak current in 600 μs and delivers nearly 8 kA to each of the 10 000 flash lamps in the system. The lamps are arrayed in 5 000 parallel circuits, each circuit consisting of two lamps in series. Table 2 summarizes features of a radial PULSAR system which would meet the requirements. An

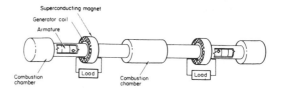

Fig. 10   Conceptual design of commercial power generator

expendable aluminium armature would be used for each pulse. Costs for the principal parts of the PULSAR system are estimated as follows:

Magnet and liquid helium system	**$1.1 M**
Generator coil	0.5 M
Muffler system	0.5 M
Armature and explosive (1000 pulses)	0.5 M
Total	**$2.6 M**

Comparing with a system based on capacitor storage alone, this 2.6 M would replace a cost of 15.3 M for capacitors, switches, and other associated hardware, and would reduce the cost of the complete power system from about 26 M to about 13 M.

Commercial power generation

Conceptual designs have been developed for PULSAR generators which could be applied to generation of commercial power. They appear to offer advantages of economy and efficiency when employed in fossil fuel power plants either as topping or as independent generators.

The generator for this application consists of a coal-powered free piston engine, the pistons of which also serve as axial mode armatures. Generator coils are incorporated into the cylinder walls of the engine, as shown in Fig. 10. In this conceptual version there are two counterpoised armatures (pistons) and two magnet-generator coil arrangements. The engine consists of a long cylinder with a single combustion chamber at each end and a double chamber at the centre. As the armatures reciprocate between combustion chambers, pulses are produced during each pass through the coils. Pulse energy is stored, converted to 60 Hz, and delivered to the grid.

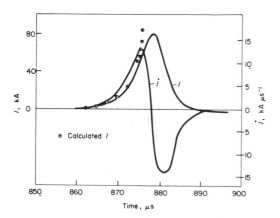

Fig. 9   Current measurements for test of plasma armature

Table 2.   PULSAR system for upgraded SHIVA laser

	Magnet	Generator coil	Armature
Radius, m	1.8	1.27	0.83 (initial) – 1.26 (final)
Length, m	2.5	2.5	2.5
Ampere turns	5.6 M	38 M (peak)	44 M (peak)
Magnetic field, T	2.2	25 (peak)	25 (peak)
Energy, MJ	49	63*	105 (kinetic)

*This energy dissipated in lamps.

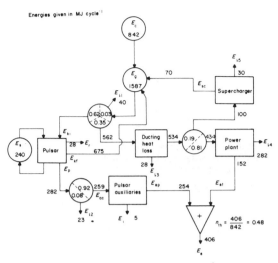

Energies given in MJ cycle⁻¹

Fig. 11   Energy flow diagram for single cycle of commercial power generator. Whole numbers are megajoules, decimal fractions refer to energy partition ratios or efficiencies

Using the PULSAR electrical operation code along with a thermo-dynamic model, a topping stage of this general design has been sized and its performance calculated for a plant of 944 MW_e output and 48% overall efficiency. Generator parameters were as follows:

Overall generator length	99 m
Cylinder diameter, generator coil and magnet lengths	2.85 m
Magnet energy	120 MJ (2 each)
Armature length	3.15 m (2 each)
Armature mass	5.4×10^4 kg (2 each)

All of the combustion for the plant was assumed to occur in the topping stage which produced about 60% of the electrical output. Heat was recovered from the exhaust gas by a conventional steam system which generated the remaining 40%, assuming a 35% efficiency which is consistent with the exhaust temperature of 1290 K. The armature (piston) velocity was required to be 135 m s⁻¹ at the coil entrance which is high compared to conventional engines, but conventional engine piston speed is limited by inertial forces on mechanical linkages (connecting rods, etc) and there are no linkages in this design.

The thermodynamic cycle used differs from conventional two-stroke cycles in that heat addition is isothermal and the volumetric compression ratio has the relatively high value of 57. The maximum values of temperature and pressure are 2050 K and 38 MPa. Total cycle time for the size given is 0.43 s, of which only 0.07 s is required for the armatures to pass through the coils.

The energy flow diagram for one cycle of this design is shown in Fig. 11. The air receives 1578 MJ per cycle from the supercharger, from compression by bringing the armatures to rest and from chemical energy released by coal combustion. Of this, 68 MJ are lost through heat transfer, 985 MJ are converted to armature kinetic energy at coil insertion and 534 MJ goes to the steam bottoming cycle. The supercharger requires

100 MJ of this and the remainder results in a net output of 152 MJ for the bottoming stage.

The PULSAR topping stage will lose 28 MJ to ohmic heating during generation of 282 MJ output. Subsequently, 23 MJ are lost in pulse-to-60 Hz conversion and 5 MJ are required for auxiliaries making a net output of 254 MJ. This gives a total output per cycle of 406 MJ. Since the chemical input was 842 MJ, the plant has a thermal efficiency of 48%. The cycle time of 0.43 s brings plant output to 944 MW.

As an independent power plant, this same PULSAR design has projected efficiency and output of 43% and 577 MW. In this application more energy may be extracted from the gas since there is no requirement for high exhaust temperatures. The exhaust is used only for supercharger power.

Conclusions

Results of analytical and experimental studies indicate that magnetic flux compression generators employing super-conducting magnets are practical as pulse power generators when the energy associated with the pulse exceeds a few tens of megajoules and when the rise time falls in a range between several microseconds and tens of milliseconds. For these conditions the energy of the pulse can range from slightly less to substantially greater than that stored in the magnet. This application is relatively straightforward since the magnet is not required to rotate or suffer significant changes in current.

Following the successful development of a suitable free piston engine, this kind of generator is also interesting for its potential applications in commercial power plants.

The authors wish to acknowledge the able assistance of the test facility personnel during the experimental phase of these studies. William P. Brooks was responsible for instrumenting and maintaining the superconducting magnet in working order and Billy W. Duggin designed and fabricated the explosive charges for the radial mode experiments. Robert A. Davis, Edward R. Ratcliff, and Larry Yellowhorse made valuable contributions while fielding the PULSAR components and in the recovery of electrical data. This work was supported by the US Energy Research and Development Administration.

References

1   Cowan, M. et al 'Multimegajoule pulsed power generation from a reusable compressed magnetic field device,' Proc Int Conf on Energy Storage, Compression, and Switching, Torino, Italy (1974)

2   Cowan, M. et al 'Electron beam power from inductive storage,' Proc Int Top Conf on Electron Beam Research & Technology (1975) 490

3   Cowan, M. et al 'PULSAR – a field compression generator for pulsed power,' Proc 6th Symp on Engineering Problems of Fusion Research (1975) 308

4   Cnare, E.C. et al 'PULSAR – the experimental program,' Proc 6th Symp on Engineering Problems of Fusion Research (1975) 312

5   Leisher, W.B. et al 'A flux compression topping stage,' Sandia Laboratories Report No SAND76-0277 (1976)

6   Cowan, M. et al 'PULSAR – a flux compression stage for coal-fired power plants,' Proc ICEC6 (forthcoming 1976)

7   Knoepfel, H. Pulsed High Magnetic Fields (North-Holland Pub Co, Amsterdam, 1970)

8   Allen, G.R., Gagnon, P.R., Rupert, P.R., Trenholme, J.B., 'Energy storage and power conditioning system for the Shiva Laser,' Proc 6th Symp on Engineering Problems of Fusion Research (1975) 325

Part III
CAPACITORS

Editor's Comments
on Papers 23 Through 27

A capacitor is a device composed of two electrical conductors separated by an electrical insulator. The classic Leyden jar, discovered in 1746 by Pieter van Musschenbrock, is perhaps the simplest form of a capacitor. It is a thin-walled glass container with a thin conductive layer on both the inside and outside. These two layers are electrically separated from each other by having the top few centimeters free of conductor. A cork with a conductor passing through it is placed in the top of jar.

A substantial spark can be drawn by discharging Leyden jars, which can be charged to several thousand volts. A typical jar of this type can store 10^{-4} to 10^{-2} J and can be discharged in a few microseconds (1 microsecond = 1 μs = one millionth of a second).

To achieve reproducible capacitance values and high-energy storage densities, manufacturers use very thin materials for both the conductor and insulator or dielectric, which are packed very tightly into as small a volume as possible. Nevertheless, capacitors can achieve only modest energy densities and are used for storage in applications where very short discharge times, between 10^{-2} and 10^{-12} s, are needed.

Various materials are used in capacitor fabrication. The dielectrics include ceramics, paper, air, mica, titanium oxide, barium titanate, Mylar, and Teflon. And the conductive plates or electrodes are made of material such as tin, aluminum, copper, and tantalum.

Special capacitors formed from a variety of exotic materials have been constructed for space application and very high- or very low-temperature environments.

There have been some proposals that the theoretical energy density of capacitors should approach 10 to 1000 MJ/m^3. These estimates are based on electric fields of 0.1 to 0.3 V/angstrom or 1 to 3×10^9 V/m and dielectric constants of 1000 to 10000, which have been separately achieved. It does not appear likely that both limits will be simultaneously reached in large capacitors in the near future.

To date, the largest capacitors, in terms of energy storage capacity, have been constructed of 0.025-mm-thick aluminum sheets separated by four or more 0.015-mm-thick sheets of kraft paper, which is impregnated with castor oil. This sandwich of materials is wound into a tight coil, electrical connections are attached to the two conductor sheets, and the entire package is enclosed in a hermetically sealed container. In this and other capacitors, the total thickness of the dielectric is determined by the maximum voltage to be applied during normal operation, and the conductor thickness is determined by the maximum current during charge or discharge.

The earliest report of the use of asymmetrical conductivity of a cell to produce a capacitor is presented here as Paper 23. The basic chemistry of electrolytic capacitors that are used to provide steady, high voltages of one polarity is developed in this paper.

Paper 24 is a set of excerpts from a book, *The ABC's of Capacitors*, by W. F. Mullin. Many of the different capacitor types used for electronic applications are presented, along with some photographs and figures showing fabrication techniques. Papers 25 and 26, both by E. L. Kemp, describe the components required in a large energy-storage capacitor bank and some of the factors that must be considered to use the fast discharge capability of capacitors on a large scale.

Paper 27 is a rather speculative assessment of the possibility of future use of capacitors for energy storage.

BIBLIOGRAPHY

Brotherton, M. 1946. *Capacitors—Their Use in Electronic Circuits*. Van Nostrand, Princeton, N.J.

Boicourt, G. P. 1968. *Problems in the Design and Manufacture of Energy Storage Capacitors*. Los Alamos Scientific Laboratory Report LA-4142-MS.

Coursey, P. R. 1927. *Electrical Condensers*. Pitman, London.

Dielectrics in Space. Symp. Westinghouse Res. Lab., Proc., 1963. Westinghouse.

Dubilier, W. 1944. Development Design and Construction of Electrical Condensers. *J. Franklin Inst.* **248**:193–204.

Dummer, G. W. A., and H. M. Nordenberg. 1960. *Fixed and Variable Capacitors*. McGraw-Hill, New York. (This book contains a comprehensive bibliography of research and development up to 1960.)

Georgiev, A. M. 1945. *The Electrolytic Capacitor*. New York: Murray Hill Books.

Kahn, L. 1955. Dielectrics in Capacitors. In *Dielectrics Materials and Applications*, ed. A. R. Von Hippel. The Technology Press of MIT, Cambridge, Mass. and Wiley, New York, pp. 221–225.

Karasz, F. E., ed. 1972. *Symp. Dielectr. Prop. Polymers, Proc.* Plenum, New York.

Reynolds, G. D. 1948. Tests for the Selection of Components for Broadcast Receivers. *Inst. Electr. Eng. (London) J.* **95**, pt. III(34):54.

Schick, W. 1944. Temperature Coefficient of Capacitance. *Wireless Eng.* **21**:65–71.

Snow, C. 1954. *Formulas for Computing Capacitance and Inductance*. National Bureau of Standards Circular 544. Govt. Printing Office, Washington, D.C.

Standard Capacitors and Their Accuracy in Practice. Notes on Applied Science No. 13., 1969. Her Majesty's Stationary Office.

23

Reprinted from *Am. Electrochem. Soc. Trans.* 5:147–162 (1904)

THE ALUMINUM ELECTROLYTIC CONDENSER.

By C. I. Zimmerman.

INTRODUCTION.—ASYMMETRICAL CELLS.

The property of asymmetrical conductivity, which is possessed by many electrolytic cells, may in some instances be developed to a remarkable degree. An illustration of such cells is one with aluminum and carbon electrodes in an aqueous solution of ammonium phosphate. This cell will allow a current to pass freely when the *carbon* is the positive terminal and will prevent the passage of a current when the *aluminum* is the positive terminal. As is well known, cells of this type (and also cells having fused salts for electrolytes) are the basis of apparatus recently placed upon the market for converting alternating currents into direct currents.

The Film.—The asymmetrical effect of the aluminum cell is located in the thin film which is formed upon the surface of the aluminum electrode by electrochemical oxidation. This film under ordinary circumstances firmly adheres to the surface of the metal, no matter in which direction the current flows through the cell. Its composition cannot be definitely determined since sufficient quantities cannot be collected to be given an accurate analysis. It is believed to be some form of oxide, or perhaps, hydroxide of aluminum.

The factors affecting the formation of a film cannot be here taken up in detail. Suffice it to state, that with fixed operating conditions, the film does not continue to grow thicker after a certain limit has been reached because it protects the metal beneath from further oxidation. The principal factors affecting the growth of a film are the kind of electrolyte, the value of the impressed pressure, the duration of the process of formation and the temperature of the electrolyte. Either alternating or direct currents may be used for the formation, since reduction of the film does not occur to an appreciable extent when current flows toward the aluminum plate.

When first formed, the film itself is ordinarily transparent and colorless. Its presence is best detected upon smooth plates by the interference colors which are produced in light reflected from the surface. The presence of these interference colors was noted by some of the earliest investigators of the "polarization phenomena of aluminum electrodes."*

After a short usage the electrode becomes grayish, but the bright colors are not necessarily destroyed. The grayish appearance is due to a slight roughness of the metallic surface beneath the film, and to the presence of numerous minute punctures in the film.

The colors to be observed as a film increases in thickness from zero are greenish-yellow, yellow, orange, red, violet, indigo, blue, green, yellow, orange, etc. The appearance of the greenish-yellow color first means that the light observed by the eye is white light minus some color which will produce greenish yellow light. The light missing is violet light, which is the light with the shortest wave-lengths. The light *blotted out* by interference follows the order violet, indigo, blue, green, yellow, orange, red, etc., as would be expected from the fact that violet light consists of the shortest waves, and red light of the longest waves.

By means of these interference color phenomena the writer has estimated the thickness of the films. They range from less than 0.000,005 centimeters to, in exceptional cases, more than 0.000,050 centimeters in thickness.†

The Condenser.—This extremely thin asymmetrical layer exists between two electrical conductors and it constitutes an asymmetrical dielectric which may be considered capable of holding

---

* *E. g.*, W. Beetz, Journal der Physik, Vol. 156, p. 456, (1875). (Poggendorff's Annalen, Vol. 244.)

† From a consideration of the interference phenomena produced by a transparent film upon the reflecting surface of a medium more dense than the film, the formula indicating the film thickness is $d = \dfrac{n\,\lambda}{4\,\beta}$; where "d" is the film thickness in centimeters; "n" is an odd number which is 1 for the first range of the spectrum observed, 3 for the second range, 5 for the third range, etc; "λ" is the wave-length in air, expressed in centimeters, of the color of light *observed to be absent in vertically reflected light*; and β is the value of the index of refraction of the film substance which is assumed in this case to be between 1.4 and 1.5, the value corresponding to the indices of refraction of other substances similar in nature to the film.

positive charges only on the side in contact with the metal, and negative charges only upon the side in contact with the liquid. The disruptive strength of this film when the metal is positive to the electrolyte is apparently greater than that of mica. Its specific inductive capacity is not unity, as has been assumed by some,[*] if we calculate it from the electrostatic capacity of a given electrode and the thickness of the film as estimated by the color phenomena. On the other hand, the specific inductive capacity seems to be abnormally high for a solid. Mr. W. R. Mott,[†] in investigating the aluminum anode properties, calculated the dielectric constant of the film to be in the neighborhood of 80.[‡]

THE ALTERNATING CURRENT CONDENSER.

An electrolytic cell having both electrodes of these coated aluminum plates will obviously prevent a direct current from flowing continuously through the cell in either direction. An alternating pressure, on the other hand, causes a comparatively large current to flow; the current being a leading or condenser current.

A cell of this type (see Fig. 1) constitutes the simplest form

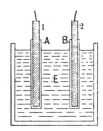

Fig. 1.- ALUMINUM CONDENSER

of an aluminum electrolytic condenser for alternating current circuits. The properties of this condenser do not appear, if we are to judge from the literature of the subject, to have been studied very thoroughly, although the presence of a condenser action has been known as long as the asymmetrical properties of the anode.

It must be borne in mind that these asymmetrical properties are properties of the chain metal-film-electrolyte, and are frequently ascribed to the film or to the electrode for convenience only. When

---

[*] *e. g.*, A. Nodon. Comptes Rendus. Vol. 136, p. 445-446, (1903).
[†] Thesis, University of Wisconsin, (1903).
[‡] See communicated discussion by Mott, p. 164, these transactions, for corrected value. —ED.

dry, the film acts like any ordinary dielectric. Again, when the aluminum electrode is immersed in the electrolyte, it is *not* a perfect asymmetrical conductor or dielectric. In the discussion to follow, perfect asymmetry, no polarization pressures, and no cell losses are assumed to exist. These conditions are only approximated, as will be shown.

In Fig. 1, 1 and 2 represent two similar cell electrodes of metallic aluminum, coated by the thin films A and B, and immersed in the electrolyte E. The electrolyte E is one which is capable of maintaining the continuity of the film by electrochemical oxidation. The films, A and B, are the two dielectrics which, it must be remembered, will allow the current to flow from the electrolyte to the electrodes, but not in the reverse direction, and which can hold positive charges only next to the metallic conductors, and negative charges only next to the electrolytic conductor.

The behavior of the cell upon an alternating current circuit is as follows, with the previous assumptions in mind:

If the cell be connected into the circuit when the alternating pressure is passing through zero and when electrode 1 is becoming positive to electrode 2, a current will flow through the entire circuit in the direction from electrode 1 toward electrode 2. Film A receives an electrostatic charge whose potential rises to that of the pressure upon the cell terminals, for film B allows the free passage of current out of the cell.

When the impressed pressure ceases to increase, the current in the circuit stops flowing. As the impressed pressure begins to decrease, film A tends to discharge itself into the circuit. Electrode 2 now becomes the anode and film B receives an electrostatic charge, positive on the side next to the metal and negative on the side in contact with the electrolyte.

Whatever may be the subsequent variation of pressure upon the cell terminals, the coulomb charge removed from either one of the films must collect upon the other; and *we have* consequently *a condenser in which the total charge in coulombs remains constant.* This statement assumes that the subsequent pressures upon the cell terminals do not rise higher than the maximum value of the pressure which caused the charge to collect in the cell.

This constant charge may be observed in the aluminum cell by disconnecting the cell from the alternating pressure circuit, short-circuiting the electrodes, and then connecting a voltmeter to the short-circuiting wire and to a carbon rod dipped into the electrolyte. A discharge always occurs through the voltmeter from the metallic connections to the electrolyte. The short circuiting of the cell is equivalent to removing the cell from the circuit when the alternating pressure passes through zero. Under similar circumstances, an ordinary condenser is completely discharged.

The condenser action of the cell will be shown later to be due to the *distribution* of the charge within the cell. The short circuiting of the electrodes of an aluminum condenser allows an equal distribution of the charges upon the two plates to occur.

A given coulomb charge, *q,* removed from one of the films as A, lowers the potential difference across the two sides of that film by a definite amount as *e.* This same charge in becoming stored up in the other film raises its potential by the same amount *e,* if the two films have equal electrostatic capacities. The *arithmetical* sum of the pressure across the two films is thus constant. The value of this sum is equal to the maximum value of the pressure applied to the cell terminals.

These two film pressures are in opposition to one another. The *algebraic* sum of their *instantaneous* values is equal to the *instantaneous* value of the impressed pressure.

This constant electrostatic charge remaining in the cell at all times sets up a difference of potential between the inside of the cell, or the electrolyte, and a point outside of the cell (in the metallic connections) which is neutral to the alternating pressure. The middle point of an inductance coil placed across the cell terminals represents such a point. In Fig. 2, an alternator G is shown con-

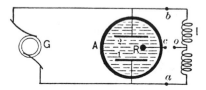

Fig. 2.- CELL PRESSURES

nected across the cell terminals of an aluminum condenser A. An inductance coil I is placed across the cell terminals. By means of a carbon rod R dipped into the electrolyte, the potential of the electrolyte may be obtained. The polarization pressure that is set up at surface of this carbon terminal will be assumed for the present to be zero, as it is small compared with the generator pressure.

The pressure between the middle point o of the inductance coil, and the electrolyte c, is a uniform direct pressure as was explained above. The pressure across one-half the inductance coil is one-half the alternating pressure. The pressure between the electrolyte c and either of the cell terminals, a or b, is the *resultant of a uniform pressure* of a magnitude equal to one-half the maximum instantaneous value of the pressure wave impressed upon the cell, *and an alternating pressure,* of a magnitude equal to one-half the effective value of the pressure impressed upon the cell. With an effective sinusoidal alternating pressure E, this resultant is theoretically equal to $\sqrt{\dfrac{3}{2}} \times$ E or .866 E.*

The average value of the pulsating pressure wave is dependent only upon the maximum instantaneous value of the impressed pressure wave. The average value for a sinusoidal impressed pressure wave is $\dfrac{E}{\sqrt{2}}$ or .707 E. These pulsating uni-directional pressures upon each film are continuous sinusoidal waves with an amplitude of one-half the alternating wave across the cell terminals, a maximum value equal to the maximum of the alternating wave and a minimum value which is zero.

The average and the effective values of these waves may be measured by means of direct and alternating current voltmeters respectively. The results obtained by such measurements are lower than the theoretical values, due, as was previously mentioned, to the film not being a perfectly asymmetrical conductor, to polarization pressures existing at the aluminum and carbon electrodes and to the small energy losses in the cell. With 100 effective alternating volts applied to the aluminum terminals the readings obtained by alternating and direct current voltmeters are about 77 and 60.5 respectively, the theoretical values being 86.6 and 70.7.

Figures 3 and 4 show these theoretical pressures by means of curves. The three curves, $b\text{-}b$, $c\text{-}c$ and $o\text{-}o$ represent the potentials of the three points, b, c and o of Fig. 2 referred to the potential of

---

* $\sqrt{(\text{eff. DC})^2 + (\text{eff. AC})^2} = \sqrt{\left(\dfrac{E}{\sqrt{2}}\right)^2 + \left(\dfrac{E}{2}\right)^2} = \dfrac{E\sqrt{3}}{2}$

one of the cell terminals *a*. Curve *d-d* will be referred to later. Horizontal distances indicate time from any assumed instant as M-M.

The instantaneous differences of potential between the middle point of an inductance coil, the electrolyte and the electrodes are represented by the vertical distances between the curves marked correspondingly. The impressed pressure is represented by *b-b*,

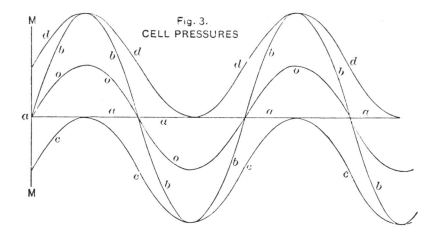

Fig. 3.
CELL PRESSURES

Fig. 4. CELL PRESSURES

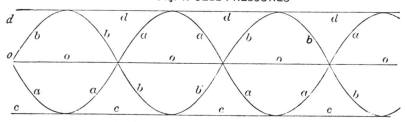

measured from the horizontal line *a-a*. The vertical distances from *a-a* to *o-o* represent the potential differences between the terminal *a* and the terminal *o*, or one-half the impressed pressure. The vertical distance between *o-o* and *c-c* represents the constant potential difference existing between the middle, *o*, of the induction coil and the electrolyte, *c*, the electrolyte remaining negative at all times to the middle point, *o*. The pressure between the cell terminal, *a*, and the electrolyte, *c*, is similarly seen to be repre-

sented by the vertical distance between *a-a* and *c-c*. This pressure
is a pulsating unidirectional pressure whose pulsations are continuous sinusoidal waves. There is one pulsation for each cycle.

Between the electrolyte and the other electrode there is a similar
pulsating pressure which, in Fig. 3, is represented by the vertical
distances between *c-c* and *b-b*. The ordnates to the curve *d-d* from
a-a represent this last pressure if one of the lines as *c-c* be considered for the moment a straight line coinciding with *a-a*.

Fig. 4 represents these same cell pressures referred to the potential of the middle point of the inductance coil, instead of one of the
electrodes.

The pulsating pressures upon the films are shown in Figs. 5 and
6 in curves taken from oscillograph records. The impressed pressures were similar in shape to these waves. Their amplitudes were

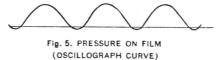

Fig. 5. PRESSURE ON FILM
(OSCILLOGRAPH CURVE)

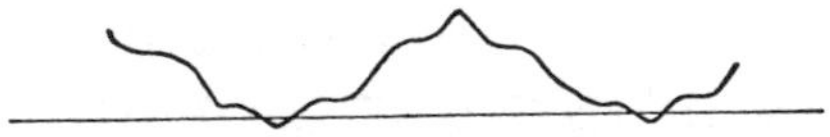

Fig. 6. PRESSURE ON FILM
(OSCILLOGRAPH CURVE)

twice as great and they were symmetrically disposed with respect
to the straight line. The curves are observed to cross the zero
line slightly, due to imperfect asymmetry, polarization pressures,
cell losses and to the effect of the oscillograph in removing a portion of the charge in the cell.

In Fig. 7 are shown oscillograph curves of all the cell pressures
and also the charging current of a 40 microfarad condenser, super-
imposed upon one another. The impressed pressure was 100 volts
and of a triangular shape. Curve *b-b* is the impressed pressure
referred to *a-a;* *c-c* is the pressure between one of the cell terminals and the electrolyte referred to *a-a;* and *d-d* is the pressure
between the other cell terminal and the electrolyte as seen by the

oscillograph. These curves correspond to curves *b-b, c-c* and *d-d* of Fig. 3 respectively. The charging current of the condenser, *i-i,* is flat-topped because the pressure wave is triangular in shape.

An electrical analogy of pressures combined to produce these

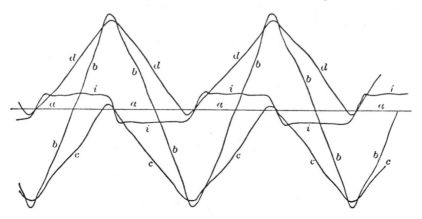

Fig. 7. OSCILLOGRAPH CURVES.

same results is shown in Fig 8. G_1 is an alternator giving a sinusoidal pressure E_1; G_2 is a constant pressure machine whose positive brush is connected to the middle point of an inductance coil I. If the constant pressure have a value of one-half the *maximum*

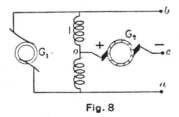

Fig. 8

value of the alternating pressure E_1 (*i. e.* $\frac{E_1}{\sqrt{2}}$) then the same pressures will be produced between the free terminal *c* of the constant pressure machine and the alternating terminals *a* and *b* as would exist in a condenser with two perfectly asymmetrical electrodes connected as shown in Fig. 2.

It is not necessary that the two asymmetrical electrodes of an aluminum condenser be placed in the same cell. If (as is shown in

Fig. 9 by the two cells to the left) two asymmetrical *cells*, each
with one aluminum and one carbon electrode be connected together
by their *carbon* terminals, and connected to the alternating pressure leads at their *aluminum* terminals, the simple condenser is
altered only by the introduction of small polarization pressures
within the cells and by additional cell resistance. The pressures
between the carbon junction, *c*, and the aluminum electrodes, *a*
and *b*, are then practically the same as in the simple condenser.

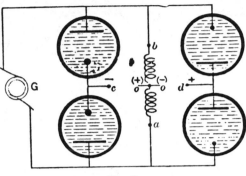

Fig. 9

If in two cells, the carbon electrodes be connected to the alternator leads and the aluminum electrodes be connected together (as
shown in Fig. 9 by the right hand pair of cells), the aluminum
junction *d* will assume a constant *positive* potential with respect to
a point, *o*, neutral to the alternating pressure. The potential difference between the wire, *c*, joining the two carbon electrodes of one
condenser and the wire *d* joining the two aluminum electrodes of
the other condenser will be twice as much as the pressure between
either of these points and the neutral point of an inductance coil.

With 100 effective alternating volts impressed upon this combination of asymmetrical cells, known as the Graetz rectifier arrangement, a pressure of approximately 121 volts is observed by
either direct or alternating voltmeters, the theoretical value being
141.4 volts. All these pressures are represented in Figs. 3 and 4,
referred to the terminals *a* and *o* respectively.

The pressures obtainable from a double-current generator correspond to these condenser pressures. As is well known, between
the positive or negative brushes (see Fig. 10) and either of the

collector rings, there may be obtained unidirectional pulsating pressures, the collector rings being positive to the negative brush, and negative to the positive brush. The middle point of an inductance coil placed across the alternating pressure terminals is at all times at a potential midway between the potentials of the constant pressure brushes and midway between the potentials of the alter-

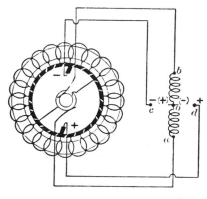

Fig. 10

nating pressure collector rings. Figs. 3 and 4 may represent the pressures obtainable from the double-current generators, as well as the pressures of the aluminum condenser.

VARIATION OF THE ENERGY WITHIN THE CELL.

The condenser action of this cell is to be found in the energy variations resulting from the variation of the distribution of the constant coulomb charge within the cell. The following discussion will illustrate the point:

Let E_m be the maximum value of the alternating pressure, and Q be the charge in coulombs which one of the films receives due to the potential difference F_m. The charge upon the second film is zero, as was shown previously, when the first film has its maximum charge.

The energy W_1 stored up in the condenser at the instant of maximum pressure is

$$W_1 = \tfrac{1}{2} Q E_m$$

When the impressed pressure passes through zero, each film holds one-half the charge, at one-half the pressure E_m and the energy W_2 stored up in the condenser is

$$W_2 = 2 \left(\tfrac{1}{2} \times \frac{Q}{2} \times \frac{E_m}{2} \right) = \tfrac{1}{4} Q E_m$$

or *one-half* as much as at the time of maximum pressure. The pressure variation is

$$W_1 - W_2 = \tfrac{1}{2} Q E_m - \tfrac{1}{4} Q E_m = \tfrac{1}{4} Q E_m$$

which is also one-half the total energy stored up when the impressed pressure is at its maximum.

COMPARISON WITH TIN-FOIL CONDENSERS.

Since, with a sinusoidal pressure wave impressed, the pressures upon the two dielectrics have been shown to vary sinusoidally, it follows that the condenser charging current is sinusoidal and likewise the energy variations in the circuit. The aluminum condenser thus has the same condenser effect upon a circuit that an ordinary tin-foil condenser has.

An aluminum condenser periodically absorbs and returns only one-half the total amount of energy it stores up. This charge is held upon one film only at the time of maximum charge. Consequently the effective electrostatic capacity of a condenser, as figured from its sinusoidal charging current, is but one-half the electrostatic capacity of one of its films. A tin-foil condenser absorbs and returns its total energy charge and its charging current is proportional to the full electrostatic capacity of its film.

In a certain sense the aluminum condenser may be considered to be like two tin-foil condensers of equal capacities placed in series, for the pressures upon each film have an alternating component equal to one-half the alternating pressure upon the condenser, and again, the effective electrostatic capacity of an aluminum condenser is one-half the electrostatic capacity of either film. However, each film of the aluminum condenser is subjected to nearly the maximum pressure upon the cell instead of but one-half the maximum as in the ordinary condensers; and either film becomes charged while the other becomes discharged, instead of being charged and discharged together as in the tin-foil components.

The charging current of both aluminum and tin-foil condensers is a function not only of the pressure and fundamental frequency of the impressed pressure wave, but a function of the irregularities or higher harmonics in the pressure wave as well. Fig. 11 shows

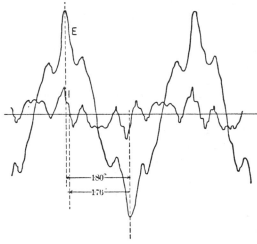

Fig. 11. CHARGING CURRENT PRODUCED BY IRREGULAR PRESSURE

the amplification of the charging current I produced in an aluminum condenser by the irregular pressure wave E. At each of the maxima and minima in the pressure wave, the condenser current passes through zero, for at these instants, the rate of change of the pressure is zero. The current is greatest where the wave is the steepest, i. e., where the rate of change of the pressure is greatest. The effective value of the current, as measured by a Hoyt instrument, was 2.4 times the value calculated for a sinusoidal pressure wave of an equal impressed pressure (114 volts) and a frequency of the fundamental wave (60 p. p. s.). The condenser upon which these curves were observed had a capacity equivalent to a 40 microfarad tin-foil condenser.

THE LOSSES IN THE ALUMINUM CELL.

The losses of the aluminum cell consist of the following:
1. Film losses.
2. C^2R losses.
3. Electrolytic decomposition losses.

309

The film losses are the principal losses in the cell and they include (a) leakage losses and (b) a loss proportional to the frequency. The leakage losses are due to minute local break-downs of the films and to an apparently uniform passage of current through the high resistance film by conduction. These losses (those due to the minute punctures of the film in particular) increase rapidly when the pressure upon a given film goes beyond a certain point which has been termed the "break-down" point. The point of break-down is dependent largely upon the kind and temperature of the electrolyte as well as upon the thickness of the film. By obtaining suitable conditions the writer has recently succeeded in operating simple two-electrode condensers directly upon effective alternating pressures, as high as 1,250 volts, without disrupting the dielectrics.

The losses proportional to the frequency are of a nature not yet clearly understood. With the higher commercial frequencies (60 to 130 cycles per second) they are in general greater than the other losses in the cell.

The C^2R losses of the electrolyte may be reduced to a negligible quantity by intermeshing sheet electrodes of opposite polarity. The electrostatic capacity per unit of surface is, fortunately, not so great as to cause the current density to be high. The enormous capacities per unit of surface in the so-called "polarization condensers" is one of their inherent practical disadvantages.

The losses caused by electrolytic decomposition are those due to the leakage current only. The wattless condenser current is induced in the electrolyte electrostatically. The products of decomposition may be made to consist chiefly of minute quantities of hydrogen and oxygen.

The presence of a break-down point in the aluminum cell, at which the losses begin to rapidly increase, is a particularly valuable property of the aluminum condenser, because the dangerous temporary rises in voltage that tend to occur through resonance are thereby practically prevented.

It is not of any serious consequence to scratch the electrode, for any break in the film is quickly repaired by electrochemical oxidization. Punctures due to abnormally high voltages are for the same reason of little consequence unless the puncturing be long continued.

In general it may be stated that the voltage limit for single cells is not much above 150 volts where high efficiencies are desired. The efficiencies may range from less than 93 per cent. to as high as 97 per cent. according to the operating conditions   It is possible to operate condensers continuously at efficiencies above 95 per cent. with pressures of 110 volts.

The weight, volume and cost per microfarad of single cell electrolytic condensers are considerably smaller than for equivalent tin-foil condensers operating at the same voltages.   This great difference is not so marked where the voltages rise much beyond those at which single condensers can operate efficiently, since the energy storing value of a condenser varies as the square of the pressure upon its terminals.

The electrostatic capacity of 110-volt aluminum condensers may be said to be roughly $\frac{1}{4}$ to $\frac{1}{2}$ microfarad per square inch of aluminum plate in the cell.

In applied electricity there is, evidently, a place for the aluminum electrolytic condenser.

SUMMARY.

The following are the principal points brought out in this paper:
The film coating the aluminum electrode varies from less than 0.000005 centimeters to more than 0.000050 centimeters as estimated by the interference color phenomena observed; this thin film acts as an asymmetrical dielectric as well as an asymmetrical conductor; a condenser with two of these asymmetrical dielectrics holds a constant coulomb charge when operating upon an alternating current circuit; this constant charge sets up a uniform pressure between the electrolyte and a point in the metallic or external circuit which is neutral to the alternating pressure; the pressures upon the films are uni-directional pressures which pulsate in continuous sinusoidal variations with one-half the amplitude and one-half the frequency of the pulsations of the impressed pressure; with the Graetz rectifier group of cells, uniform pressures of approximately 1.2 times the impressed alternating pressure may be observed; the condenser action of the cell is due to the energy variations accompanying the variations of the distribution of the constant coulomb charge within the cell; the energy varia-

tions are one-half the maximum energy held by the cell; the ordinary condenser behavior upon AC circuits is possessed by the aluminum condenser; the losses in the cell are small and consist of film losses, C^2R losses and decomposition losses, the film losses being due to leakage and to a loss directly proportional to the frequency.

The writer hopes to continue this subject at later meetings of the Society, taking up among other points, the properties of the aluminum cell as shown by curves and other quantitative measurements; the peculiarities of cells having electrodes with films of different electrostatic capacities; and polyphase condensers, with their relation to polyphase asymmetrical rectifiers.

In concluding, the writer wishes to express his deep feeling of obligation to the electrochemical and electrical departments of the University of Wisconsin, whose instructional forces and whose laboratory equipments enabled him to take up the investigation of the aluminum condenser under most favorable surroundings. To Prof. C. F. Burgess in particular is he deeply indebted for inspiration and suggestions received. The thanks of the writer are tendered to Messrs. W. R. Mott and J. G. Zimmerman for their co-operation and assistance in some of the work, and to Mr. Budd Frankerfield for suggestions relative to the treatment of the subject in this and future papers.

University of Wisconsin.

[*Editor's Note:* The discussion has been omitted.]

24

Copyright © 1978 by Howard W. Sams & Co., Inc.

Reprinted from pages 38, 40–41, 46, 47, 52, 62–63, 67, 71–72, and 73 of *The ABC's of Capacitors*, Howard W. Sams & Co., 1978, 127pp.

CAPACITOR CONSTRUCTION

W. F. Mullin

AIR CAPACITORS

Although air is a poor insulator, it is still highly useful as the dielectric in a capacitor because its power factor is almost nil and its stability is excellent. Furthermore, it doesn't cost anything.

The capacitance ratings of air-dielectric capacitors range from about 3 pF, to above 330 pF. Voltage ratings reach a practical limit at about 30,000 Vdc. Air capacitors may be either fixed or variable, but their primary advantage is their relative simplicity.

Fig. 3-3 shows the basic arrangement of a simple air capacitor. The plates of the fixed unit (Fig. 3-3A) are insulated from the supporting frame by an appropriate material. Fig. 3-3B shows a simple, variable air capacitor. A screw thread varies the distance between the two plates and thus

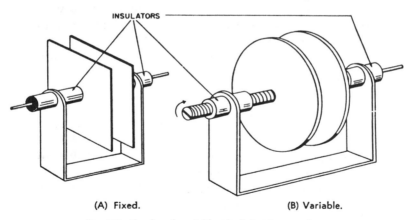

(A) Fixed. (B) Variable.

Fig. 3-3. Fixed and variable air-dielectric capacitors.

increases or decreases the capacitance. This arrangement has a disadvantage, however. The resistance to voltage breakdown increases or decreases inversely with capacitance. Therefore, its application is limited.

The more common variable air capacitor is shown in Fig. 3-4. This is the constant-gap type, in which the capacitance is varied by exposing more or less of the plate surfaces to change the area ratio between plates. Since the plates remain the same distance apart at all times, the resistance to voltage breakdown never changes.

[Editor's Note: Material has been omitted at this point.]

Fig. 3-5. Variable air capacitors.

[*Editor's Note:* Material has been omitted at this point.]

MICA CAPACITORS

Mica was used commercially long before the advent of capacitors. More popularly known as *isinglass*, its most common application was for lanterns, the doors of stoves, etc. This indicates one of the important characteristics of mica: its ability to operate at very high temperatures (up to 500°C). An additional characteristic is that the material is almost totally inert and will not change with age, either chemically or physically.

The development of mica capacitors can be attributed to the fact that mica is usable as a dielectric in its natural state. Mica blocks are mined in conjunction with granite and similar igneous rock formations. The blocks exhibit almost perfect cleavage. That is, they can readily be split into very thin leaves or sheets, often as thin as 0.0001 inch. The sheets are quite uniform in thickness, which is important in capacitor construction. Furthermore the dielectric constant (K) averages about 6.85.

Mica also has some disadvantages. Since it is a natural material, mica is subject to all of the variations of nature.

Great care must be exercised in the selection of mica sheets to ensure that they are as nearly uniform as possible. And, as might be expected, higher-quality mica is also more expensive. As a result, man-made materials often have both a design and an economic advantage.

From a construction standpoint, the mica capacitor is a perfect model of the classic flat-plate type. A thin sheet of mica is sandwiched between two layers of foil (Fig. 3-6).

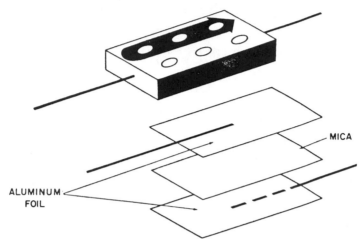

ALUMINUM FOIL

MICA

Fig. 3-6. Basic mica capacitor construction.

The most common foil material is a tin-lead compound. The tin-lead foil readily conforms to microscopic variations in the surface of the mica. Capacitance is a function of the area of the conductors and the thickness of the mica. It should be noted here that the mica is slightly larger than the foil sheets to prevent contact between the sheets. To show the size/thickness ratio, let us calculate the value of a mica capacitor 1-inch square made from 0.0001-inch mica having a dielectric constant (K) of 7. The formula is:

$$C \text{ (pF)} = 0.2235 \frac{KA}{d} (N - 1)$$

$$= 0.2235 \times \frac{7 \times 1}{0.0001} \times (1)$$

$$= 15{,}645 \text{ pF or } 0.015 (+) \ \mu F$$

[*Editor's Note:* Material has been omitted at this point.]

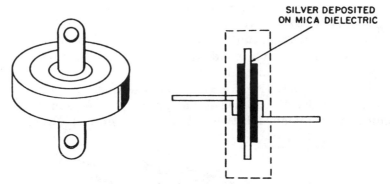

Fig. 3-8. Button style silver-mica capacitor construction.

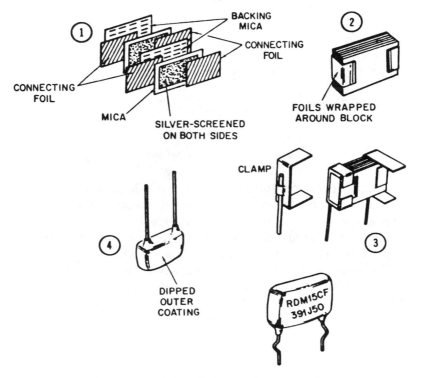

Fig. 3-9. Typical dipped-mica capacitor construction.

GLASS CAPACITORS

The glass capacitor is actually the original Leyden jar modernized after two centuries of much neglect. The main reason for the interest in glass as a dielectric came about because the sources for high-quality mica were endangered during World War II. It wasn't until the 1950s that all of the problems of producing a practical glass dielectric were solved.

The foremost problem was in producing an ultrathin ribbon of highly stable glass of sufficient width and length. As the problems were solved, volume production came about. Glass is superior to mica in many ways. The quality of the dielectric can be controlled, and there are no voids or natural impurities.

Construction is almost exactly the same as that of the mica capacitor (Fig. 3-12). Aluminum foil is used instead

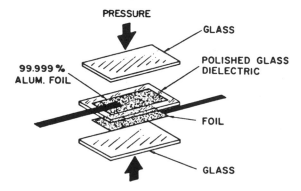

Fig. 3-12. Basic glass capacitor construction.

of tin-lead foil because the surface of glass is substantially flatter than mica, and because the melting point of aluminum foil is much higher than that of tin-lead foil. Layers of foil and glass are interleaved, and the entire structure is fused at high temperature to create an essentially monolithic structure of great strength and superb resistance to moisture. Of the known capacitor types, no other is more resistant to destructive moisture than the glass type.

Capacitance values for glass capacitors range from 0.5 pF to 10,000 pF (0.01 μF), very much the same as mica.

[*Editor's Note:* Material has been omitted at this point.]

CERAMIC CAPACITORS

No other dielectric is as versatile as ceramic. This man-made material can be mixed in many different ways to produce a wide variety of results. One mixture yields a capacitor with characteristics nearly the same as mica. On the other hand, another mixture produces paperlike characteristics.

In configuration, ceramic capacitors are generally either disc-shaped, tubular, or rectangular. The rectangular configuration is widely used in solid-state electronics, particularly in the newer *monolithic* ceramics which will be discussed in more detail later on in this chapter. Both tubular and disc types are made in adjustable as well as in fixed capacitance values. Fig. 3-17 illustrates several common types.

Capacitance values range from as low as 0.5 pF to as high as 3 μF. Operating voltages vary from a mere 3 Vdcw to over 30,000 Vdcw. Ceramics offer good size-to-capacitance ratios and outstanding temperature and frequency characteristics.

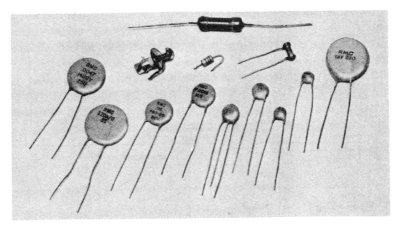

Fig. 3-17. Ceramic capacitors.

Ceramic capacitors have similar capacitance values but lower inductances than mica types of corresponding values. The four types are general-purpose, temperature-compensating, temperature-stable, and frequency-stable. Each of these four broad types has characteristics which suit it to particular applications.

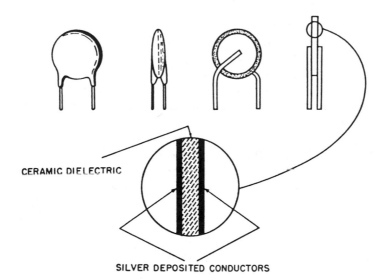

CERAMIC DIELECTRIC

SILVER DEPOSITED CONDUCTORS

Fig. 3-18. Disc ceramic capacitor construction.

[*Editor's Note:* Material has been omitted at this point.]

319

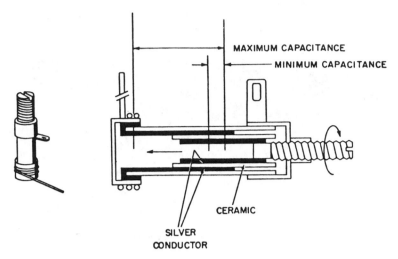

MAXIMUM CAPACITANCE

MINIMUM CAPACITANCE

CERAMIC

SILVER
CONDUCTOR

Fig. 3-22. Variable tubular ceramic capacitor construction.

[*Editor's Note:* Material has been omitted at this point.]

ELECTROLYTIC CAPACITORS

Electrolytic capacitors provide more capacitance for their size than any other type. From the service technician's viewpoint, they are also the most confusing. They seem to defy all the rules applicable to capacitors, yet electrolytics are able to perform jobs which no other type can. Fig. 3-26 shows only a few of the wide range of values and sizes that are available.

Electrolytics all have one thing in common, however— they are made differently from other capacitors (see Fig. 3-27). Instead of the usual plates separated by a dielectric, the electrolytic has a metallic anode coated with an oxide film. This outer covering is the dielectric, and a liquid electrolyte acts as the cathode. A second metallic conductor serves primarily as the connection to the liquid cathode, providing an external termination.

Fig. 3-26. Typical electrolytic capacitors.

In actual practice, porous paper is wrapped around the anode and saturated with the electrolyte to eliminate the spillage problem.

There are two common types of electrolytic capacitors: aluminum and tantalum. Both employ the same basic principles. The aluminum or tantalum anode is covered with an oxide film. A suitable liquid (or solid) electrolyte is the cathode. The aluminum type is by far the most popular because of its lower cost.

The aluminum-oxide film has a very high resistance to current in one direction and a very low resistance in the opposite direction. In other words, the film acts as a dielectric in the first instance and as a plate in the second. Because of this, electrolytics are polarized. If the designated polarity is not observed, the oxide film on the anode will break down and migrate to the cathode connection, resulting in prompt failure of the capacitor.

[*Editor's Note:* Material has been omitted at this point.]

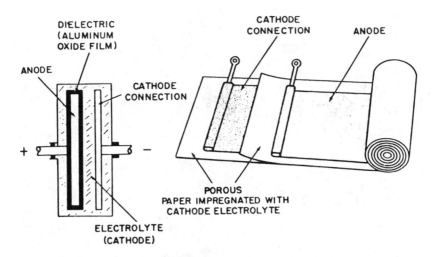

Fig. 3-27. Basic electrolytic capacitor construction.

[*Editor's Note:* Material has been omitted at this point.]

25

Copyright © 1962 by the Institute of Electrical and Electronics Engineers, Inc.
Reprinted by permission from Electr. Eng. 81:681–685 (1962)

ELEMENTS OF ENERGY STORAGE CAPACITOR BANKS

E. L. Kemp

Large magnetic fields are common laboratory tools today mainly because of the increased interest in thermonuclear research and plasma propulsion. Fields are usually generated by passing high currents through solenoid coils, which require large amounts of power during the time of interest. The capacitor bank, the most universal system for storing and delivering this power, can deliver very high currents for short periods and is cheap compared to a high-current generator

The primary components in a capacitor bank are the capacitor, switch, transmission system, and charging system. The pertinent characteristics of each of these will be discussed.

COMPONENTS

Capacitors • Energy storage capacitors are rated from 1 to 100 µf at voltages between 3 and 100 kv. They usually store between 1 and 3 kilojoules at rated voltage. Their internal inductance is less than 0.1 µh so that they may routinely produce peak discharge currents of more than 100 kiloamperes. A typical guaranteed life is 1,000 shots at the design rating.

Fig. 1 shows three typical capacitors. They are a 100-µf 3-kv capacitor, a 15-µf 20-kv capacitor, and an 0.8-µf 120-kv unit.

When these capacitors fail they usually develop an internal short circuit. If several units are connected in parallel, the unit which short-circuits will receive all the energy stored in that module. Most energy storage capacitor cans are able to absorb a maximum of 25 kilojoules without rupturing.

Switches • The capacitor bank switch must hold off voltage during the charge cycle, close with a small jitter, and pass large currents. Historically, spark gaps were used for capacitor bank switching. A single 2-ball spark gap was connected in series with the load, and the bank was charged until the gap was overvolted, discharging the capacitors into the load. Today, a triggering pin has been added to the 2-element gap, and under proper conditions, the 3-element "triggertron" is an inexpensive, reliable bank switch. Its disadvantage is its limited operating voltage range, 10 to 15 per cent of the maximum. Triggertrons are operated at atmospheric pressure at

voltages up to 25 kv. Some pressurized models operate at 40 kv.

A vacuum spark gap[1] has been developed which will operate from 100 volts to 20 kv. The vacuum gap requires constant pumping and the usual vacuum technology, but the reliability and wide voltage range make it attractive for some low-inductance banks.

Spark gaps can also be designed to operate at

Fig. 1. Three typical energy storage capacitors

higher voltages. Fig. 2 shows a 4-element spark gap which operates from 60 to 100 kv with a maximum jitter of 10⁻⁸ seconds.[2,3] It resembles the common 3-electrode gap with an additional triggering electrode. The center electrode is biased at a potential midway between the upper and lower electrodes. The fourth electrode is the irradiating spark plug. The gap is triggered by driving the center electrode to the breakdown potential of one of the gaps. This trigger pulse also causes the spark plug gap to break down and irradiate both halves of the gap, actually before the center electrode reaches the breakdown point. When

E. L. Kemp is with the Los Alamos Scientific Laboratory, University of California, Los Alamos, N.M.

The author gratefully acknowledges the assistance and criticisms of the following people: G. P. Boicou.t, H. K. Jennings, J. P. Mize, T. M. Putnam, A. E. Schofield, Elizabeth J. Pohlmann, and Mary E. Smith.

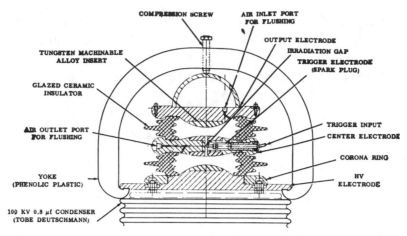

Fig. 2. Cross section of a 4-element spark gap

one-half of the gap breaks down, the other half is immediately overvolted and follows.

Spark gaps are reliable and have low jitter, so many of them can be used in parallel. Usually, each new bank design requires a unique spark gap design, and such designs are often unavailable from commercial sources.

Ignitrons are the only commercially available switches suitable for general capacitor bank switching. The size A (left) and size D ignitrons shown in Fig. 3 are the most popular. The size A is relatively inexpensive, has a low inductance, and has recently been improved for reliable operation above 20 kv.[4] The size D is more rugged and will handle large currents for milliseconds.

The size A ignitron is available with a carbon anode or a metal anode. The carbon anode model is reliable at voltages up to 15 kv, passing peak currents of 15 ka (kiloamperes). The metal anode ignitron will operate above 20 kv, passing peak currents greater than 100 ka.

The size A ignitron has less than 0.25-μsecond jitter when properly triggered. The trigger pulse should be over 3 kv and have a rise time comparable to the required jitter. The peak trigger current should be at least 250 amperes. After 1,000 shots or more, the resistance of the ignitor in a metal anode tube will drop from its nominal value of 25 to 50 ohms to perhaps 0.1 ohm and make the tube fail to fire when the ignitor is triggered. The ignitor resistance can be increased to nearly the original value by applying a 500- to 1,000-ampere short pulse to the ignitor for a few times when there is no voltage on the anode.

The operating characteristics of a bank will usually determine the choice of the switch. Spark gaps are indicated for operating voltages above 30 kv. They can be designed for low inductance and low jitter, but they must be designed for each application.

Ignitrons will operate reliably below 30 kv. They are commercially available and economical. They must be mounted vertically and usually require a water-cooling clamp. Their inductance is low if the mounting is coaxial. Ignitrons are more rugged than spark gaps and will sustain accidental short-circuit currents without requiring subsequent maintenance.

Transmission Systems • The transmission system for a capacitor bank may be bus work, a coaxial cable array, or a parallel plate transmission line, depending on the inductance and flexibility of the bank.

Open bus work is satisfactory where the inductance of the bank is not critical. The voltage hold-off distance should be conservative. One accidental discharge in air will make a true believer of any designer.

The bus should be designed for the maximum short-circuit current available from the bank. The electrical engineer is aware that parallel currents attract and opposing currents repel. This should be kept firmly in mind when designing bus work for capacitor banks. Kiloampere currents will generate magnetic fields that play havoc with a bus that is not sturdily made.

Fig. 4 shows a test bank using a bus transmission system. This bank stores 25,000 joules at 20 kv and delivers 60,000-ampere peak current into an inductive load.

Coaxial cables are well suited for capacitor banks where the load is flexible and when the transmission system must have low inductance. The standard types of RG-8/U, RG-14/U, RG-17/U, and RG-19/U can be used for voltages between 20 and 100 kv, respectively. They will pass peak currents of 30,000 amperes at a repetition rate of two shots per minute. These cables have a nominal high-frequency inductance of approximately 0.08 μh per foot.

Some special low-inductance cable has been developed for capacitor bank application. It is now commercially available and is usually designated by its component diameters, type 17/14 being a typical case. This cable has RG-17/U braid as the outer conductor and RG-14/U braid as the inner conductor. The central core is foamed polyethylene. The dielectric between conductors is normal polyethylene. It is covered with the usual RG-17/U vinyl jacket.

This cable has a measured low-frequency inductance of 0.036 µh per foot with a resistance of approximately 1 milliohm per foot. Combinations of RG-14/8 and RG-19/14 and other nonstandard designations are also available.

Parallel plate transmission lines can be designed for any desired inductance if space is available.

The inductance for a parallel plate transmission line is

$$L = 4\pi \cdot 10^{-7} \frac{t}{w} \; h/m \tag{1}$$

where t is the separation distance and w is the width, both dimensions in meters.

The current flowing in opposite directions in the two parallel plate transmission lines creates a magnetic pressure which tends to separate the plates. The pressure generated by the magnetic field is

$$P = 2\pi \left(\frac{I}{W}\right)^2 10^{-12} \; atm \tag{2}$$

I is current in amperes and W is the width of the plates in meters.

Although this pressure will last for only a short time, its amplitude may be quite large. The problem of holding parallel plate lines together often proves formidable.

Charging Systems • The simplest charging scheme for a capacitor bank is an *R-C* circuit. When this method is used, the charge resistor will dissipate an amount of power equal to the power stored in the bank. The wattage rating for the charge resistor can be computed at one watt per 8 joules of stored energy for a repetition rate of 2 shots per minute. The power dissipated in the charge resistor and the long charge time make the *R-C* method unsuitable for charging large capacitor banks.

A constant-current charging system is more desirable. It is efficient in power transfer, while the charge time is limited only by the current capacity of the power supply. A power supply whose output voltage is changing at a constant rate will deliver a constant current into a capacitor bank.

A motor driven autotransformer in the primary of a power supply will provide the dv/dt characteristic. The system has one disadvantage: if the bank switches fire during the charge cycle, the bank will discharge and present a low impedance to the power supply. The resulting high current will destroy the power supply rectifiers unless they are protected by a fast circuit breaker in the primary.

Fig. 5 shows another constant-current circuit often called a monocyclic network. The inductor and capacitor are in series resonance at the line frequency. The network is connected to the primary of the power supply and maintains a constant output current independently of the impedance connected to it. It can be shown[5] that at resonance

$$i_{ac} = j \frac{V}{X} \tag{3}$$

where

$$X = \omega L = \frac{1}{\omega C} \tag{4}$$

In this application X is chosen to equal the reflected impedance of the power supply. The value of X will be

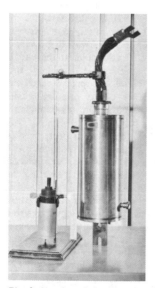

Fig. 3. Size A and size D ignitrons

Fig. 4. A capacitor bank connected with open bus

$$X = \frac{V_{ac}^{\cdot 2}}{V_{dc}I_{do}K} \tag{5}$$

The constant K, sometimes called the "line power factor," is a characteristic of the power supply circuit and can be found in a power supply handbook.[6] The inductance and capacitor are chosen to present this impedance at the line frequency. This circuit is applicable in a single-phase or a 3-phase circuit.

The major disadvantage of the circuit is its tendency to deliver a constant current into any load, regardless of the load impedance. This characteristic will produce output voltages that are several times the rated value if the output impedance is high. For example, it will charge a capacitor bank to several times the rated voltage if the voltage detector system fails. This circuit will likewise produce a high voltage if the power supply output sees an open circuit.

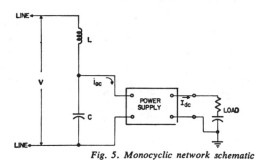

Fig. 5. Monocyclic network schematic

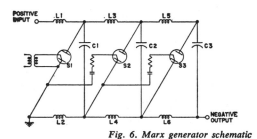

Fig. 6. Marx generator schematic

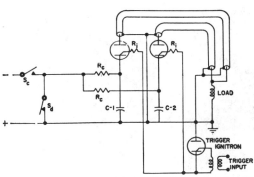

Fig. 7. Low-inductance bank schematic

When a monocyclic network is used, the bank should be protected with an overvoltage detector. It is also necessary to turn off the power supply before it is isolated from the bank at completion of charge.

Surge charging has several advantages for charging capacitor banks operating at high voltages. It provides high voltage from a modest supply, and it charges a bank quickly, keeping the voltage on the components for a minimum time.

Modified Marx generators are usually used to surge-charge capacitor banks. The modified circuit uses inductors rather than resistors to minimize power dissipation. The switches are reversed from the usual Marx circuit. This keeps the load capacitor bank grounded except when it is being charged, an important safety feature when the bank may store many joules at high voltage. The output polarity is opposite the input polarity.

Fig. 6 is a schematic of a 3-stage modified Marx generator which can be used to charge a capacitor bank. This circuit uses ignitrons as switches. With the polarities shown, the triggering of the switches is automatic after $S1$ triggers. The voltage on $C1$ is impressed across L_2 and on the ignitor of $S2$. The resistor and capacitor in the ignitor circuit limit the ignitor current. When $S2$ fires, it fires $S3$ in the same way. More stages can be used as necessary. If a positive output is required, each switch can be inverted and the bank charged negative. The ignitor capacitor must then hold off the charge voltage during charge. A pulse transformer to each ignitor will also operate in this circuit. Spark gaps are also well suited for Marx generator switches.

CIRCUITS

The capacitor bank circuit is usually determined by the size of the bank, its operating characteristics, and the switching system. In large banks, it is necessary to isolate each capacitor in some way to prevent it from exploding when it short-circuits at the end of its life. Exploding wire fuses[7] have been used successfully in some applications to protect each capacitor, but the inductance of the fuse will sometimes prevent the use of fuses in low-inductance banks.

The circuit in Fig. 7 can be used for protecting the capacitors in a low-inductance bank. An ignitron is mounted directly on each capacitor in a coaxial housing. The size A ignitron in such a housing has an inductance of approximately 0.02 μh. One or more coaxial cables are used to carry current to the load. Each capacitor has its own charging resistor. If one capacitor fails during the charge cycle, current from the other capacitors is limited by the charging resistors. If a capacitor fails during discharge, it is in parallel with the load, but it is isolated by its cable. The impedance of the cable is high compared to the load, so most of the bank energy goes into the load.

326

This circuit has performed satisfactorily in systems having over 100 units in parallel.

Fig. 7 also shows a simple triggering method that has proved reliable. The capacitors are charged negative. When the trigger ignitron is fired, it grounds the ignitors, making them positive with respect to their cathodes. The firing energy is taken from the storage capacitors. Ignitor current is limited by the resistor R_i in each ignitor lead. A large number of ignitrons can be fired in this manner from one trigger ignitron.

BANK PARAMETERS

Since most capacitor bank circuits are simple *RLC* circuits, their performance can be calculated accurately. The energy stored in the bank will be

$$U = 1/2 \ CV^2 \text{ joules} \qquad (6)$$

In most cases, the bank will be designed to produce a maximum current from the energy stored.

The peak current will be

$$I_p = V \sqrt{\frac{C}{L}} \ e^{-(Rt/2L)} \qquad (7)$$

where t is the time to the first quarter cycle.

Expressions for the ringing frequency, the charge time, and other incidental characteristics of a bank are well known.

CAPACITOR BANK SAFETY

All energy storage banks are lethal! Adequate personnel safety must be the first specification for any bank design. There are several areas where thoughtful design will eliminate potential hazards.

The bank should be placed in unconfined quarters. Visual and manual access to all parts is an operating necessity. The bank should be barricaded behind strong walls. Many times when a capacitor fails or a bus arcs over, there will be shrapnel flying with considerable velocity.

All entrances to the bank should be interlocked. All banks should have reliable fail-safe short-circuits on them except when in use. Short-circuiting hooks should be available to reach each capacitor. An unshort-circuited energy storage capacitor will recover a lethal charge. They should always be short-circuited during handling or while in storage.

The fire danger in a bank is acute. Energy storage capacitors may be impregnated with mineral oil, castor oil, or chlorinated diphenyl, the first two of which are inflammable. If a capacitor short-circuits during discharge, it sometimes ruptures and spews impregnant over a considerable area. If the impregnant is flammable, a highly noxious fire may start. The fire-fighting personnel should be informed of the hazards in spraying water on a bank which may be charged.

Ground currents are always a problem in bank design. Banks should be grounded at only one point to a counterpoise ground with wide, low-inductance straps.

REFERENCES

1. Some Properties of a Graded Vacuum Spark Gap, J. W. Mather. A. H. Williams. *Review of Scientific Instruments*, New York, N. Y., vol. 31, Mar. 1960, pp. 297-303.
2. J. L. Tuck. *Proceedings*, Second International Conference on the Peaceful Uses of Atomic Energy, United Nations, Geneva, Switzerland, vol. 32, 1958, p. 19.
3. "High Powered Spark Gap Switches," E. L. Kemp. Report *TID-6670*, Oak Ridge National Laboratory, Oak Ridge, Tenn., 1960, p. 10.
4. Report *LAMS-2416*, G. P. Boicourt, E. L. Kemp, F. K. Tallmadge. Los Alamos Scientific Laboratory, Los Alamos, Calif., Dec. 1959.
5. Charging Large Capacitor Banks in Thermonuclear Research, H. K. Jennings. *Electrical Engineering*, June 1961, pp. 419-21.
6. Reference Data for Radio Engineers. International Telephone and Telegraph Corp., New York, N. Y., fourth edition, 1956, p. 306.
7. Report *UCRL-4733*, H. B. McFarlane. Lawrence Radiation Laboratory, Livermore, Calif., Aug. 9, 1956.

26

Copyright © 1976 by the Institute of Electrical and Electronics Engineers, Inc.

Reprinted by permission from pages IIC1-1–IIC1-6 and IIC1-10 of *IEEE Int. Pulsed Power Conf., 1976, Proc.*, Institute of Electrical and Electronics Engineers, Inc., 1976

PRINCIPAL CONSIDERATIONS IN LARGE ENERGY-STORAGE CAPACITOR BANKS

E. L. Kemp

ABSTRACT

Capacitor banks storing one or more megajoules and costing more than one million dollars have unique problems not often found in smaller systems. Two large banks, Scyllac at Los Alamos and Shiva at Livermore, are used as models of large, complex systems. Scyllac is a 10-MJ, 60-kV theta-pinch system while Shiva is a 20-MJ, 20-kV energy system for laser flash lamps.

A number of design principles are emphasized for expediting the design and construction of large banks.

The sensitive features of the charge system, the storage system layout, the switching system, the transmission system, and the design of the principal bank components are presented.

Project management and planning must involve a PERT chart with certain common features for all the activities. The importance of the budget is emphasized.

Introduction

The capacitor bank is the most universal system for producing the pulse power requirements for several types of fusion experiments. Large capacitor banks, here defined as banks storing more than 1 MJ and costing more than one million dollars, require special attention because of their cost and complexity.

Three of the most important factors in the design and construction of large banks are performance, cost, and reliability. Performance is essential and obviously most important. Cost can be a determinant, but is more subtle and may become a critical issue after the project is well under way. Reliability is a necessary and worthy objective, but it can jeopardize both performance and cost if it is given too much priority. The appropriate balancing of performance, cost, and reliability requires considerable attention and mature judgement.

Two capacitor banks, Scyllac at the Los Alamos Scientific Laboratory[1] and Shiva at the Lawrence Livermore Laboratory,[2] will be used to illustrate certain characteristics of large systems. Scyllac, shown in Fig. 1 during construction, is a 60-kV, 10-MJ toroidal theta-pinch experiment. The bank was built in 1970 for 60 cents per joule. The Shiva system, shown in Fig. 2, is a 20-kV, 20-MJ laser fusion experiment. This

Fig. 1. Scyllac capacitor banks during construction.

bank was built in 1974 for about 20 cents per joule. It must be emphasized that the criteria for these two systems were considerably different.
The Scyllac system drives a few-nanohenry load coil delivering over 100 MA
in 3-4 µs and is crowbarred. The principal components for Scyllac were
developed during the design phase and were at the edge of the state-of-
the-art. Reliability was compromised for performance. The Shiva bank
drives a large number of flash lamps in a slow risetime, non-reversing circuit. Reliability was a major consideration so conservative designs were
used whenever possible. Both systems meet their criteria specifications
adequately.

Design Tenets

     The design of a large capacitor bank often takes one to two years.
Recognizing a few basic tenets can make this process go more smoothly and
minimize the design time and cost.

     The criteria should be completely specified at the very beginning.
As in all system design, everything affects almost everything else. New
requirements late in the design phase can impact or nullify many completed
designs. Careful review of the criteria is a cost-effective exercise in
the early phase of the project.

Fig. 2.   Shiva capacitor banks.

The system design should be directed to meet the specified criteria. While many designs may meet the criteria, cost and time considerations must limit the design options in order to keep the project within the budget and on schedule.

Large systems must use conservative designs. Even modest design changes can be devastating. The simple addition of one charge resistor in the Scyllac circuit during construction cost over $10,000 for the components and several man-weeks of installation. All new and novel designs should be tested in a substantial prototype system before they are approved for the main system.

Installation and maintainability are two requirements that must be incorporated into all designs. It must be appreciated that the installation cost of most electrical components is typically 20% of the component cost. Installation cost can be minimized by prefabricating and

330

preassembling many components before they are installed. For instance, the capacitors in the Shiva system were assembled into the racks and connected before the module was installed in the system. Much of the air distribution system of Scyllac was also prefabricated before installation.

Maintenance is often ignored by designers. All design reviews should address the operating and maintenance features of the design.

Finally, remember the control system. Often the majority of the energy system is designed before the control system is considered. After a number of the energy subsystems are installed and ready to check out, it is awkward to discover there is no control system available. The design and installation of the control system should have a high priority in the program.

Circuits

The system circuit will depend upon the application so no particular circuit will be analyzed here. Rather, a few general comments will be addressed to various subsystems of the circuit, including the charge system, the storage system layout, the switching system, and the transmission system.

Large banks should be charged at a constant current. An RC charge system takes a long time or requires a very large power supply, and the energy dissipated in the resistor equals the energy stored in the bank. There are a variety of constant-current charging systems. Scyllac uses monocyclic networks, shown in Fig. 3, for the main charging system. Monocyclic networks are one of the least expensive methods for controlling large KVA power supplies, but they must always be connected to a load or they will generate extremely high voltages with accompanying spectacular arcs. The Shiva bank uses the three-phase voltage doubling circuit shown in Fig. 4. Like the monocyclic network, it can operate continuously into a short circuit. Both circuits use only passive components which enhance their reliability.

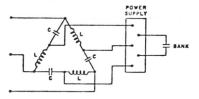

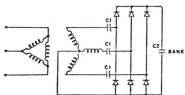

Fig. 3.   Scyllac monocyclic network circuit.   Fig. 4.   Shiva voltage doubler circuit.

Scyllac uses vacuum-tube rectifiers in the power supplies while Shiva uses solid-state rectifiers. Although vacuum tubes require filament power and have a limited life, they are more forgiving to transients which always occur during system checkout, and occasionally during operations. Modern solid-state power supplies are now operating satisfactorily in Scylla IV-P, which is similar to Scyllac, after the rectifiers were changed to have an adequate PIV rating. The capacitors will ring occasionally and the rectifiers must accommodate the reverse load voltage, as well as the normal PIV.

The principal safety features of the entire bank are incorporated in the charge system. These include the dump circuit for passively discharging the bank when a shot aborts and the manual shorting system to short out a particular capacitor before it is physically touched. Convenient shorting places should be built into the charge system with shorting sticks permanently located at these locations.

There are a host of options for bank layout. The Shiva system, shown in Fig. 2, places the capacitors on their edge so that the aisle can be used for high-voltage isolation, as well as maintenance access. Six capacitors are also connected solidly together. The typical energy-storage capacitor can absorb 25 to 30 kJ without rupture so the Shiva design of 18 kJ is conservative.

The switching system is often the most critical item in the entire bank. Prefires and failures to fire may generate transients that destroy or injure many other components. In the Scyllac system a crowbar failure to fire allowed the bank to ring and broke the plasma discharge tube which shut the experiment down for three weeks. Although the crowbar trigger system was cross triggered and redundant, this event happened twice. Exhaustive diagnosis eventually revealed that an intermittent transient was getting back to the master clock and resetting it before it generated the crowbar trigger signal. This problem was overcome by taking a back-up signal from the bank to fire the crowbar system slightly later than the normal clock signal. Failure analysis should be performed on all large systems before the final design is approved.

When possible the switches should be physically grouped together. Such designs minimize installation costs, simplify the trigger system, and enhance operations and maintenance.

There are two generic transmission systems: solid conductors and coaxial cables. Solid conductor systems are often attractive because they can be built for almost any reasonable inductance and resistance characteristic, and you can do it yourself. However, in practice a few problems arise. The physical system must be located exactly as designed, there is very little flexibility with solid conductors. Small particles of foreign matter can penetrate between the conductors and eventually cause electrical failure. Finally, the magnetic force between conductors can be a very significant problem.

Coaxial cable has certain advantages for transmission systems. It is commercially available and fairly inexpensive. It is mechanically flexible, contains the magnetic forces, and also the magnetic field which minimizes electromagnetic noise problems. A coaxial cable system can be designed for low inductance by paralleling cables, so the net transmission line inductance competes well with a solid conductor system.

Components

The principal components of an energy storage capacitor bank are the capacitors, the switches, and the coaxial transmission cable. Some features of these components will be discussed. Figure 5 shows two energy-storage capacitors. The 1.85-μF, 60-kV capacitor has 22-nH self-inductance

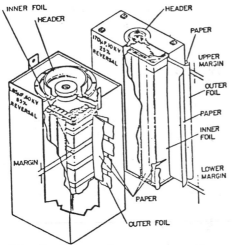

Fig. 5. Energy storage capacitors.

and has a life of about 50,000 shots with 85% voltage reversal. This capacitor stores 3300 J at 0.85 J/in.[3]. Recently a 2.8-μF unit has been developed with a life of about 50,000 shots at 15% voltage reversal. It stores 5100 J at 1.3 J/in.[3]. The 170-μF, 10-kV capacitor shown has a life exceeding 250,000 shots at 15% voltage reversal. This capacitor stores 8500 J at 3.5 J/in.[3]. Both of these capacitors are made with Kraft paper and impregnated with castor oil. Recently the 10-kV design has been made with paper and film and impregnated with another type of impregnant. It has an even higher energy density and lifetimes exceeding 50,000 shots at 15% reversal.

[*Editor's Note:* Material has been omitted at this point.]

References

[1]E. L. Kemp, "The Final Design of Scyllac," Proc. of·Symposium on Engineering Problems of Fusion Research, Los Alamos Scientific Laboratory Report LA-4250 (1969).

[2]G. R. Allen, W. L. Gagnon, P. R. Rupert, and J. B. Trenholme, "Engineering Storage and Power Conditioning System for the Shiva Laser," Sixth Symposium on Engineering Problems of Fusion Research, IEEE Pub. 75CH1097-5-NPS.

[3]W. H. Harris, E. L. Kemp, W. E. Quinn, and F. L. Ribe, "Project Management of Scyllac Construction," Fifth Symposium on Engineering Problems of Fusion Research, IEEE Pub. 73CH0843-3-NPS.

27

Reprinted from *Report ORNL/TM-5529*, Oak Ridge National Laboratory
(operated by Union Carbide for the Dept. of Energy), 1976, 8pp.

ELECTROSTATIC ENERGY STORAGE
R. L. Macklin*

Abstract

The advantages of capacitor storage of energy, particularly for
automotive use are reviewed and compared with alternatives. A few suggestions for research possibilities for possibly suitable organic dielectrics are given.

Introduction

With shortages of oil and gasoline clearly facing the world, it is
time to consider all other possibilities for storing energy effectively.
Leading possibilities for automotive use include advanced electric batteries[1] and specially designed flywheels.[2] Some interest is also shown
in superconducting magnet storage although the cryogenic problem seems
formidable for automotive use.

The object of this paper is to urge the reconsideration of electrostatic or capacitor storage for such an application. First, note that
there are several practical advantages to such a system if a suitable
organic chemical polymer can be developed as the dielectric.

The efficiency in converting the stored electric energy to mechanical energy to turn wheels can confidently be expected to be very high;
close to 90%. This means that substantially less energy need be stored in
a vehicle (about 1/7) for equal range, speed, acceleration, etc., compared
to a gasoline burning system. There is no exhaust to pollute the air and

*Research sponsored by the U. S. Energy Research and Development Adminis-
tration under contract with Union Carbide Corporation.

even thermal pollution by wasted heat is reduced by a factor of 9 or
so at the vehicle. Of course, heat must be carefully dissipated back
at the power house that generates the electricity but an overall gain
of a factor of nearly 3 is still to be expected. The real but modest
efficiency gains from regenerative braking have been emphasized previously vis a vis the gasoline driven system.[2]

The storage system can be expected to operate at a normal range
of temperature though some form of protection against fire and collision
would be needed. Operation from sea level to mountain altitudes should
present no problems as any air pressure sensitive components could be
hermetically sealed at the factory.

Recharging can be done rapidly and efficiently compared with
conventional battery charging. A few minutes should suffice at a service station equipped with suitable rectifying equipment and industrial
capacity electric supply lines. Alternatively, residential electric
supply is generally adequate for overnight charging at home (at off peak
electric rates too!)

There should be no need for raw materials in short supply unless
the organic dielectric were based on a rare element as a major constituent.
The solid state switches for control and series-parallel interconnection
of individual capicitor "cells" require silicon, a very abundant element.
The organic material for the dielectric should be comparable in amount
to the synthetic rubber currently used in tires. Aluminum could be used
for wiring if a shortage of copper became critical. Only in the newest
magnetic materials for wheel motors is a possible shortage of raw materials
anticipated, and there, of course, the older materials based on iron and
the iron group elements would be satisfactory substitutes.

Capital costs for the dielectric should be comparable to those already known for specialized plastic sheet.[3] The paucity of moving parts should provide a significant improvement in both capital and maintenance costs. For instance, the relative economic advantages of DC vs AC drive motors would no longer be clear and the choice might lead to even fewer parts such as brushes and commutators to wear out.

Material Basis of Energy Storage Systems

The combustion engine systems (gasoline, Diesel fuel) exploit the availability of free air to achieve a high capacity for chemical energy storage; $\sim 4 \times 10^7$ joules/kg^{-1}. Conversion efficiency from thermal energy to automotive propulsion reduces this nearly to 4×10^6 joules/kg^{-1}. On a molecular basis, carbon to hydrogen (and carbon to carbon) chemical bonds are exchanged through combustion for the stronger carbon-oxygen and hydrogen-oxygen bonds, freeing the difference in energy as heat.

The lead storage battery exploits the overall reaction $Pb+PbO_2 \rightarrow PbO + PbO$, effected through ion (and electron!) migration, to produce electric power directly. The stored energy is only about 4.4×10^4 joules/kg^{-1} but there is no (further) thermal cycle inefficiency involved in its use. It is notable however that even with its drawbacks the lead-acid storage battery is being revived to operate cars for city driving, and battery operated delivery trucks have had a long history.

Advanced batteries such as the lithium-sulfur cell[1] promise a gain of several fold in energy storage density, but share some of the battery drawbacks such as slow charging rate and will add new ones, especially the need for high temperature operation.

A great deal of work has gone into the development of fuel cells but automotive application seems to have been abandoned. Through the use of free air, the potential for energy storage of such systems rivalled the gasoline-combustion engine system. Technical problems presumably have proved intractable.

Another recent development has turned to mechanical energy storage in flywheels.[2] By the use of ultra strong fiber composites and optimized structural design, capacities comparable to gasoline combustion in useful automotive energy are expected.

One of the possibilities considered in superconducting magnet research is electric energy storage. At temperatures below the boiling point of hydrogen (near 20°K) many materials have been found to support persistent ring currents. The loss of energy in continuously refrigerating the material far outweighs inherent ohmic losses due to imperfections. The limits of magnetic field that can be sustained have risen substantially in the past decade, particularly for niobium based composites. For practical energy storage densities one will again be pressing the strength of materials of construction as in a flywheel storage.

All of these approaches can be viewed in a unified way in terms of ionic, molecular and intermolecular bond strengths or bond energies. These are typically a few electron volts per bond. Examples are Li-Li (1.0 eV), Li-Cl (4.9 eV), C-H (3.5 eV), C-C (6.25 eV), C-N (7.55 eV), C-O (11.1 eV), C-F (4.6 eV), Si-O (8.3 eV), Pb-Pb (1.0 eV), Pb-O (4.3 eV). In terms of power storage density, other factors such as overall density and molecular weight per bond will also be important. In general, though, all the approaches considered earlier will experience similar limits in storage capacity. Factors of a more practical nature will then determine the best system for a particular use. These include materials availability, manufacturing techniques, operating conditions such as temperature range and speed of recharge, etc.

Many years ago Locke[4] suggested that new dielectrics might provide the best hope for the ultimate solution to the electric energy storage problem. The stored energy in a flat sheet capacitor is given as

$$E = CV^2/2$$

with units: joules or watt seconds (E), farads (C), and volts (V). In practice thin sheets of dielectric material, coated with thin metal electrodes to apply the voltage, are rolled up and put into cylindrical tubes. Thus it is convenient to express the energy storage in the dielectric "meat" of such a sandwich on a volume basis, recognizing that there will be some additional volume required for the thin electrodes and an outer casing

$$D = 0.4427 \times 10^6 \ K \ a^2$$

with D in kilojoules per cubic meter and a in volts per Angstrom. K is the low frequency (static) dielectric "constant" measured as the ratio of the capacitance using a particular material (at a particular temperature) to that of a vacuum occupying the same space in a test condenser. Many commonly used condenser dielectrics have dielectric constants below ten, corresponding to strong ionic or more often to non-polar interatomic bonds. A class of metal titanate ceramics have shown dielectric constants up to 12000. Such behavior is correlated with weakening some titanium-oxygen bonds (over a fraction of an Angstrom), in this case by maximum thermal stretching of the Perovskite lattice structure.

The dielectric "strength" a is sharply reduced by impurities and imperfections in a material and thus rather ill defined. At a slightly higher voltage gradient "breakdown" occurs, resembling miniature lightning whereby conducting pathways through the material are formed. Presumably the process must start by breaking at least one bound electron loose or by breaking an ionic bond. To avoid this the bonding must remain uniformly strong at atomic dimensions ($\sim$2 Angstrom). In paper thin sheets, appropriate to capacitor construction, high purity can be achieved in mica and organic polymers. Adhesive electric insulating tape can be guaranteed for use at 1000 volts per 2.54×10^{-5} m (mil), and mica approaches 225,000 volts per millimeter (a = 0.0225). Other organic polymers (polyamides) rate as well as mica in dielectric strength (a = 0.028) and probably exceed it in still thinner sheets. Using the energy storage formula for mica with a dielectric constant of $\sim$5 to 7 we might expect a storage capacity of 1000-1500 kj m^{-3} of mica. Commercial energy storage capacitors were found listed up to 600 kj m^{-3}, based on a waxed paper dielectric,[5] so this level of energy density is close to commercial practice. We can expect that no materials could appreciably exceed a dielectric strength of one volt per Angstrom as bond strengths and lengths are of this order. Extremely thin layers formed by oxidation of silicon and aluminum surfaces may approach this limit.

If we take the lead-acid storage battery at 44,000 j kg^{-1} as an aiming point, we would need to develop a material with the dielectric strength of the best plastic, incorporating enough polarizable sites to raise the dielectric constant well above 130. This is a reasonable aim for economic impact as the lead-acid battery already provides marginal performance for electric cars and trucks. At the same time it is clearly unreasonable to hope for materials allowing a thousand times greater energy storage density, exceeding the heat from gasoline burned with free air.

Although many promising materials are beginning to be reported, perhaps I can add a few of my own speculations. First, simple dipolar molecules dispersed in an insulating liquid seems short of the mark. They could be largely aligned (polarized) by an applied field but the restoring force of thermal agitation would provide less than 1/40 electron volt per dipole. Of course, if the dipole could be significantly stretched without breakdown, that would be another matter. If the dipoles were several atoms long (like TTF-TCNQ) and the liquid could be polymerized to long chain molecules, the restoring force could be increased. In this technique, a thin polymer sheet[3b] is stretched in two directions during manufacture so the dipoles tend to be trapped with their axes within a few degrees of the plane of the sheet.

More sophisticated approaches envisage specially tailored organic molecules incorporating ionic bonds such as polyelectrolytes, metal-organics, ion exchange resins, etc. Flat molecules like the porphyrin rings in haemoglobin can bind polyvalent metals in the plane of the rings, leaving the perpendicular direction free. Layers of such substituents might provide the basis for a material of high dielectric constant in the one dimension.

Conclusion

A search for materials (particularly polymers) with high dielectric strength and static dielectric constant might lead to economically attractive electric storage systems, applicable to mobile as well as stationary installations.

References

1. For instance Argonne National Laboratory is proceeding to an automotive test development with their lithium-sulfur storage battery. M. Warshay and L. O. Wright in NASA TM X-3192 (February 1975) have emphasized the economic attractiveness of a pumped iron chloride-titanium chloride electrochemical storage system for power load leveling in electric utility systems.

2. R. F. Post and S. F. Post, Sci. Amer. <u>229</u>, 17-23 (1973).

3. "Mylar", "Polaroid", Polyethylene and Polyvinyl Fluoride are examples of plastics produced in thin sheets.

4. E. L. Locke, "New Dielectrics" A.S.F. 72-86 (∿1955).

5. ELMAG Corporation 0.01 mfd, 160 kv, Rapid Discharge Energy Storage Capacitor rated "10 joules per cubic inch".

AUTHOR CITATION INDEX

SUBJECT INDEX

About the Editor

WILLIAM V. HASSENZAHL is a senior scientist at the Lawrence Berkeley Laboratory. He received the B.S. in physics from the California Institute of Technology in 1962 and the Ph.D in elementary particle physics from the University of Illinois in 1967.

During 1967 to 1980 Dr. Hassenzahl was a physicist at the Los Alamos Scientific Laboratory, where he was manager of the superconducting magnetic energy storage program from its inception in 1972 to 1978. After a sabbatical at the Centre d'Etudes Nucleaires de Saclay he returned briefly to Los Alamos before transferring to the Lawrence Berkeley Laboratory in 1980.

Dr. Hassenzahl's principal research interest has been in the development of superconducting magnets including investigations of static and transient heat transfer to liquid helium, conductor stability, and magnet performance in superfluid helium. He has authored or edited 40 papers on superconducting magnets.

He is a member of the American Physical Society and a consultant to the Electric Power Research Institute on superconducting magnetic energy storage.